Nonlinear Optimization in Electrical Engineering with Applications in MATLAB®

Nonlinear Optimization in Electrical Engineering with Applications in MATLAB®

Mohamed Bakr

The Institution of Engineering and Technology

Published by The Institution of Engineering and Technology, London, United Kingdom

The Institution of Engineering and Technology is registered as a Charity in England & Wales (no. 211014) and Scotland (no. SC038698).

First published 2013

The Institution of Engineering and Technology
Michael Faraday House
Six Hills Way, Stevenage
Herts, SG1 2AY, United Kingdom

www.theiet.org

British Library Cataloguing in Publication Data
A catalogue record for this product is available from the British Library

ISBN 978-1-84919-543-0 (hardback)
ISBN 978-1-84919-544-7 (PDF)

Typeset in India by MPS Limited
Printed in the UK by CPI Group (UK) Ltd, Croydon

To my wife Mahetab, my children Jannah, Omar, and Youssef, and to my parents to whom I am indebted for as long as I live

Contents

Preface

In 2008, I was asked to give a sequence of lectures about optimization to graduate electrical engineering students in University of Waterloo, Canada. These lectures were attended by over 40 students, postdoctoral fellows, and industry CAD people. I tried in this course to teach optimization, which is a highly mathematical subject, with more focus on applications. The course was well received by all attendees. The feedback I received from them regarding other optimization courses, in comparison to my course, was surprising. Most of them attended other courses in optimization that contributed very little to their understanding of the subject. Most of these courses were too abstract with very little connection to practical examples. There were so many confusing mathematical proofs that distracted them from the actual meanings and implementations. Engineering students, both graduate and undergraduate, would like to get a better focus on the meaning hidden behind the lengthy proofs. They are interested in real-life engineering applications rather than abstract material that add very little to their global understanding of the subject.

I was encouraged by this experience to write an optimization book dedicated to electrical engineers; a book that combines simplicity in introducing the subject with mathematical rigor; a book that focuses more on examples and applications rather than on lengthy mathematical proofs. I use MATLAB® as a tool to show simplified implementations of the discussed algorithms. This should help in removing the abstractness of the theorems and improve understanding of the different subjects.

This book is intended as an initial book in optimization for undergraduate students, graduate students, and industry professionals. Third-year engineering students should have the necessary background to understand different concepts. Special focus is put on giving as many examples and applications as possible. Only absolutely necessary mathematical proofs are given. These proofs, usually short in nature, serve the purpose of increasing the depth of understanding of the subject. For lengthy proofs, the interested reader is referred to other more advanced books and published research papers in the subject.

For the applications given, I try to show a MATLAB implementation as much as possible. In more advanced applications that require non-MATLAB commercial codes, I just report on the problem formulation and the results. All the MATLAB codes given in this book are only for educational purposes. The students should be able to fully understand these codes, change their parameters, and apply them to different problems of interest. These codes can also be developed further for research- and industry-related applications. The full MATLAB codes can be downloaded through the link www.optimizationisfun.com.

Even though most of the addressed examples and applications are related to electrical engineering, students, researchers, and technical people from other engineering areas will find the material also useful. Electrical engineering has been integrated with other areas resulting in innovative programs in, for example, mechatronics and biomedical engineering. I give as much explanation to the theory behind every application so that the unfamiliar reader will grasp the basic concepts.

This book is organized as follows. In Chapter 1, we review the basic mathematical background needed in this book. Most third-year electrical engineering students will have good familiarity with the content of this chapter. We discuss in this chapter basic rules of differentiation of vectors and matrices. I added more content related to linear algebra that is useful in understanding many of the discussed optimization subjects.

Chapter 2 gives an overview of the linear programming approach. As well known in the area of engineering, nonlinear problems can be solved by converting them to a sequence of linear problems. Linear programming addresses the issue of optimizing a linear objective function subject to linear constraints. I explain in this chapter the basic Simplex method and the theory behind it. I give several examples and applications relevant to the area of electrical engineering.

Chapter 3 reviews some classical optimization techniques. These techniques found wide applications before the advent of the digital computer era. The mathematical foundations of these techniques form the bases for many of the numerical optimization techniques.

In Chapter 4, we address the one-dimensional line search problem. This problem is very important in optimization theory as it is routinely utilized by other linear and nonlinear optimization techniques. I introduce different approaches for solving this problem including derivative-free techniques, first-order line search techniques, and Newton-based techniques.

Derivative-free nonlinear optimization techniques for unconstrained problems are discussed in Chapter 5. These techniques do not require any derivative information for solving general multidimensional problems. They are useful in problems where sensitivity information may not be available or costly to obtain. Only objective function values are used to guide the optimization iterations.

Chapter 6 discusses gradient-based techniques for unconstrained optimization. These are techniques that assume the availability of first-order sensitivity information. We discuss the basic steepest descent method and a number of conjugate optimization techniques. These techniques have robust convergence proofs under certain assumptions on the objective function.

Quasi-Newton techniques for the unconstrained optimization of general nonlinear problem are discussed in Chapter 7. These techniques aim at approximating second-order sensitivity information and using them in guiding the optimization iterations. These techniques are known for their good convergence rate. I illustrate these techniques through examples and applications.

Chapter 8 reviews some constrained optimization techniques. Some of these techniques build on different concepts of unconstrained optimization. Other techniques sequentially approximate the nonlinear problem by a corresponding linear problem.

All the techniques discussed in Chapters 1 through 8 are local optimization techniques. They obtain a minimum of the optimization problem that satisfies certain optimality conditions only locally. Chapter 9 introduces a number of techniques for obtaining the global minimum of an optimization problem. This is the minimum that has the lowest value of the objective function over all other minima. A number of nature-inspired techniques are discussed and illustrated through examples and applications.

Chapter 10 addresses adjoint-based techniques for estimating the gradient of an objective function. I show in this chapter how, by using only one extra simulation, the sensitivities of an objective function with respect to all parameters are estimated regardless of their number. I illustrate this approach for the adjoint network method, the frequency-domain case, and the time-domain case. Several examples and applications are presented.

Finally, I should say that the area of optimization is a huge field with so many techniques and applications. No one book can cover all material by its own. I just attempted here to help the readers scratch the surface of this area. I encourage them to build on the material in this book and continue to seek more advanced learning in this area.

Dr. Mohamed Bakr
Professor, Department of Electrical and Computer Engineering,
McMaster University, Hamilton, Ontario, Canada
July 2013

Acknowledgments

I would like to acknowledge the mentorship of Drs. Hany L. Abdel Malek and Abdel Karim Hassan from Cairo University, Egypt. They were the first ones to introduce me as a Master student to the fascinating area of optimization. Being electrical engineers themselves, they have always attempted to focus on applications without abandoning the mathematical rigor.

I would like also to thank Drs. John Bandler from McMaster University and Radek Biernacki (currently with Agilent technologies) for their guidance during my graduate studies in Canada. I have learnt a lot from my Ph.D. supervisor Dr. John Bandler. He is a dedicated person with high commitment to excellence. His work in computer aided design (CAD) of microwave structures has helped shape the field and motivated all his students. It was also my pleasure that I helped teach EE3KB3 course with Dr. Biernacki. Dr. Biernacki gave me the opportunity to conduct the tutorials for this applied optimization course. This was an amazing experience to learn about applications of optimization theory to the design of microwave circuits and antennas. I include the material of this course as a reference in some of the chapters because they helped inspire several of the presented examples.

I would like to thank all the good researchers who worked with me over the years and helped shape my experience in the areas of optimization and computational electromagnetics. These excellent researchers include Dr. Payam Abolghasem, Dr. Peter Basl, Dr. Ezzeldin Soliman, Dr. Mahmoud ElSabbagh, Dr. Ahmed Radwan, Dr. Mohamed Swillam, Dr. Osman Ahmed, Peipei Zhao, Kai Wang, Laleh Kalantari, Yu Zhang, Mohamed Negm, Harman Malhi, and Mohamed Elsherif.

I would also like to thank my wife and children for their patience during the development of this book. Last but foremost, I would like to thank God for giving me the strength and energy to continue this work to its end.

Chapter 1
Mathematical background

1.1 Introduction

Mathematics is the language we use to express different concepts and theorems in optimization theory. Like all languages, one has to master the basic building blocks of a language such as the letters, the words, and the symbols in order to gain some fluency in a language.

The main target of this chapter is to briefly review some of the mathematical concepts needed for subsequent chapters. We will review some properties of vectors and matrices. I will introduce some of the vocabulary used in optimization theory. We will also discuss the properties of the solution of a system of linear equations.

1.2 Vectors

A vector expresses a number of variables or parameters combined together. It offers a more compact way for manipulating these parameters. A general n-dimensional real vector $\boldsymbol{x} \in \Re^n$ is given by:

$$\boldsymbol{x} = \begin{bmatrix} x_1 \\ x_2 \\ \vdots \\ x_n \end{bmatrix} = [x_1 \quad x_2 \quad \ldots \quad x_n]^T \tag{1.1}$$

where x_i, $i = 1, 2, \ldots, n$ is the ith component of the vector $\boldsymbol{x}$. We utilize a column vector notation where all vectors are assumed to be column vectors. The symbol "T" in the superscript abbreviates the transpose operator which converts a row vector to a column vector and vice versa. In all MATLAB® code outputs, we use the MATLAB symbol "′" to also denote the transpose of a column vector.

In optimization theory, the components of a vector are usually used in an abstract way. They are not usually given a physical meaning. In practice, however, these components have a meaning. They may represent the values of inductors, capacitors, and resistors of an electric circuit. They may alternatively represent the values of the different dimensions of discontinuities and their electromagnetic properties in high-frequency structures. The meaning depends on the optimization problem at hand.

Consider, for example, an electric circuit with two resistors, two inductors, and two capacitors. We assume that the optimization parameters of this circuit are:

$$\boldsymbol{x} = [R_s \quad C_1 \quad L_1 \quad C_2 \quad L_2 \quad R_L]^T \tag{1.2}$$

The parameters in (1.2) can have completely different orders of magnitude. The resistors can be in the KΩ range, the capacitors are in the μF range, while the inductors may be in the mH range. This difference in magnitudes can cause ill-conditioning in optimization algorithms that usually assume that all parameters have the same order of magnitude. To fix this issue, we use scaling. For example, within the optimization algorithm the used values of the parameters can be $\boldsymbol{x} = [1.0 \quad 3.0 \quad 1.5 \quad 2.0 \quad 2.0 \quad 1.5]^T$. These are scaled values of the parameters with a different scale factor for each parameter. Every time it is required to evaluate the response of the circuit, the parameters are scaled back to their physical values. This approach allows us to use the same optimization algorithm for parameters with different orders.

During an optimization algorithm, the values of the parameters are iteratively adjusted until an optimal design is reached. We denote by $\boldsymbol{x}^{(k)}$ the value of the optimizable parameters at the kth iteration. The optimal design is denoted by $\boldsymbol{x}^*$. An optimization algorithm aims at obtaining a good approximation of $\boldsymbol{x}^*$ within a number of iterations. We denote the final design achieved by an optimization algorithm as $\bar{\boldsymbol{x}}$ which is expected to be very close to $\boldsymbol{x}^*$.

The iterative adjustment done to the variables at the kth iteration involves determining a perturbation $\Delta\boldsymbol{x}^{(k)}$ of the parameters. The parameters at the $(k+1)$st iteration are given by:

$$\boldsymbol{x}^{(k+1)} = \boldsymbol{x}^{(k)} + \Delta\boldsymbol{x}^{(k)} \tag{1.3}$$

where vector addition implies adjusting every component of the vector $\boldsymbol{x}^{(k)}$ by the corresponding component in the vector $\Delta\boldsymbol{x}^{(k)}$ to give the new parameters $\boldsymbol{x}^{(k+1)}$.

Another important vector operation is the inner product of two vectors. If $\boldsymbol{x}, \boldsymbol{y} \in \Re^n$, the n-dimensional parameter space, then their inner product is given by:

$$a = \boldsymbol{x}^T\boldsymbol{y} = [x_1 \quad x_2 \quad \ldots \quad x_n]^T \begin{bmatrix} y_1 \\ y_2 \\ \vdots \\ y_n \end{bmatrix} = \sum_{i=1}^{n} x_i y_i \tag{1.4}$$

where $a \in \Re^1$ is a scalar. The Euclidean norm of a vector is defined as the square root of the inner product between the vector and itself. The square of the Euclidean norm is given by:

$$\|\boldsymbol{x}\|^2 = \boldsymbol{x}^T\boldsymbol{x} = x_1^2 + x_2^2 + \cdots + x_n^2 \tag{1.5}$$

This norm evaluates the length of the vector. The inner product between two vectors is a measure of the angle between the two vectors as it can be written as:

$$\boldsymbol{x}^T\boldsymbol{y} = \|\boldsymbol{x}\|\,\|\boldsymbol{y}\|\cos(\theta) \tag{1.6}$$

where θ is the angle between the two vectors. If the inner product between two vectors is positive, then the angle between them is less than 90°. If the inner product is negative then this angle is greater than 90°.

Another type of vector product can create a matrix. Consider two vectors $\boldsymbol{x}$ and $\boldsymbol{y}$ that do not necessarily have the same number of components. If $\boldsymbol{x} \in \Re^n$ and $\boldsymbol{y} \in \Re^m$, then the following matrix can be constructed:

$$\boldsymbol{x}\boldsymbol{y}^T = \begin{bmatrix} x_1 \\ x_2 \\ \vdots \\ x_n \end{bmatrix} [y_1 \quad y_2 \quad \dots \quad y_m] = \begin{bmatrix} x_1y_1 & x_1y_2 & \dots & x_1y_m \\ x_2y_1 & x_2y_2 & \dots & x_2y_m \\ \vdots & \vdots & \vdots & \vdots \\ x_ny_1 & x_ny_2 & \dots & x_ny_m \end{bmatrix} \tag{1.7}$$

This matrix has n rows and m columns. Notice that all the rows of the matrix in (1.7) are multiples of the vector $\boldsymbol{y}$. Such a matrix is called a rank-one matrix because it has only one linearly independent row. In later chapters, I will show that such matrices play a role in optimizing nonlinear problems. The following example illustrates some concepts in vector operations.

Example 1.1: Consider the two vectors $\boldsymbol{x} = [1.0 \quad 2.0 \quad -1.0]^T$ and $\boldsymbol{y} = [-3.0 \quad 2.0 \quad 2.0]^T$. Evaluate $\boldsymbol{x}+\boldsymbol{y}$, $\boldsymbol{x}-\boldsymbol{y}$, $\boldsymbol{x}^T\boldsymbol{y}$, $\boldsymbol{x}\boldsymbol{y}^T$, and the angle between these two vectors.

Solution: The results for this example are as follows:

Addition: $\boldsymbol{x}+\boldsymbol{y} = \begin{bmatrix} 1 \\ 2 \\ -1 \end{bmatrix} + \begin{bmatrix} -3 \\ 2 \\ 2 \end{bmatrix} = \begin{bmatrix} -2 \\ 4 \\ 1 \end{bmatrix}$

Subtraction: $\boldsymbol{x}-\boldsymbol{y} = \begin{bmatrix} 1 \\ 2 \\ -1 \end{bmatrix} - \begin{bmatrix} -3 \\ 2 \\ 2 \end{bmatrix} = \begin{bmatrix} 4 \\ 0 \\ -3 \end{bmatrix}$

Inner product: $\boldsymbol{x}^T\boldsymbol{y} = [1 \quad 2 \quad -1] \begin{bmatrix} -3 \\ 2 \\ 2 \end{bmatrix} = -1$, and

Multiplication: $\boldsymbol{x}\boldsymbol{y}^T = \begin{bmatrix} 1 \\ 2 \\ -1 \end{bmatrix} [-3 \quad 2 \quad 2] = \begin{bmatrix} -3 & 2 & 2 \\ -6 & 4 & 4 \\ 3 & -2 & -2 \end{bmatrix}$

Notice that in the last result, the matrix is a rank-one matrix as all rows are multiples of the vector $\boldsymbol{y}$. Because the inner product of the two vectors is negative, these two vectors make an obtuse angle. To determine this angle we use (1.6) to get:

$$\begin{aligned} \cos(\theta) &= \frac{\boldsymbol{x}^T\boldsymbol{y}}{\|\boldsymbol{x}\|\|\boldsymbol{y}\|} \\ &= \frac{-1}{\sqrt{(1)^2+(2)^2+(-1)^2}\sqrt{(-3)^2+(2)^2+(2)^2}} = \frac{-1}{\sqrt{6}\sqrt{17}} = -0.099 \\ &\Downarrow \\ \theta &= \cos^{-1}(-0.099) = 1.67\,\text{radians} = 95.68^\circ \end{aligned} \tag{1.8}$$

1.3 Matrices

Matrices are another means of manipulating a group of components together. A matrix can be looked at as a number of column vectors arranged in columns or as the transpose of column vectors arranged in rows. These two different points of view are useful in different manipulations. A matrix $\boldsymbol{A} \in \Re^{m\times n}$ has the form:

$$\boldsymbol{A} = \begin{bmatrix} a_{11} & a_{12} & \ldots & a_{1n} \\ a_{21} & a_{22} & \ldots & a_{2n} \\ \vdots & \vdots & \vdots & \vdots \\ a_{m1} & a_{m2} & \ldots & a_{mn} \end{bmatrix} \tag{1.9}$$

This matrix may be looked at as a number of row vectors:

$$\boldsymbol{A} = \begin{bmatrix} \boldsymbol{a}_1^T \\ \boldsymbol{a}_2^T \\ \vdots \\ \boldsymbol{a}_m^T \end{bmatrix}, \quad \text{with } \boldsymbol{a}_i = \begin{bmatrix} a_{i1} \\ a_{i2} \\ \vdots \\ a_{in} \end{bmatrix}, \quad i = 1, 2, \ldots, m \tag{1.10}$$

Alternatively, it may be looked at as the following column arrangement of column vectors:

$$\boldsymbol{A} = [\boldsymbol{a}_1 \quad \boldsymbol{a}_2 \quad \ldots \quad \boldsymbol{a}_n], \quad \text{with } \boldsymbol{a}_i = \begin{bmatrix} a_{1i} \\ a_{2i} \\ \vdots \\ a_{mi} \end{bmatrix}, \quad i = 1, 2, \ldots, n \tag{1.11}$$

The matrix is a square matrix if the number of rows and the number of columns are equal ($m = n$). Otherwise, it is a rectangular matrix. An important matrix that is used often in this book is the identity matrix $\boldsymbol{I}$ given by:

$$\boldsymbol{I}_m = \underbrace{\begin{bmatrix} 1 & 0 & 0 & 0 \\ 0 & 1 & 0 & 0 \\ 0 & 0 & \ddots & 0 \\ 0 & 0 & 0 & 1 \end{bmatrix}}_{m \text{ columns}} \tag{1.12}$$

The identity is a square diagonal matrix with only 1s on the diagonal and zeros elsewhere. Multiplying a matrix or a vector with the identity matrix does not change them. The ith column of the identity matrix $\boldsymbol{e}_i$ is a vector that has only 1 in the ith row and zero elsewhere. We use such vectors very often in optimization theory to express changes made to only one parameter at a time.

The rank of the matrix is the number of linearly independent rows or columns of the matrix. By linear independence I mean that none of these vectors can be written as a linear sum of the other vectors. A matrix is said to be full rank if its rank equals min(m,n). A square matrix has full rank if all its rows and columns are linearly independent. The inverse of a square full rank matrix $\boldsymbol{A}$ is another matrix $\boldsymbol{A}^{-1}$ that satisfies:

$$\boldsymbol{A}\boldsymbol{A}^{-1} = \boldsymbol{A}^{-1}\boldsymbol{A} = \boldsymbol{I} \tag{1.13}$$

The inverse is not defined for a square matrix with linearly dependent columns or rows. Right multiplying a matrix with a vector is defined by:

$$b = Ax = \begin{bmatrix} a_{11} & a_{12} & \dots & a_{1n} \\ a_{21} & a_{22} & \dots & a_{2n} \\ \vdots & \vdots & \vdots & \vdots \\ a_{m1} & a_{m2} & \dots & a_{mn} \end{bmatrix} \begin{bmatrix} x_1 \\ x_2 \\ \vdots \\ x_n \end{bmatrix} = \begin{bmatrix} a_{11}x_1 + a_{12}x_2 + \dots + a_{1n}x_n \\ a_{21}x_1 + a_{22}x_2 + \dots + a_{2n}x_n \\ \vdots \\ a_{m1}x_1 + a_{m2}x_2 + \dots + a_{mn}x_n \end{bmatrix} \tag{1.14}$$

The product of a matrix $A \in \Re^{m \times n}$ with a vector $x \in \Re^n$ is a vector $b \in \Re^m$. The number of columns in the matrix A and the number of rows in the vector x must be equal for multiplication to be possible. It is obvious from the formula (1.14) that a matrix-vector multiplication is equivalent to taking the inner product between every row of the matrix and the vector. Another way of looking at (1.14) is to consider this product as a linear combination of the columns of the matrix with coefficients equal to the vector components:

$$\begin{bmatrix} a_{11}x_1 + a_{12}x_2 + \dots + a_{1n}x_n \\ a_{21}x_1 + a_{22}x_2 + \dots + a_{2n}x_n \\ \vdots \\ a_{m1}x_1 + a_{m2}x_2 + \dots + a_{mn}x_n \end{bmatrix} = \begin{bmatrix} a_{11} \\ a_{21} \\ \vdots \\ a_{m1} \end{bmatrix} x_1 + \begin{bmatrix} a_{12} \\ a_{22} \\ \vdots \\ a_{m2} \end{bmatrix} x_2 + \dots + \begin{bmatrix} a_{1n} \\ a_{2n} \\ \vdots \\ a_{mn} \end{bmatrix} x_n \tag{1.15}$$

This way of looking at matrix-vector product is useful in many applications. We call all possible linear combinations of the columns of a matrix the range of the matrix.

Multiplying a square matrix $A \in \Re^{m \times m}$ by a vector $x \in \Re^m$ results in a vector $y \in \Re^m$ that is usually different from x unless it is an eigenvector of the matrix A. In other words, the matrix A "operates" on the vector x and causes a change in its magnitude and direction. If the matrix A rotates any vector x by less than 90°, we call this matrix a positive definite matrix. It follows that a matrix A is positive definite if it satisfies:

$$x^T A x > 0, \quad \forall x \neq 0 \tag{1.16}$$

To prove that a matrix is indeed positive definite, one does not have to check (1.16) for all non-zero vectors. Rather, checking the sign of the eigenvalues of the matrix A and showing that they are all positive are sufficient to prove positive definiteness.

Positive definiteness is an important concept in optimization theory. It is used for checking optimality of obtained solutions. A positive semi-definite matrix is the one that can rotate a vector by up to 90°. In this case, the vector and its rotated counterpart can be orthogonal to each other. In this case, (1.16) is relaxed to satisfy:

$$x^T A x \geq 0, \quad \forall x \tag{1.17}$$

A diagonally dominant symmetric matrix with positive diagonal elements can be shown to be positive semi-definite. A negative definite matrix is the negative of a positive definite matrix. The same applies for a negative semi-definite matrix.

The following example illustrates some of the discussed properties of matrix operations.

Example 1.2: Consider the matrix:

$$A = \begin{bmatrix} 2 & -1 & 0 \\ -1 & 2 & -1 \\ 0 & -1 & 2 \end{bmatrix}$$

and the vectors:

$$\boldsymbol{x} = \begin{bmatrix} 1 \\ -1 \\ 2 \end{bmatrix} \quad \text{and} \quad \boldsymbol{b} = \begin{bmatrix} 2 \\ 3 \\ -2 \end{bmatrix}$$

Evaluate the quadratic forms $\boldsymbol{x}^T \boldsymbol{A}\boldsymbol{x}$ and $\boldsymbol{b}^T \boldsymbol{A}\boldsymbol{b}$.

Solution: In this example, we check the definiteness of the matrix by considering two vector samples. For the vector $\boldsymbol{x}$, we have:

$$\begin{aligned} \boldsymbol{y} = \boldsymbol{A}\boldsymbol{x} &= \begin{bmatrix} 2 & -1 & 0 \\ -1 & 2 & -1 \\ 0 & -1 & 2 \end{bmatrix}\begin{bmatrix} 1 \\ -1 \\ 2 \end{bmatrix} = \begin{bmatrix} 3 \\ -5 \\ 5 \end{bmatrix} \Rightarrow \boldsymbol{x}^T\boldsymbol{y} = \boldsymbol{x}^T\boldsymbol{A}\boldsymbol{x} \\ &= \begin{bmatrix} 1 & -1 & 2 \end{bmatrix}\begin{bmatrix} 3 \\ -5 \\ 5 \end{bmatrix} = 18 \end{aligned} \tag{1.18}$$

For the vector $\boldsymbol{b}$ we have:

$$\begin{aligned} \boldsymbol{c} = \boldsymbol{A}\boldsymbol{b} &= \begin{bmatrix} 2 & -1 & 0 \\ -1 & 2 & -1 \\ 0 & -1 & 2 \end{bmatrix}\begin{bmatrix} 2 \\ 3 \\ -2 \end{bmatrix} = \begin{bmatrix} 1 \\ 6 \\ -7 \end{bmatrix} \Rightarrow \boldsymbol{b}^T\boldsymbol{c} = \boldsymbol{b}^T\boldsymbol{A}\boldsymbol{b} \\ &= \begin{bmatrix} 2 & 3 & -2 \end{bmatrix}\begin{bmatrix} 1 \\ 6 \\ -7 \end{bmatrix} = 34 \end{aligned} \tag{1.19}$$

For both cases, it is evident that both vectors are rotated by less than 90°. For this matrix, it can be shown that its eigenvalues are all positive and given by 0.5858, 2.0000, 3.4142. It follows that indeed this matrix is a positive definite matrix.

1.4 The solution of linear systems of equations

One of the main problems that frequently appears in electrical engineering is the solution of a linear system of equations. When writing the nodal equations of a circuit, we solve a linear system of equations for the unknown currents. When writing the loop equations of an electric circuit we solve a linear system of equations for the different unknown voltages. Similar applications can be found in many other electrical applications. A linear system of n equations in n unknowns is given by:

$$\begin{aligned} a_{11}x_1 + a_{12}x_2 + \cdots + a_{1n}x_n &= b_1 \\ a_{21}x_1 + a_{22}x_2 + \cdots + a_{2n}x_n &= b_2 \\ &\vdots \\ a_{n1}x_1 + a_{n2}x_2 + \cdots + a_{nn}x_n &= b_n \end{aligned} \tag{1.20}$$

These equations impose n constraints on the vector of unknowns $\boldsymbol{x}$. In matrix form, (1.20) is given by:

$$\boldsymbol{Ax} = \boldsymbol{b} \tag{1.21}$$

If the n equations in (1.20) are independent from one another, then the rows of the matrix $\boldsymbol{A}$ are linearly independent. In this case, the matrix $\boldsymbol{A}$ is invertible and the solution of the linear system of equations is obtained by multiplying both sides of (1.21) by the inverse $\boldsymbol{A}^{-1}$ to get:

$$\boldsymbol{x}^* = \boldsymbol{A}^{-1}\boldsymbol{b} \tag{1.22}$$

Another approach for solving the system (1.20) that does not require direct evaluation of the inverse matrix is to apply elementary row operations on (1.20). The basic concept of these operations is that a system of equations does not change by dividing an equation by a constant or by adding a multiple of one equation to another equation. Using these operations we make one variable responsible for only one equation. We scale this equation to make the coefficient of this variable in this equation unity. We add to all other equations multiple of this equation to make the coefficients of that variable in all other equations zero. Repeating this for all variables one by one we end up with the diagonalized system of equations:

$$\begin{aligned} x_1 \qquad\qquad\qquad &= x_1^* \\ x_2 \qquad\qquad &= x_2^* \\ \ddots \qquad & \\ x_n &= x_3^* \end{aligned} \tag{1.23}$$

The elimination technique is illustrated through the following example.

Example 1.3: Utilize basic row operations to solve the system of equations:

$$\begin{aligned} 2x_1 + 3x_2 + x_3 + x_4 &= 4 \\ x_1 + x_2 + 2x_3 - x_4 &= 1 \\ 3x_1 - 2x_2 + 2x_3 + 3x_4 &= 18 \\ 2x_1 - x_2 - 2x_3 + 2x_4 &= 5 \end{aligned} \tag{1.24}$$

Solution: We first build the augmented matrix by adding the vector $\boldsymbol{b}$ as the $(n+1)$st column of the matrix $\boldsymbol{A}$ to get Tableau 1.

The variable x_1 is assigned to the first equation. We thus make the coefficient of this variable in this equation unity by dividing the first equation by 2. We then force the coefficients of x_1 in all other equations to zero by multiplying the scaled first equation by -1 and adding it to the second equation, multiplying it by -3 and

Tableau 1

x_1	x_2	x_3	x_4	RHS
2	3	1	1	4
1	1	2	−1	1
3	−2	2	3	18
2	−1	−2	2	5

adding it to the third equation, and multiplying it by −2 and adding it to the fourth equation. This gives Tableau 2.

Tableau 2

x_1	x_2	x_3	x_4	RHS
1	3/2	1/2	1/2	2
0	**−1/2**	3/2	−3/2	−1
0	−13/2	1/2	3/2	12
0	−4	−3	1	1

We then assign the variable x_2 to the second equation. We make the coefficient of this variable in this equation unity by multiplying the second equation by −2. We then force the coefficients of x_2 in all other equations to zero by multiplying the scaled second equation by −3/2 and adding it to the first equation, multiplying it by 13/2 and adding it to the third equation, and multiplying it by 4 and adding it to the fourth equation. This gives Tableau 3.

Tableau 3

x_1	x_2	x_3	x_4	RHS
1	0	5	−4	−1
0	1	−3	3	2
0	0	**−19**	21	25
0	0	−15	13	9

Similarly, we assign the variable x_3 to the third equation and the variable x_4 to get the Tableaus 4 and 5:

Tableau 4

x_1	x_2	x_3	x_4	RHS
1	0	0	29/19	106/19
0	1	0	−6/19	−37/19
0	0	1	−21/19	−25/19
0	0	0	**−68/19**	−204/19

Tableau 5

x_1	x_2	x_3	x_4	RHS
1	0	0	0	1
0	1	0	0	−1
0	0	1	0	2
0	0	0	1	3

It follows that the solution is $x^* = [1 \quad -1 \quad 2 \quad 3]^T$. This means that the coefficients in the last column give the expansion of the vector $\boldsymbol{b}$ in the initial tableau in terms of the columns of the matrix in the final tableau, i.e.:

$$\begin{bmatrix} 4 \\ 1 \\ 18 \\ 5 \end{bmatrix} = 1 \times \begin{bmatrix} 2 \\ 1 \\ 3 \\ 2 \end{bmatrix} - 1 \times \begin{bmatrix} 3 \\ 1 \\ -2 \\ -1 \end{bmatrix} + 2 \times \begin{bmatrix} 1 \\ 2 \\ 2 \\ -2 \end{bmatrix} + 3 \times \begin{bmatrix} 1 \\ -1 \\ 3 \\ 2 \end{bmatrix} \tag{1.25}$$

Actually, if the original matrix is augmented by more columns other than $\boldsymbol{b}$, the components of each column at the final tableau are the expansion of that column in terms of the columns corresponding to the identity matrix in the initial tableau. This property is important in the Simplex algorithm which will be discussed in Chapter 2.

The same concept applies as well to rectangular full-rank matrices. For example, consider the system of m equations in n unknowns with $m < n$:

$$\begin{aligned} a_{11}x_1 + a_{12}x_2 + \cdots + a_{1n}x_n &= b_1 \\ a_{21}x_1 + a_{22}x_2 + \cdots + a_{2n}x_n &= b_2 \\ &\vdots \\ a_{m1}x_1 + a_{m2}x_2 + \cdots + a_{mn}x_n &= b_m \end{aligned} \tag{1.26}$$

Pivoting can be applied to this system similar to the square matrix case. If we assign the first m variables to the first m equations, one gets:

$$\begin{array}{llllllll} x_1 + & & & & a_{1,m+1}x_{m+1} & a_{1,m+2}x_{m+2} & \cdots & a_{1,n}x_n = x_1^* \\ & x_2 + & & & \vdots & \ddots & & \vdots \\ & & \ddots & & \vdots & & \ddots & \vdots \\ & & & x_m + & a_{m,m+1}x_{m+1} & a_{m,m+2}x_{m+2} & & a_{m,n}x_n = x_m^* \end{array} \tag{1.27}$$

This system has infinite number of solutions. We can obtain some of these solutions by assigning values to $(n-m)$ variables and then solving for the remaining m variables. For example, in (1.27) if we set the last $(n-m)$ variables to zero, we obtain the solution $\boldsymbol{x} = \begin{bmatrix} x_1^* & x_2^* & \ldots & x_m^* & \boldsymbol{0}_{n-m}^T \end{bmatrix}^T$, where $\boldsymbol{0}_{n-m}$ is a vector of zeros of size $(n-m)$.

In (1.23) and (1.27), we assume that the ith variable is assigned to the ith equation resulting in an identity in the first m rows. However, this is not a limitation. The pth variable can be assigned to the qth equation. Pivoting in this case results in a unity coefficient at the location a_{qp} and zero in all other components a_{jp}, $j \neq q$. This results in a permuted identity matrix, but all the concepts discussed earlier still hold.

The following example illustrates the solution of rectangular systems of equations.

Example 1.4: Obtain a solution of the underdetermined system:

$$\begin{aligned} 3.0x_1 - x_2 + 3x_3 \quad\quad &= 25 \\ 0.4x_1 + 0.6x_2 \ + x_4 \quad &= 8.5 \\ 3.0x_1 + 6.0x_2 \quad\quad + x_5 &= 70 \end{aligned}$$

through basic row operations.

Solution: This problem has three equations and five unknowns. It thus has infinite number of solutions. Tableau 1 of this problem is first constructed as shown:

Tableau 1

x_1	x_2	x_3	x_4	x_5	**RHS**
3.0	**−1.0**	3.0	0	0	25
0.4	0.6	0	1.0	0	8.5
3.0	6.0	0	0	1.0	70

First, we assign x_2 ($p=2$) to the first equation ($q=1$) to get Tableau 2:

Tableau 2

x_1	x_2	x_3	x_4	x_5	**RHS**
−3.0	1.0	−3.0	0	0	−25
2.2	0	1.8	1.0	0	23.5
21.0	0	18	0	1.0	220

By setting $x_1 = x_3 = 0$, we have the solution $\boldsymbol{x} = [0 \quad -25 \quad 0 \quad 23.5 \quad 220]^T$. As stated earlier, the coefficients of the first column in the last tableau represent the expansion coefficients of this column in the initial tableau in terms of the columns corresponding to the identity in the first tableau, i.e.:

$$\begin{bmatrix} 3.0 \\ 0.4 \\ 3 \end{bmatrix} = -3\begin{bmatrix} -1 \\ 0.6 \\ 6 \end{bmatrix} + 2.2\begin{bmatrix} 0 \\ 1 \\ 0 \end{bmatrix} + 21\begin{bmatrix} 0 \\ 0 \\ 1 \end{bmatrix}$$

Similarly for the third and the RHS columns, we have, respectively:

$$\begin{bmatrix} 3.0 \\ 0 \\ 0 \end{bmatrix} = -3\begin{bmatrix} -1 \\ 0.6 \\ 6 \end{bmatrix} + 1.8\begin{bmatrix} 0 \\ 1 \\ 0 \end{bmatrix} + 18\begin{bmatrix} 0 \\ 0 \\ 1 \end{bmatrix}$$

$$\begin{bmatrix} 25 \\ 8.5 \\ 70 \end{bmatrix} = -25\begin{bmatrix} -1 \\ 0.6 \\ 6 \end{bmatrix} + 23.5\begin{bmatrix} 0 \\ 1 \\ 0 \end{bmatrix} + 220\begin{bmatrix} 0 \\ 0 \\ 1 \end{bmatrix}$$

These properties of the basic row operations will be utilized later in the Simplex method for solving linear programs.

1.5 Derivatives

The concepts associated with derivatives of scalars, vectors, and matrices are important in optimization theory and they find applications in many proofs. In this book, we consider, in general, differentiable and continuous functions of n variables $\boldsymbol{x} = [x_1 \ \ x_2 \ \ \ldots \ \ x_n]^T$. These functions are denoted as $f(\boldsymbol{x})$ or $f(x_1, x_2, \ldots, x_n)$. The function f may be the gain of an amplifier, the return loss of a microwave circuit, or the real part of the input impedance of an antenna. Changing one of the parameters affects the value of f. The derivative with respect to the ith parameter tells us how sensitive the function is with respect to that parameter.

1.5.1 Derivative approximation

Mathematically speaking, the first-order derivatives are obtained by perturbing the parameters one at a time and determining the corresponding relative change of the function. The derivative with respect to the ith parameter is given by:

$$\frac{\partial f}{\partial x_i} = \underset{\Delta x_i \to 0}{\text{limit}} \frac{f(x_1, x_2, \ldots, x_i + \Delta x_i, \ldots, x_n) - f(x_1, x_2, \ldots, x_i, \ldots, x_n)}{\Delta x_i} \tag{1.28}$$

The formula (1.28) means that the perturbation should be as small as possible to get an accurate result. The formula (1.28) can be approximated by the forward difference formula:

$$\frac{\partial f}{\partial x_i} \approx \frac{f(x_1, x_2, \ldots, x_i + \Delta x_i, \ldots, x_n) - f(x_1, x_2, \ldots, x_i, \ldots, x_n)}{\Delta x_i} \tag{1.29}$$

where a finite small perturbation Δx_i is used. This formula is denoted as the forward difference finite difference (FFD) formula. Other formulas that can be utilized for approximating the first-order derivatives include the backward finite difference (BFD) and central finite difference (CFD) given by, respectively:

$$\frac{\partial f}{\partial x_i} \approx \frac{f(x_1, x_2, \ldots, x_i, \ldots, x_n) - f(x_1, x_2, \ldots, x_i - \Delta x_i, \ldots, x_n)}{\Delta x_i} \tag{1.30}$$

$$\frac{\partial f}{\partial x_i} \approx \frac{f(x_1, x_2, \ldots, x_i + \Delta x_i, \ldots, x_n) - f(x_1, x_2, \ldots, x_i - \Delta x_i, \ldots, x_n)}{2\Delta x_i} \tag{1.31}$$

The FFD approach requires one extra function evaluation per parameter in the forward direction. The BFD formula requires one extra function evaluation per parameter in the backward direction. The CFD approximation is the average of both the FFD and BFD approximations. It offers a better accuracy for derivative estimation at the expense of using two extra function evaluations per parameter. To estimate derivatives relative to all parameters, the FFD and the BFD require n extra function evaluations while the CFD requires $2n$ extra function evaluations.

In many practical engineering problems, the function is evaluated through a numerical simulation with a finite accuracy. It follows that the finite difference formulas (1.29)–(1.31) should be used with a carefully selected perturbation. Using a too small perturbation may result in an inaccurate result because of the numerical noise of the simulator. Using a large perturbation will also result in an inaccurate result. Also, the computational cost of one simulation may be too expensive. In that case the less accurate FFD or BFD is preferable over the costly but more accurate CFD. We will discuss in Chapter 10 the adjoint sensitivity approaches that offer more efficient approaches for sensitivity estimation.

1.5.2 The gradient

The gradient of a function $\nabla f(\boldsymbol{x})$ is the column vector of its derivatives relative to all parameters. It is given by:

$$\nabla f = \begin{bmatrix} \frac{\partial f}{\partial x_1} & \frac{\partial f}{\partial x_2} & \cdots & \frac{\partial f}{\partial x_n} \end{bmatrix}^T \tag{1.32}$$

The gradient varies in general from one point in the parameter space to another. This vector plays an important role in optimization theory as will become clear in the following chapters.

A differentiable nonlinear function can always be approximated within a small neighborhood of a point $\boldsymbol{x}_o$ by the first-order Taylor expansion:

$$\begin{aligned} f(\boldsymbol{x} + \Delta\boldsymbol{x}) &= L(\Delta\boldsymbol{x}) = f(\boldsymbol{x}_o) + \nabla f^T \Delta\boldsymbol{x} \\ &= f(\boldsymbol{x}_o) + \frac{\partial f}{\partial x_1}\Delta x_1 + \frac{\partial f}{\partial x_2}\Delta x_2 + \cdots + \frac{\partial f}{\partial x_n}\Delta x_n \end{aligned} \tag{1.33}$$

where $\Delta\boldsymbol{x} = \boldsymbol{x} - \boldsymbol{x}_o = [\Delta x_1 \quad \Delta x_2 \quad \ldots \quad \Delta x_n]^T$. All the first derivatives in (1.33) are evaluated at the point $\boldsymbol{x}_o$. The linearization $L(\Delta\boldsymbol{x})$ approximates the original nonlinear function only over a small neighborhood of the point $\boldsymbol{x}_o$. We will discuss in Chapter 3 in detail the different properties of the gradient.

The following example illustrates the concepts of the gradient and linearizations of nonlinear functions.

Example 1.5: Consider the function:

$$f(\boldsymbol{x}) = (x_1 - 2)^2 + (x_2 - 1)^2 - x_1 x_2 \tag{1.34}$$

Evaluate the gradient of the function at the point $\boldsymbol{x}_o = [1.0 \quad 1.0]^T$ both analytically and numerically. Construct a linearization of the function $f(\boldsymbol{x})$ at this point and test its accuracy.

Solution: The gradient of this function at any point $\boldsymbol{x}$ is given by:

$$\nabla f = \begin{bmatrix} \frac{\partial f}{\partial x_1} \\ \frac{\partial f}{\partial x_2} \end{bmatrix} = \begin{bmatrix} 2(x_1 - 2) - x_2 \\ 2(x_2 - 1) - x_1 \end{bmatrix} \tag{1.35}$$

At the point $\boldsymbol{x}_o = [1.0 \quad 1.0]^T$, the gradient is obtained by setting $x_1 = 1.0$ and $x_2 = 1.0$ in (1.35) to get:

$$\nabla f = \begin{bmatrix} 2(1-2) - 1 \\ 2(1-1) - 1 \end{bmatrix} = \begin{bmatrix} -3 \\ -1 \end{bmatrix} \tag{1.36}$$

The same answer can be obtained through FFD, BFD, or CFD approximations. We utilize a perturbation of $\Delta x = 0.05$ for both parameters. The function values at the original point and the perturbed points are given by:

$$\begin{aligned} f(1.0, 1.0) &= 0 \\ f(\boldsymbol{x}_o + \Delta x \boldsymbol{e}_1) = f(1.05, 1) &= -0.1475 \\ f(\boldsymbol{x}_o - \Delta x \boldsymbol{e}_1) = f(0.95, 1) &= 0.1525 \\ f(\boldsymbol{x}_o + \Delta x \boldsymbol{e}_2) = f(1.0, 1.05) &= -0.0475 \\ f(\boldsymbol{x}_o - \Delta x \boldsymbol{e}_2) = f(1.0, 0.95) &= 0.0525 \end{aligned} \tag{1.37}$$

where $\boldsymbol{e}_1$ and $\boldsymbol{e}_2$ are the first and second column of the identity matrix $\boldsymbol{I}_2$. The forward, backward, and central finite difference approximations of the gradient are thus given by:

$$\begin{aligned} \text{FFD} &= \begin{bmatrix} \dfrac{f(\boldsymbol{x}_o + \Delta x \boldsymbol{e}_1) - f(\boldsymbol{x}_o)}{\Delta x} \\ \dfrac{f(\boldsymbol{x}_o + \Delta x \boldsymbol{e}_2) - f(\boldsymbol{x}_o)}{\Delta x} \end{bmatrix} = \begin{bmatrix} \dfrac{-0.1475 - 0}{0.05} \\ \dfrac{-0.0475 - 0}{0.05} \end{bmatrix} = \begin{bmatrix} -2.95 \\ -0.95 \end{bmatrix} \\ \text{BFD} &= \begin{bmatrix} \dfrac{f(\boldsymbol{x}_o) - f(\boldsymbol{x}_o - \Delta x \boldsymbol{e}_1)}{\Delta x} \\ \dfrac{f(\boldsymbol{x}_o) - f(\boldsymbol{x}_o - \Delta x \boldsymbol{e}_2)}{\Delta x} \end{bmatrix} = \begin{bmatrix} \dfrac{0 - 0.1525}{0.05} \\ \dfrac{0 - 0.0525}{0.05} \end{bmatrix} = \begin{bmatrix} -3.05 \\ -1.05 \end{bmatrix} \\ \text{CFD} &= \begin{bmatrix} \dfrac{f(\boldsymbol{x}_o + \Delta x \boldsymbol{e}_1) - f(\boldsymbol{x}_o - \Delta x \boldsymbol{e}_1)}{2\Delta x} \\ \dfrac{f(\boldsymbol{x}_o + \Delta x \boldsymbol{e}_2) - f(\boldsymbol{x}_o - \Delta x \boldsymbol{e}_2)}{2\Delta x} \end{bmatrix} = \begin{bmatrix} \dfrac{-0.1475 - 0.1525}{0.1} \\ \dfrac{-0.0475 - 0.0525}{0.1} \end{bmatrix} \\ &= \begin{bmatrix} -3.0 \\ -1.0 \end{bmatrix} \end{aligned} \tag{1.38}$$

It can be seen the gradient estimated through the CFD approximation is more accurate than the other approaches. Actually, in this example, it is identical to the analytical gradient. This comes at the expense of requiring four extra function evaluations.

At the point $\boldsymbol{x}_o$, a linearization of the function is given by (1.33) as follows:

$$\begin{aligned} L(\boldsymbol{x}) = L(\Delta \boldsymbol{x}) &= f(\boldsymbol{x}_o) + \nabla f^T \Delta \boldsymbol{x} = 0 - 3\Delta x_1 - \Delta x_2 \\ &= -3(x_1 - 1) - 1(x_2 - 1) \end{aligned} \tag{1.39}$$

Notice that $\Delta \boldsymbol{x} = \boldsymbol{x} - \boldsymbol{x}_o$ is the deviation from the expansion point. Notice also that $L(\boldsymbol{x} = \boldsymbol{x}_o) = 0$ implying that linearized function has the same value as the original function at the expansion point. Also we have $L(\boldsymbol{x} = [1.1 \quad 1.1]^T) = -0.4$. The original function value at this point is $f(\boldsymbol{x} = [1.1 \quad 1.1]^T) = -0.39$. It follows that the linearization approximates the original function well at this point. At the point $\boldsymbol{x} = [2.0 \quad 2.0]^T$, both the linearization and the original function have the values $L(\boldsymbol{x} = [2.0 \quad 2.0]^T) = -4$ and $f(\boldsymbol{x} = [2.0 \quad 2.0]^T) = -3$. It is obvious that as the perturbation Δx becomes larger, the accuracy of the linearization starts to drop. Linearizations are usually used within a "trust region" around the expansion point to guarantee reasonable accuracy.

1.5.3 The Jacobian

In many engineering applications, we consider at the same time a number of functions. These functions may, for example, represent the response of a circuit at a number of frequencies. In this case, we have a vector of functions where each function depends on the parameters. A vector of m functions is given by:

$$\boldsymbol{f}(\boldsymbol{x}) = [f_1 \quad f_2 \quad \dots \quad f_m]^T \tag{1.40}$$

Every one of these functions has its gradient with respect to the parameters $\boldsymbol{x}$. The Jacobian matrix is the matrix combining all these gradients together and is given by:

$$\boldsymbol{J}(\boldsymbol{x}) = \begin{bmatrix} \nabla f_1^T \\ \nabla f_2^T \\ \vdots \\ \nabla f_m^T \end{bmatrix} = \begin{bmatrix} \dfrac{\partial f_1}{\partial x_1} & \dfrac{\partial f_1}{\partial x_2} & \cdots & \dfrac{\partial f_1}{\partial x_n} \\ \dfrac{\partial f_2}{\partial x_1} & \dfrac{\partial f_2}{\partial x_2} & \cdots & \dfrac{\partial f_2}{\partial x_n} \\ \vdots & \vdots & \vdots & \vdots \\ \dfrac{\partial f_m}{\partial x_1} & \dfrac{\partial f_m}{\partial x_2} & \cdots & \dfrac{\partial f_m}{\partial x_n} \end{bmatrix} \tag{1.41}$$

As evident from (1.41), the Jacobian matrix is a rectangular matrix of m rows and n columns. The ith row of the Jacobian is the transpose of the gradient of the function f_i. The linearization (1.33) can be generalized for the vector $\boldsymbol{f}$ to get the expansion:

$$\boldsymbol{f}(\boldsymbol{x} + \Delta\boldsymbol{x}) \approx \boldsymbol{f}(\boldsymbol{x}) + \boldsymbol{J}(\boldsymbol{x})\Delta\boldsymbol{x} \tag{1.42}$$

The following example illustrates the calculation of the Jacobian and its utilization in approximating functions.

Example 1.6: Consider the vector of functions $\boldsymbol{f}(\boldsymbol{x}) = [3x_1^2 + x_2^3 \quad x_1^3 - 2x_3]^T$. Evaluate the Jacobian at the point $\boldsymbol{x}_o = [2.0 \quad 1.0 \quad 1.0]^T$. Utilize this Jacobian to estimate the value of this vector function at the point $\boldsymbol{x}_1 = [2.1 \quad 1.2 \quad 0.9]^T$.

Solution: For this problem, we have two functions ($m = 2$) and three variables ($n = 3$). It follows that the Jacobian matrix has two rows and three columns. Each row is the transpose of the gradient of the corresponding function. The value of vector of functions at the expansion point is $f(\boldsymbol{x}_o) = [13.0 \quad 6.0]^T$. The Jacobian at a general point $\boldsymbol{x}$ is thus given by:

$$\boldsymbol{J}(\boldsymbol{x}) = \begin{bmatrix} \nabla f_1^T \\ \nabla f_2^T \end{bmatrix} = \begin{bmatrix} 6x_1 & 3x_2^2 & 0 \\ 3x_1^2 & 0 & -2 \end{bmatrix} \tag{1.43}$$

At the point $\boldsymbol{x}_o$, the Jacobian is given by:

$$\boldsymbol{J}(\boldsymbol{x}_o) = \begin{bmatrix} \nabla f_1^T \\ \nabla f_2^T \end{bmatrix} = \begin{bmatrix} 12 & 3 & 0 \\ 12 & 0 & -2 \end{bmatrix} \tag{1.44}$$

Utilizing the formula (1.42), the value of the function vector at the new point is given by:

$$\begin{gathered} \boldsymbol{f}(\boldsymbol{x}_1) \approx \boldsymbol{f}(\boldsymbol{x}_o) + \boldsymbol{J}(\boldsymbol{x}_o)(\boldsymbol{x}_1 - \boldsymbol{x}_o) \\ \Downarrow \\ \boldsymbol{f}(\boldsymbol{x}_1) \approx \begin{bmatrix} 13 \\ 6 \end{bmatrix} + \begin{bmatrix} 12 & 3 & 0 \\ 12 & 0 & -2 \end{bmatrix} \left(\begin{bmatrix} 2.1 \\ 1.2 \\ 0.9 \end{bmatrix} - \begin{bmatrix} 2.0 \\ 1.0 \\ 1.0 \end{bmatrix} \right) = \begin{bmatrix} 14.8 \\ 7.4 \end{bmatrix} \end{gathered} \tag{1.45}$$

The exact value of the function at the new point is $\boldsymbol{f}(\boldsymbol{x}_1) = [14.958 \quad 7.46]^T$. The accuracy of the linearization formula (1.42) continues to deteriorate as we move further away from the expansion point as expected.

1.5.4 Second-order derivatives

The derivatives discussed so far are first-order derivatives. They require only perturbing one parameter at a time. The second-order derivatives are, by definition, the rate of change of the first-order derivatives. For a function $f(\boldsymbol{x})$ with n-parameters, the mixed second-order derivative with respect to the ith and jth parameters is given by:

$$\frac{\partial^2 f}{\partial x_i \partial x_j} = \underset{\Delta x_j \to 0}{\text{limit}} \frac{\dfrac{\partial f(x_1, x_2, \ldots, x_j + \Delta x_j, \ldots, x_n)}{\partial x_i} - \dfrac{\partial f(x_1, x_2, \ldots, x_i, \ldots, x_n)}{\partial x_i}}{\Delta x_j} \tag{1.46}$$

In other words, the derivative $\partial f/\partial x_i$ is calculated for two different values of the parameter x_j and their relative difference is calculated. The same value can be obtained if the derivative $\partial f/\partial x_j$ was differentiated relative to x_i as the order of differentiation does not matter. These second-order derivatives play an important role in some optimization algorithms. They are also used in deriving optimality conditions as will become clear in following chapters.

All the second-order derivatives of a function $f(\boldsymbol{x})$ are combined in the compact Hessian matrix:

$$\boldsymbol{H} = \begin{bmatrix} \dfrac{\partial^2 f}{\partial x_1^2} & \dfrac{\partial^2 f}{\partial x_1 \partial x_2} & \cdots & \dfrac{\partial^2 f}{\partial x_1 \partial x_n} \\ \dfrac{\partial^2 f}{\partial x_2 \partial x_1} & \dfrac{\partial^2 f}{\partial x_2^2} & \cdots & \dfrac{\partial^2 f}{\partial x_2 \partial x_n} \\ \vdots & \vdots & & \vdots \\ \dfrac{\partial^2 f}{\partial x_n \partial x_1} & \dfrac{\partial^2 f}{\partial x_n \partial x_2} & \cdots & \dfrac{\partial^2 f}{\partial x_n^2} \end{bmatrix} \tag{1.47}$$

The Hessian matrix is a symmetric matrix because the order of differentiation does not make a difference. The following example illustrates the calculation of the Hessian matrix.

Example 1.7: Evaluate the gradient and the Hessian of the function $f(\boldsymbol{x}) = 3x_1^2 + 2x_1x_2 + x_1x_3 + 2.5x_2^2 + 2x_2x_3 + 2x_3^2 - 8x_1 - 3x_2 - 3x_3$ at the point $\boldsymbol{x}_o = [1.0 \quad -1.0 \quad 1.0]^T$.

Solution: Utilizing the gradient definition (1.32) we get the gradient at an arbitrary point:

$$\nabla f(\boldsymbol{x}) = \begin{bmatrix} 6x_1 + 2x_2 + x_3 - 8 \\ 2x_1 + 5x_2 + 2x_3 - 3 \\ x_1 + 2x_2 + 4x_3 - 3 \end{bmatrix} \Rightarrow \nabla f(\boldsymbol{x}_o) = \begin{bmatrix} -3 \\ -4 \\ 0 \end{bmatrix} \tag{1.48}$$

The Hessian can be obtained using (1.47) which is equivalent to differentiating each component of the transpose of gradient with respect to all parameters. It follows that we have:

$$\boldsymbol{H} = \frac{\partial \nabla f^T}{\partial \boldsymbol{x}} = \begin{bmatrix} 6 & 2 & 1 \\ 2 & 5 & 2 \\ 1 & 2 & 4 \end{bmatrix} \tag{1.49}$$

In this case, the Hessian does not change with parameters as the function is quadratic.

1.5.5 Derivatives of vectors and matrices

In most undergraduate courses, basic rules for differentiating products or summations of functions are discussed. The same rules can be extended to scalars, vectors, and matrices that are functions of optimization variables. I show here the way the derivatives are calculated for a number of important cases.

The derivatives of a scalar function $f(\boldsymbol{x})$ with respect to the vector of parameters $\boldsymbol{x}$ is a column vector given by the gradient (1.32). If $f = \boldsymbol{a}^T\boldsymbol{x}$, a linear function of $\boldsymbol{x}$, then its gradient is obtained by applying (1.32) and is given by:

$$\frac{\partial f}{\partial \boldsymbol{x}} = \frac{\partial(\boldsymbol{a}^T\boldsymbol{x})}{\partial \boldsymbol{x}} = \frac{\partial(\boldsymbol{x}^T\boldsymbol{a})}{\partial \boldsymbol{x}} = \boldsymbol{a} \tag{1.50}$$

where $\boldsymbol{a}$ is a constant column vector. Similarly, the derivative of a row vector $\boldsymbol{g}^T$ of m components with respect to the variables $\boldsymbol{x}$ is defined as:

$$\begin{gathered} \frac{\partial \boldsymbol{g}^T}{\partial \boldsymbol{x}} = \frac{\partial}{\partial \boldsymbol{x}}[g_1 \quad g_2 \quad \ldots \quad g_m] = \begin{bmatrix} \frac{\partial g_1}{\partial \boldsymbol{x}} & \frac{\partial g_2}{\partial \boldsymbol{x}} & \cdots & \frac{\partial g_m}{\partial \boldsymbol{x}} \end{bmatrix} \\ \Downarrow \\ \frac{\partial \boldsymbol{g}^T}{\partial \boldsymbol{x}} = [\nabla g_1 \quad \nabla g_2 \quad \ldots \quad \nabla g_m] \end{gathered} \tag{1.51}$$

The derivative is thus a matrix with n rows and m columns. Notice that a column vector cannot be differentiated with respect to another column vector. A byproduct of this result is the derivative of a row vector that is a linear function of the variables of the form:

$$\frac{\partial(\boldsymbol{x}^T\boldsymbol{A})}{\partial \boldsymbol{x}} = \boldsymbol{A} \tag{1.52}$$

where $\boldsymbol{A}$ is a constant matrix. The rule (1.52) can be derived by differentiating each component of the row vector separately relative to all parameters and putting the result in a matrix form. A useful formula that can be derived using the previously discussed rules is the derivative of the quadratic term $\boldsymbol{x}^T\boldsymbol{A}\boldsymbol{x}$ with $\boldsymbol{A}$ being a square matrix. This derivative is obtained by applying differentiation and multiplication rules:

$$\frac{\partial(\boldsymbol{x}^T\boldsymbol{A}\boldsymbol{x})}{\partial\boldsymbol{x}} = \frac{\partial(\boldsymbol{x}^T\boldsymbol{A})}{\partial\boldsymbol{x}}\boldsymbol{x} + \frac{\partial(\boldsymbol{x}^T)}{\partial\boldsymbol{x}}(\boldsymbol{x}^T\boldsymbol{A})^T = \boldsymbol{A}\boldsymbol{x} + \boldsymbol{A}^T\boldsymbol{x} \tag{1.53}$$

For a symmetric matrix $\boldsymbol{A}$, (1.41) gives $2\boldsymbol{A}\boldsymbol{x}$.

Differentiating a matrix with respect to a variable implies differentiating all elements of the matrix. The derivative of a matrix $\boldsymbol{A}$ with respect to the ith parameter x_i is given by:

$$\frac{\partial A}{\partial x_i} = \begin{bmatrix} \frac{\partial a_{11}}{\partial x_i} & \frac{\partial a_{12}}{\partial x_i} & \cdots & \frac{\partial a_{1n}}{\partial x_i} \\ \frac{\partial a_{21}}{\partial x_i} & \frac{\partial a_{22}}{\partial x_i} & \cdots & \frac{\partial a_{2n}}{\partial x_i} \\ \vdots & \vdots & \vdots & \vdots \\ \frac{\partial a_{m1}}{\partial x_i} & \frac{\partial a_{m2}}{\partial x_i} & \cdots & \frac{\partial a_{mn}}{\partial x_i} \end{bmatrix} \tag{1.54}$$

The following example illustrates differentiation rules of vectors and matrices.

Example 1.8: Repeat Example 1.7 using vector differentiation rules.

Solution: The objective function in Example 1.7 can be written as:

$$\begin{aligned} f(\boldsymbol{x}) &= 0.5[x_1 \quad x_2 \quad x_3]\begin{bmatrix} 6 & 2 & 1 \\ 2 & 5 & 2 \\ 1 & 2 & 4 \end{bmatrix}\begin{bmatrix} x_1 \\ x_2 \\ x_3 \end{bmatrix} + [-8 \quad -3 \quad -2]\begin{bmatrix} x_1 \\ x_2 \\ x_3 \end{bmatrix} \\ &= 0.5\boldsymbol{x}^T\boldsymbol{A}\boldsymbol{x} + \boldsymbol{b}^T\boldsymbol{x} \end{aligned} \tag{1.55}$$

Applying the vector differentiation rules, the gradient at an arbitrary point is given by:

$$\begin{aligned} \nabla f &= \frac{\partial f}{\partial \boldsymbol{x}} = \boldsymbol{A}\boldsymbol{x} + \boldsymbol{b} = \begin{bmatrix} 6 & 2 & 1 \\ 2 & 5 & 2 \\ 1 & 2 & 4 \end{bmatrix}\begin{bmatrix} x_1 \\ x_2 \\ x_3 \end{bmatrix} + \begin{bmatrix} -8 \\ -3 \\ -2 \end{bmatrix} \\ &= \begin{bmatrix} 6x_1 + 2x_2 + x_3 - 8 \\ 2x_1 + 5x_2 + 2x_3 - 3 \\ x_1 + 2x_2 + 4x_3 - 2 \end{bmatrix} \end{aligned} \tag{1.56}$$

which is identical to the result obtained earlier. The Hessian (1.47) can be evaluated as:

$$\boldsymbol{H}(\boldsymbol{x}) = \frac{\partial \nabla f^T}{\partial \boldsymbol{x}} = \frac{\partial}{\partial \boldsymbol{x}}(\boldsymbol{x}^T\boldsymbol{A}^T + \boldsymbol{b}^T) = \boldsymbol{A}^T = \begin{bmatrix} 6 & 2 & 1 \\ 2 & 5 & 2 \\ 1 & 2 & 4 \end{bmatrix} \tag{1.57}$$

which is identical to (1.49).

1.6 Subspaces

Linear algebra concepts appear in some theorems or proofs of convergence of optimization techniques. We cover here some of the essentials of these concepts for good understanding of the rest of the material.

A fundamental concept in linear algebra is the concept of a subspace. A subset S of the n-dimensional space is a subspace if $\forall \boldsymbol{x}, \boldsymbol{y} \in S$, $\alpha\boldsymbol{x} + \beta\boldsymbol{y} \in S$, where α and β are any scalars. In other words, any combination of two vector members of the subset must also belong to this subset. Notice that implies that the origin must also belong to this subset. There are many examples of such subsets. For example, consider the hyperplane $\boldsymbol{a}^T\boldsymbol{x} = 0$ in a general n-dimensional space where $\boldsymbol{a}$ is a constant vector. The points $\boldsymbol{x}_1$ and $\boldsymbol{x}_2$ are on this hyperplane satisfying $\boldsymbol{a}^T\boldsymbol{x}_1 = 0$ and $\boldsymbol{a}^T\boldsymbol{x}_2 = 0$. If we consider any point $\boldsymbol{x}_3$ that is a linear combination of these two points with $\boldsymbol{x}_3 = \alpha\boldsymbol{x}_1 + \beta\boldsymbol{x}_2$, then it also satisfies:

$$\boldsymbol{a}^T\boldsymbol{x}_3 = \boldsymbol{a}^T(\alpha\boldsymbol{x}_1 + \beta\boldsymbol{x}_2) = \alpha\boldsymbol{a}^T\boldsymbol{x}_1 + \beta\boldsymbol{a}^T\boldsymbol{x}_2 = 0 \tag{1.58}$$

It follows that $\boldsymbol{x}_3$ also belongs to this hyperplane. This hyperplane is thus a subspace. Notice that the hyperplane $\boldsymbol{a}^T\boldsymbol{x} = b$, $b \neq 0$ is not a subspace because it does not include the origin.

An important subspace that is utilized frequently in optimization problems with constraints, linear or nonlinear, is the null subspace of a matrix $\boldsymbol{A}$, denoted by Null($\boldsymbol{A}$). To define this subspace, we start by considering a set of m given vectors $\boldsymbol{a}_i \in \Re^n$, $i = 1, 2, \ldots, m$. The set of vectors normal to all these vectors is a subspace defined by:

$$s = \{\boldsymbol{x} | \boldsymbol{a}_i^T\boldsymbol{x} = 0, \quad i = 1, 2, \ldots, m\} \tag{1.59}$$

The proof of this property is very similar to the one given in (1.58) for the case of a single vector. This property can be extended to the case of matrices. Consider a matrix $\boldsymbol{A} \in \Re^{m\times n}$. The set of all vectors satisfying $\boldsymbol{A}\boldsymbol{x} = \boldsymbol{0}$ is a subspace called the null subspace of the matrix $\boldsymbol{A}$. Using the definition of matrix vector multiplication, this set is normal to all the rows of the matrix $\boldsymbol{A}$. If $\boldsymbol{x}_1$ and $\boldsymbol{x}_2$ are vectors satisfying $\boldsymbol{A}\boldsymbol{x}_1 = \boldsymbol{0}$ and $\boldsymbol{A}\boldsymbol{x}_2 = \boldsymbol{0}$ then $\boldsymbol{x}_3 = \alpha\boldsymbol{x}_1 + \beta\boldsymbol{x}_2$ satisfies:

$$\boldsymbol{A}\boldsymbol{x}_3 = \boldsymbol{A}(\alpha\boldsymbol{x}_1 + \beta\boldsymbol{x}_2) = \alpha\boldsymbol{A}\boldsymbol{x}_1 + \beta\boldsymbol{A}\boldsymbol{x}_2 = \boldsymbol{0} \tag{1.60}$$

It follows that the set of all vectors normal to all the rows of a matrix is a subspace. Obviously, the origin belongs to this subspace. Notice that if $m = n$ and the matrix is invertible, the equation $\boldsymbol{A}\boldsymbol{x} = \boldsymbol{0}$ has only the trivial solution $\boldsymbol{x} = \boldsymbol{0}$. In this case, the null subspace contains only the origin.

Another important subspace is the range subspace of a matrix. It is defined as the set of all vectors that can be written as a linear combination of the columns of that matrix. This subspace is defined by:

$$\text{Range}(\boldsymbol{A}) = \{\boldsymbol{x} | \boldsymbol{x} = \boldsymbol{A}\boldsymbol{w}\} \tag{1.61}$$

where $\boldsymbol{w}$ is a vector of coefficients. The definition (1.61) includes the origin by setting $\boldsymbol{w} = \boldsymbol{0}$. Also, if $\boldsymbol{x}_1 = \boldsymbol{A}\boldsymbol{w}_1$ and $\boldsymbol{x}_2 = \boldsymbol{A}\boldsymbol{w}_2$ then $\boldsymbol{x}_3 = \alpha\boldsymbol{x}_1 + \beta\boldsymbol{x}_2 = \boldsymbol{A}(\alpha\boldsymbol{w}_1 + \beta\boldsymbol{w}_2) = \boldsymbol{A}\boldsymbol{w}_3$. It follows that the set (1.61) satisfies all the characteristics of a subspace.

The fundamental theory of linear algebra states that any vector in $\Re^n$ can be uniquely written as the summation of two orthogonal vectors, one in the range of $\boldsymbol{A}^T$ and the other one in the null subspace of $\boldsymbol{A}$. In other words:

$$\text{Range}(A^T) \oplus \text{Null}(A) = R^n \tag{1.62}$$

It can be shown that all vectors in the Null(**A**) are orthogonal to all vectors in the Range(A^T). In other words, (1.62) shows a decomposition of a vector in $\Re^n$ in two orthogonal components.

If the matrix **A** has a full rank, it can be shown that the projection of a vector into the range of A^T is given by:

$$x_p = Px = A^T(AA^T)^{-1}Ax \tag{1.63}$$

where **P** is a projection matrix. The component of **x** in the null subspace of **A** is given by:

$$x_n = x - x_p = x - Px = (I - A^T(AA^T)^{-1}A)x \tag{1.64}$$

The concepts of null subspace and range are illustrated by the following examples.

Example 1.9: Consider the matrix:

$$A = \begin{bmatrix} 1 & 1 & 3 \\ 1 & -1 & 1 \\ 1 & -3 & -1 \end{bmatrix}$$

Show that the vector $v = [2 \quad 1 \quad -1]^T$ lies in the null space of this matrix. Show that this vector is normal to all vectors in Range(A^T). Get the projection of the vector $w = [3 \quad 2 \quad -1]^T$ in the null subspace of **A**.

Solution: Multiplying the vector **v** by the matrix **A** we get:

$$Av = \begin{bmatrix} 1 & 1 & 3 \\ 1 & -1 & 1 \\ 1 & -3 & -1 \end{bmatrix} \begin{bmatrix} 2 \\ 1 \\ -1 \end{bmatrix} = \begin{bmatrix} 0 \\ 0 \\ 0 \end{bmatrix}$$

It follows that **v** is normal to all the rows of the matrix **A** and thus belongs to Null(**A**). All vectors in the range of A^T can be written as a linear combination of the rows of **A** as follows:

$$d = \alpha_1 \begin{bmatrix} 1 \\ 1 \\ 3 \end{bmatrix} + \alpha_2 \begin{bmatrix} 1 \\ -1 \\ 1 \end{bmatrix} + \alpha_3 \begin{bmatrix} 1 \\ -3 \\ -1 \end{bmatrix}$$

$$\Downarrow$$

$$v^T d = [2 \quad 1 \quad -1] \left(\alpha_1 \begin{bmatrix} 1 \\ 1 \\ 3 \end{bmatrix} + \alpha_2 \begin{bmatrix} 1 \\ -1 \\ 1 \end{bmatrix} + \alpha_3 \begin{bmatrix} 1 \\ -3 \\ -1 \end{bmatrix} \right) = 0$$

regardless of the values of the coefficients α_1, α_2, and α_3. As the matrix **A** has only two independent rows, we consider only the first two rows to construct the projection matrix. Using (1.64), the matrix projection in the null space of the matrix $\boldsymbol{A}$ is given by:

$$\boldsymbol{P}_n = \begin{bmatrix} 1 & 0 & 0 \\ 0 & 1 & 0 \\ 0 & 0 & 1 \end{bmatrix} - \begin{bmatrix} 1 & 1 \\ 1 & -1 \\ 3 & 1 \end{bmatrix} \left(\begin{bmatrix} 1 & 1 & 3 \\ 1 & -1 & 1 \end{bmatrix} \begin{bmatrix} 1 & 1 \\ 1 & -1 \\ 3 & 1 \end{bmatrix} \right)^{-1} \begin{bmatrix} 1 & 1 & 3 \\ 1 & -1 & 1 \end{bmatrix}$$

$$\boldsymbol{P}_n = \begin{bmatrix} 1 & 0 & 0 \\ 0 & 1 & 0 \\ 0 & 0 & 1 \end{bmatrix} - \begin{bmatrix} 1/3 & -1/3 & 1/3 \\ -1/3 & 5/6 & 1/6 \\ 1/3 & 1/6 & 5/6 \end{bmatrix} = \begin{bmatrix} 2/3 & 1/3 & -1/3 \\ 1/3 & 1/6 & -1/6 \\ -1/3 & -1/6 & 1/6 \end{bmatrix}$$

The projection of the vector $\boldsymbol{w}$ in the null space of the vector $\boldsymbol{A}$ is thus given by:

$$\boldsymbol{P}_n\boldsymbol{w} = \begin{bmatrix} 2/3 & 1/3 & -1/3 \\ 1/3 & 1/6 & -1/6 \\ -1/3 & -1/6 & 1/6 \end{bmatrix} \begin{bmatrix} 3 \\ 2 \\ -1 \end{bmatrix} = \begin{bmatrix} 3 \\ 3/2 \\ -3/2 \end{bmatrix}$$

It can be easily verified that this vector is indeed in the null space of the matrix $\boldsymbol{A}$.

1.7 Convergence rates

Optimization algorithms carry out a sequence of iterations. Starting with an initial guess of the optimal parameters $\boldsymbol{x}^{(0)}$, the algorithm creates a sequence of points $\boldsymbol{x}^{(k)}$ that should converge to an optimal set of variables $\boldsymbol{x}^*$ satisfying certain optimality conditions. The rate at which the iterations approach the optimal point is called the convergence rate of the algorithm. The faster the algorithm approaches the solution, the better the algorithm is.

An algorithm is said to have linear rate of convergence if it satisfies:

$$\|\boldsymbol{x}^{(k+1)} - \boldsymbol{x}^*\| \le r\|\boldsymbol{x}^{(k)} - \boldsymbol{x}^*\| \tag{1.65}$$

where r is a positive scalar with $r < 1$. This means that at the kth iteration, the new point $\boldsymbol{x}^{(k+1)}$ is closer to the optimal solution but the distance reduces in a linear way. Linear rate of convergence is considered slow in reaching the optimal solution. An example of a sequence that converges linearly is $x^{(k)} = 1 + (0.5)^k$. The limit of this sequence is $x^* = 1.0$. The ratio between the distances of two consecutive points to the optimal point is 0.5.

Algorithms with improved rate of convergence satisfy:

$$\|\boldsymbol{x}^{(k+1)} - \boldsymbol{x}^*\| \le r\|\boldsymbol{x}^{(k)} - \boldsymbol{x}^*\|^p \tag{1.66}$$

with $1 < p < 2$ is said to have superlinear convergence. This is the rate of convergence of the quasi-Newton optimization techniques to be covered in Chapter 7. If $p = 2$, the algorithm is said to have quadratic convergence which is considered a fast rate of convergence. At one iteration, if the distance from the current point to the optimal solution is 0.1 then in the following iteration, this distance is below 0.01. The following iteration will have a distance below 0.0001 and so on. Newton's method, which requires exact second-order sensitivities, enjoys this fast rate of convergence.

1.8 Functions and sets

Throughout this book, we make use of some properties of functions and sets. Convexity is one of the important concepts used routinely in optimization theory. A function is said to be convex if it satisfies:

$$f(\lambda \boldsymbol{x}_1 + (1-\lambda)\boldsymbol{x}_2) \leq \lambda f(\boldsymbol{x}_1) + (1-\lambda) f(\boldsymbol{x}_2) \tag{1.67}$$

with $0 \leq \lambda \leq 1.0$. In other words, the value of a function at the line connecting two points is less than the convex combination of their function values. Figure 1.1 shows an example of a convex function and a non-convex function for the one-dimensional case.

Convexity can also be defined for sets. If $\boldsymbol{x}_1$ and $\boldsymbol{x}_2$ are points in a convex set S, then any point on the line connecting them, $\boldsymbol{x} = \lambda \boldsymbol{x}_1 + (1-\lambda)\boldsymbol{x}_2$, $0 \leq \lambda \leq 1.0$, is also in that set. Interesting properties can be proven for problems with convex objective functions and convex constraints. Figure 1.2 shows an illustration of a convex and non-convex sets for the two-dimensional case.

An extreme point of a set is a point that cannot be written as a convex combination of any two other points. An example of that are the vertices of a triangular area.

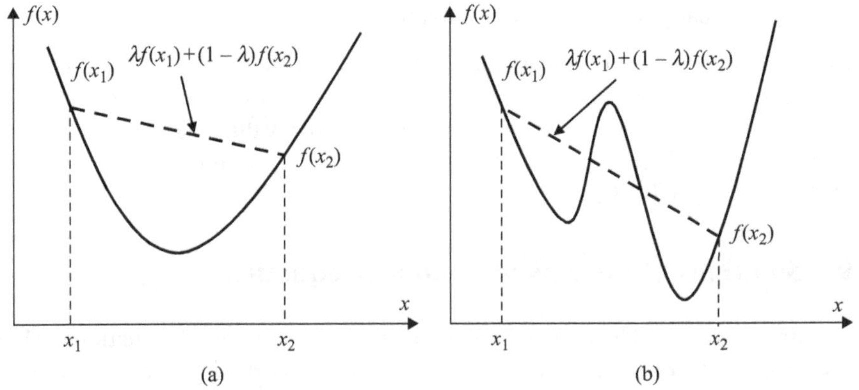

Figure 1.1 An illustration of convex and non-convex functions. (a) A convex function is shown where the function value at any point $x = \lambda x_1 + (1-\lambda)x_2$ between x_1 and x_2 ($0 \leq \lambda \leq 1$) is below the line connecting their function values $\lambda f(x_1) + (1-\lambda) f(x_2)$ and (b) a non-convex function where some of the points have function values higher than the line $\lambda f(x_1) + (1-\lambda) f(x_2)$. Notice how the convex function has only one minimum while the non-convex function has two minima

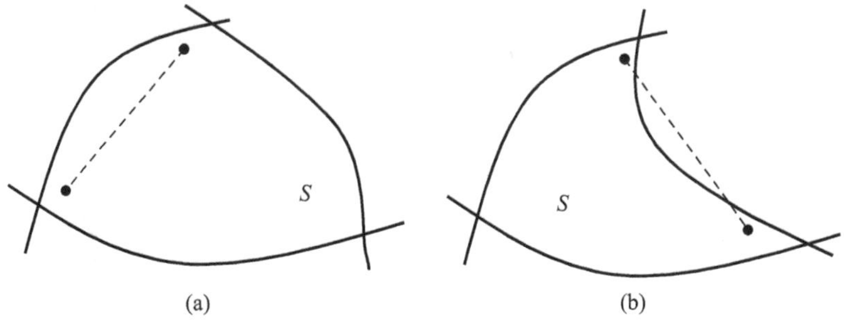

Figure 1.2 An illustration of convex and non-convex sets. (a) A convex set S that is defined by three constraints $y_i(\boldsymbol{x}) \leq 0$, $i = 1, 2, 3$. The line connecting any two points belonging to S lies entirely in this set and (b) a non-convex where the line connecting two points have some points outside the set

All points inside the triangle can be written as a convex combination of two other points except the vertices. This definition of extreme points will be utilized in Chapter 2 when discussing linear programs.

Another important property of functions is the concept of a constant value surface. For an n-dimensional function $f(\boldsymbol{x})$, the surface $f(\boldsymbol{x}) = C$, with C being a constant, defines a surface in the n-dimensional space. All points on that surface have the same objective function value (C). For two-dimensional problems, this surface becomes a contour. These contours are usually useful in obtaining graphical solutions of some problems. As will be seen later, the constant surface contours are related to the gradient at every point in the parameter space.

The following example helps illustrate the concepts of convexity and constant value surfaces.

Example 1.10: Consider the two-dimensional function $f(x_1, x_2) = x_1^2 + x_2^2$. Obtain the constant value contours of this function. Determine the minimum of this function by visually inspecting the contours.

Solution: This function represents the square of the L_2 norm of a point. All norms can be shown to be convex. The contour plot of this function is shown in Figure 1.3. We see that as we move away from the origin, the values associated with the contours increase. It can be seen from the contours that the minimum, as expected, is at the origin $\boldsymbol{x} = [0 \quad 0]^T$.

1.9 Solutions of systems of nonlinear equations

In Section 1.4, we addressed the solution of linear systems of equations. Many engineering problems may, however, result in a system of nonlinear equations of the form $\boldsymbol{f}(\boldsymbol{x}) = \boldsymbol{0}$, where $\boldsymbol{f} \in \Re^m$ is a vector of nonlinear functions. The target is to find the point in the parameter space $\boldsymbol{x} \in \Re^n$ at which all these functions have zero value.

If $m < n$, the system is underdetermined and we do not have sufficient information to solve for a unique solution. Infinite number of solutions may exist.

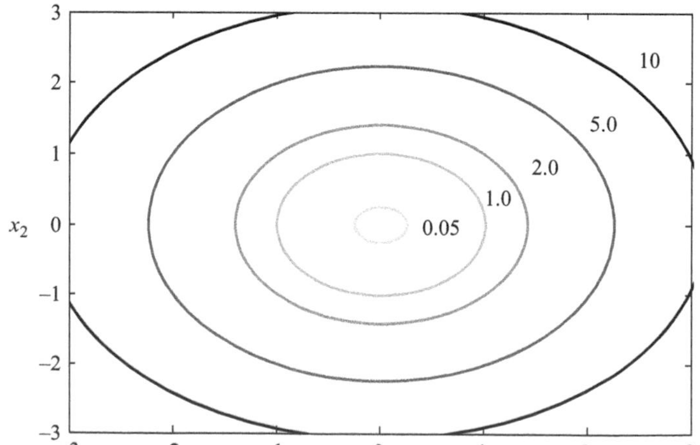

Figure 1.3 The contours of the function $f(x_1, x_2) = x_1^2 + x_2^2$

If $m > n$, the system is overdetermined and an exact solution may not exist. The problem should then be treated as an optimization problem in which we aim at minimizing $\|f\|$ at the solution. We will address the overdetermined case in Chapter 7.

We address here the Newton–Raphson technique for solving the case $m = n$. The algorithm assumes that at the kth iteration, we have the approximation to the solution $\boldsymbol{x}^{(k)}$. Writing the Taylor expansion (1.42) at this point, we have:

$$\boldsymbol{f}(\boldsymbol{x}^{(k)} + \Delta x) \approx \boldsymbol{f}(\boldsymbol{x}^{(k)}) + \boldsymbol{J}(\boldsymbol{x}^{(k)})\Delta \boldsymbol{x} \tag{1.68}$$

The algorithm selects a perturbation $\Delta\boldsymbol{x}^{(k)}$ such that the function at the new $\boldsymbol{x}^{(k)} + \Delta\boldsymbol{x}^{(k)}$ satisfies $\boldsymbol{f}(\boldsymbol{x}^{(k)} + \Delta\boldsymbol{x}^{(k)}) = \boldsymbol{0}$. Imposing this condition in (1.68) gives the equations:

$$\boldsymbol{f}(\boldsymbol{x}^{(k)}) + \boldsymbol{J}(\boldsymbol{x}^{(k)})\Delta\boldsymbol{x}^{(k)} = \boldsymbol{0} \Rightarrow \Delta\boldsymbol{x}^{(k)} = -\boldsymbol{J}(\boldsymbol{x}^{(k)})^{-1}\boldsymbol{f}(\boldsymbol{x}^{(k)}) \tag{1.69}$$

It is assumed here that the Jacobian matrix is $\boldsymbol{J} \in \Re^n$ has full rank and is thus invertible. This will be the case if the equations are independent from each other. It should be noted though that this algorithm may not converge if the starting point is far from the solution. A reasonable starting point usually guarantees convergence of this method.

The step taken by (1.69) is repeated at every iteration until the algorithm converges to the solution $\boldsymbol{x}^*$. A possible termination condition is to check that $\|\boldsymbol{f}(\boldsymbol{x}^{(k)})\|$ is small enough. The following example illustrates the application of the Newton–Raphson method.

Example 1.11: Apply the Newton–Raphson method to the solution of the system of nonlinear equations $x_2 - x_1^2 - 2 = 0$ and $x_1 - (x_2 - 2)^2 + 1 = 0$ starting from the point $\boldsymbol{x}^{(0)} = [-2.0 \quad 4.0]^T$. Carry out only three iterations.

Solution: Our target is to solve the two nonlinear equations.

$$\boldsymbol{f}(\boldsymbol{x}) = \begin{bmatrix} x_2 - x_1^2 - 2 \\ x_1 - (x_2 - 2)^2 + 1 \end{bmatrix} = \begin{bmatrix} 0 \\ 0 \end{bmatrix}$$

The Jacobian of this nonlinear function at an arbitrary point is given by:

$$\boldsymbol{J}(\boldsymbol{x}) = \begin{bmatrix} -2x_1 & 1 \\ 1 & -2(x_2 - 2) \end{bmatrix}$$

At the initial point $\boldsymbol{x}^{(0)}$, we have:

$$\boldsymbol{f}(\boldsymbol{x}^{(0)}) = \begin{bmatrix} -2 \\ -5 \end{bmatrix}, \quad \boldsymbol{J}(\boldsymbol{x}^{(0)}) = \begin{bmatrix} 4 & 1 \\ 1 & -4 \end{bmatrix}$$

The step taken is thus given by:

$$\Delta\boldsymbol{x}^{(0)} = -\begin{bmatrix} 4 & 1 \\ 1 & -4 \end{bmatrix}^{-1}\begin{bmatrix} -2 \\ -5 \end{bmatrix} = -\begin{bmatrix} 0.23529 & 0.05882 \\ 0.05882 & -0.23529 \end{bmatrix}\begin{bmatrix} -2 \\ -5 \end{bmatrix}$$

$$= \begin{bmatrix} 0.76471 \\ -1.05882 \end{bmatrix}$$

The new point is thus given by:

$$x^{(1)} = x^{(0)} + \Delta x^{(0)} = \begin{bmatrix} -2 \\ 4 \end{bmatrix} + \begin{bmatrix} 0.76471 \\ -1.05882 \end{bmatrix} = \begin{bmatrix} -1.23529 \\ 2.94118 \end{bmatrix}$$

At the point $x^{(1)}$, we have:

$$f(x^{(1)}) = \begin{bmatrix} -0.58478 \\ -1.12111 \end{bmatrix}, \quad J(x^{(1)}) = \begin{bmatrix} 2.47059 & 1.0000000 \\ 1.000000 & -1.88235 \end{bmatrix}$$

Notice that the algorithm is converging as $\|f(x^{(1)})\| \leq \|f(x^{(0)})\|$. The step taken in the second iteration is given by:

$$\begin{aligned} \Delta x^{(1)} &= -\begin{bmatrix} 2.47059 & 1.0000000 \\ 1.000000 & -1.88235 \end{bmatrix}^{-1} \begin{bmatrix} -0.58478 \\ -1.12111 \end{bmatrix} \\ &= \begin{bmatrix} 0.33313 & 0.17697 \\ 0.17697 & -0.43723 \end{bmatrix} \begin{bmatrix} -0.58478 \\ -1.12111 \end{bmatrix} = \begin{bmatrix} 0.39321 \\ -0.38669 \end{bmatrix} \end{aligned}$$

The new point is thus given by:

$$x^{(2)} = x^{(1)} + \Delta x^{(1)} = \begin{bmatrix} -1.23529 \\ 2.94118 \end{bmatrix} + \begin{bmatrix} 0.39321 \\ -0.38669 \end{bmatrix} = \begin{bmatrix} -0.84208 \\ 2.55448 \end{bmatrix}$$

Notice how the step taken is getting smaller which is also a characteristic of converging iterations. At the point $x^{(2)}$, we have:

$$f(x^{(2)}) = \begin{bmatrix} -0.15462 \\ -0.14953 \end{bmatrix}, \quad J(x^{(2)}) = \begin{bmatrix} 1.684161 & 1.00000 \\ 1.00000 & -1.10897 \end{bmatrix}$$

The step taken in the third iteration is given by:

$$\begin{aligned} \Delta x^{(2)} &= -\begin{bmatrix} 1.684161 & 1.00000 \\ 1.00000 & -1.10897 \end{bmatrix}^{-1} \begin{bmatrix} -0.15462 \\ -0.14953 \end{bmatrix} \\ &= \begin{bmatrix} 0.38671 & 0.34871 \\ 0.34871 & -0.58729 \end{bmatrix} \begin{bmatrix} -0.15462 \\ -0.14953 \end{bmatrix} = \begin{bmatrix} 0.11194 \\ -0.03390 \end{bmatrix} \end{aligned}$$

The last point is given by:

$$x^{(3)} = x^{(2)} + \Delta x^{(2)} = \begin{bmatrix} -0.84208 \\ 2.55448 \end{bmatrix} + \begin{bmatrix} 0.11194 \\ -0.03390 \end{bmatrix} = \begin{bmatrix} -0.73014 \\ 2.52058 \end{bmatrix}$$

The value of the nonlinear equations at this point is:

$$f(x^{(3)}) = \begin{bmatrix} -0.01253 \\ -0.00115 \end{bmatrix}$$

It follows that the norm of the error has dropped to only 0.01258 at the last iteration. The exact solution of this problem is $x^* = [-0.72449 \quad 2.52489]^T$.

1.10 Optimization problem definition

For the remainder of this book, we are concerned with techniques for solving optimization problems of the form:

$$\begin{aligned} & x^* = \min_x f(x) \\ \text{subject to:} \quad & y_j(x) \leq 0, \quad j = 1, 2, \ldots, m \\ & h_k(x) = 0, \quad k = 1, 2, \ldots, p \end{aligned} \tag{1.70}$$

We are searching for an optimal point x^* that has the minimum value of the objective function over all points satisfying the given constraints. The constraints in (1.70) consist of m inequality constraints and p equality constraints. The nature of the constraints and the objective function determine how the problem should be solved. We will focus mainly on problems with nonlinear differentiable objective functions. Constrained and unconstrained problems will be addressed.

Most of the algorithms discussed in this book are local optimization techniques. They obtain a minimum x^* satisfying $f(x^*) \leq f(x)$, $x \in N(x^*)$, the neighborhood of the point x^*. In other words, the optimal point has the best value of the objective function over all close points. Other minima may also exist and may have a better value of the objective function. A global optimization technique aims at finding the point x^* that has the best objective function value over all points satisfying the constraints not only within a small neighborhood. Such a point is called the global minimum of the problem. Convex functions discussed in Section 1.8 enjoy the characteristic that a local minimum of a convex function is also a global minimum.

As will be shown in later chapters, local techniques require much less number of function evaluations than global techniques.

References

1. Paul McDougle, *Vector Algebra*, Wadsworth Publishing Company, Belmont, CA, 1971
2. John W. Bandler and R.M. Biernacki, *ECE3KB3 Courseware on Optimization Theory*, McMaster University, Canada, 1997
3. Singiresu S. Rao, *Engineering Optimization Theory and Practice*, Third Edition, John Wiley & Sons Inc, New York, 1996

Problems

1.1 Obtain two different solutions of the underdetermined system of equations:

$$\begin{aligned} x_1 - 2x_2 + x_3 - x_4 &= 4 \\ 2x_1 + x_2 - x_3 + 2x_4 &= 1 \\ -x_1 + x_2 + 2x_3 + x_4 &= 3 \end{aligned}$$

through basic row operations.

1.2 Evaluate analytically the gradient of the function $f(x_1,x_2) = x_1 + x_2^2 + e^{x_1}\sin(x_2)$ at the point $\boldsymbol{x} = [1.0 \quad 0.5]^T$. Confirm your answer by using FFD, BFD, and CFD approximations with $\Delta x = 0.05$ for both parameters.

1.3 Evaluate analytically the gradient and the Hessian of the function $f(x_1,x_2,x_3) = 3x_1^2 + 6x_2^2 + 2x_3^2 - 2x_1 + x_2 + x_3 - 5$ at the point $\boldsymbol{x} = [2.0 \quad -2.0]^T$. Confirm your result using FFD approximation with $\Delta x = 0.02$.

1.4 Repeat Problem 1.3 by putting the function in a matrix form similar to (1.55). Apply vector differentiation rules.

1.5 Consider the circuit shown in Figure 1.4.
By applying Kirchhoff's voltage and current laws, express the power delivered to the load resistor R_L as a function of the other circuit parameter. Obtain analytically the gradient of power relative to the parameters $\boldsymbol{x} = [R_1 \quad R_s \quad R_L]^T$. Evaluate the gradient for the shown values. Confirm your answer by using FFD with a perturbation of $\Delta x = 2.0\ \Omega$ for all resistors.

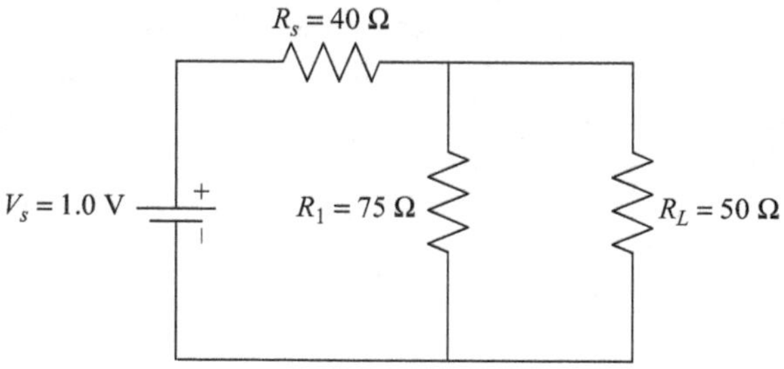

Figure 1.4 The circuit of Problem 1.5

1.6 Consider the matrix:

$$A = \begin{bmatrix} 1 & 3 & 5 \\ 3 & 10 & 8 \\ 5 & 8 & -7 \end{bmatrix}$$

Evaluate the quadratic forms $\boldsymbol{x}^T\boldsymbol{A}\boldsymbol{x}$ and $\boldsymbol{y}^T\boldsymbol{A}\boldsymbol{y}$ for $\boldsymbol{x} = [1 \quad 0 \quad 0]^T$ and $\boldsymbol{y} = [0 \quad 0 \quad 1]^T$. What is your conclusion of the definiteness of this matrix?

1.7 Using elementary row operations, obtain two solutions of the underdetermined system:

$$\begin{aligned} 2x_1 + x_2 - 4x_3 &= 8 \\ 3x_1 - x_2 + 2x_3 &= -1 \end{aligned}$$

1.8 Project the vector $\boldsymbol{x} = [1.0 \quad -1 \quad 1.0]^T$ in the null subspace of the matrix $A = \begin{bmatrix} 1 & 3 & 4 \\ 1 & -1 & 1 \end{bmatrix}$. Find its component in the Range($\boldsymbol{A}^T$). Show that both vectors are orthogonal to each other.

1.9 Express analytically the Jacobian of the vector function $\boldsymbol{f} = [3x_1^2 + 2x_2^4 \quad 3x_1^3 - 2x_2]^T$. Evaluate the Jacobian at the point $\boldsymbol{x} = [2 \quad -2.0]^T$. Write the linearization of this vector function at this point. Compare its value at the points $[2.1 \quad -1.9]^T$ and $[3.0 \quad -3.0]^T$ with that obtained directly from the vector function.

1.10 Consider the nonlinear circuit of Figure 1.5. The circuit elements have the values $V_s = 5.0$ V, $R_s = 1.0$ KΩ, $R_1 = 300$ Ω, $R_2 = 500$ Ω, $R_3 = 200$ Ω, and

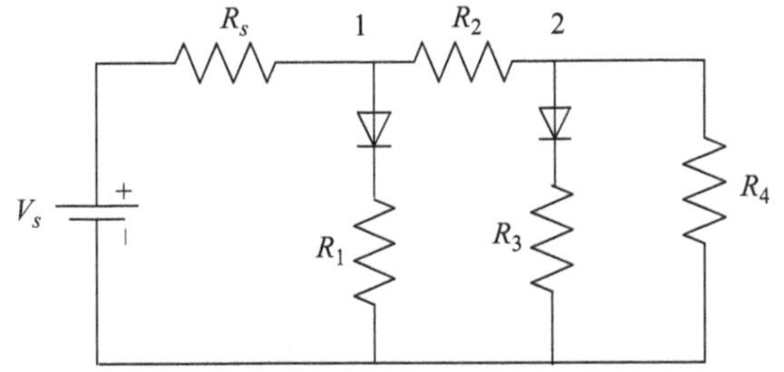

Figure 1.5 The circuit of Problem 1.10

$R_4 = 1.0$ KΩ. The two diodes follow the model $i_D = I_s \exp(v_D/V_T)$, with $I_s = 1.0e-13$ A and $V_T = 0.025$ V.

(a) Write the two nonlinear nodal equations at nodes 1 and 2 in terms of the two unknown diode voltages.

(b) Derive the Jacobian of the two nonlinear equations with respect to $\boldsymbol{x} = [v_{D1} \quad v_{D2}]^T$.

(c) Write a MATLAB program that implements the Newton–Raphson method to obtain the solution of this circuit starting with initial solution $\boldsymbol{x}^{(0)} = [0.4 \text{ V} \quad 0.4 \text{ V}]^T$. Does the algorithm converge? Repeat for the starting point $\boldsymbol{x}^{(0)} = [0.5 \text{ V} \quad 0.5 \text{ V}]^T$. What is $\boldsymbol{x}^*$?

Chapter 2

An introduction to linear programming

2.1 Introduction

Many engineering optimization problem can be cast as a linear Program. By a Linear Programming (LP) I mean an optimization problem where the objective function and the constraints are linear functions of the optimization variables. In addition, several nonlinear optimization problems can be solved by iteratively solving linearized versions of the original problem. Several examples of such problems will be illustrated in this chapter and in the following chapters. The word "programming" in this context does not mean writing computer codes using programming languages but rather it means "scheduling."

It was not until the end of the Second World War that significant progress in the solution of linear programs was achieved. Dantzig developed his famous Simplex approach. Following his pioneering work, several variations of the Simplex method were introduced. The more efficient interior-point methods (IPMs) for solving linear programs were developed later.

In this chapter, we focus on the solution of linear programs. Different statements of the LP problem are introduced. The Simplex method is explained in both the tabular and matrix forms. Several approaches for starting the Simplex approach are discussed.

2.2 Examples of linear programs

Linear programs arise in many fields when there is a needed compromise between the linear cost or resources in a problem and the expected profit. I give here few simple examples.

2.2.1 A farming example

A farmer owns a piece of land of area 5.0 km^2. He wants to grow in his land either wheat or barley or a combination of both. He has only 100.0 kg of fertilizer and 20.0 kg of pesticides. Each square kilometer of barely requires 30.0 kg of fertilizer and 4.0 kg of pesticide, while each square kilometer of wheat requires 25.0 kg of fertilizer and 6.0 kg of pesticide. He is expected to sell his crop for \$1.0 per kilogram of barely and \$1.25 per kilogram of wheat. Each square kilometer of land should deliver 1 ton of barely or 0.75 ton of wheat. Our target is to formulate this problem as a linear program.

The farmer aims at maximizing his profit within the limited resources he has. We denote the area assigned to barely by x_b and the area assigned to wheat as x_w. The expected income is thus $f = 1.00 \times 1000 \times x_b + 1.25 \times 750 \times x_w$. As the farmer cannot use more than 100 kg of fertilizer, it follows that $(30 \times x_b) + (25 \times x_w) \leq 100$.

For the pesticide, the physical constraint $(4.0 \times x_b) + (6.0 \times x_w) \leq 20$ applies. Also, the sum of the two areas should not exceed the area of the piece of land (5 km^2). It follows that the corresponding linear program is given by:

$$\begin{aligned} \text{maximize} \quad & 1.00 \times 1000 \times x_b + 1.25 \times 750 \times x_w \\ \text{subject to} \quad & (30 \times x_b) + (25 \times x_w) \leq 100 \\ & (4 \times x_b) + (6 \times x_w) \leq 20 \\ & x_b + x_w \leq 5 \end{aligned} \tag{2.1}$$

In matrix form, (2.1) can be written as:

$$\begin{aligned} & \min_{x} \boldsymbol{c}^T \boldsymbol{x} \\ \text{subject to} \quad & \boldsymbol{A}\boldsymbol{x} \leq \boldsymbol{b}, \quad \boldsymbol{x} \geq 0 \end{aligned} \tag{2.2}$$

where $\boldsymbol{x} = [x_b \quad x_w]^T$, $\boldsymbol{b} = [100 \quad 20 \quad 5]^T$, and $\boldsymbol{c} = -1.0 \times [1000 \quad (1.25 \times 750)]^T$. The matrix $\boldsymbol{A}$ is given by:

$$\boldsymbol{A} = \begin{bmatrix} 30 & 25 \\ 4 & 6 \\ 1 & 1 \end{bmatrix} \tag{2.3}$$

Notice that maximizing a quantity is equivalent to minimizing its negative. Also, the non-negativity constraints in (2.2) are needed to ensure a feasible answer with non-negative areas.

2.2.2 A production example

A company produces two types of products. Projections indicate an expected demand of at least 75 units for the first product and 50 units for the second product per day. Because of physical limitations, no more than 150 units of the first product and 100 units of the second product can be produced daily. The company should ship at least 200 units per day of both types to fulfill its commitments. The profit per unit of the first product is \$3.00 while the profit per unit of the second product is \$5.00. How many units of each type should be produced per day to maximize profit?

Similar to the previous example, the target is to maximize the profit or alternatively minimize the negative of the profit subject to the constraints. We denote by x_1 the number of produced units of the first product per day and by x_2 the number of produced units of the second type. Our target is to maximize the objective function $f(x_1, x_2) = 3 \times x_1 + 5 \times x_2$ subject to the constraints $75 \leq x_1 \leq 150$ and $50 \leq x_2 \leq 100$. Also, the shipments' constraint implies that $200 \leq (x_1 + x_2)$. Collecting all these equations in one linear program, we get:

$$\begin{aligned} & \min_{x} \boldsymbol{c}^T \boldsymbol{x} \\ & x_1 \leq 150 \\ & -x_1 \leq -75 \\ \text{subject to} \quad & x_2 \leq 100 \\ & -x_2 \leq -50 \\ & -x_2 - x_2 \leq -200 \\ & \boldsymbol{x} \geq \boldsymbol{0} \end{aligned} \tag{2.4}$$

where $\boldsymbol{x} = [x_1 \quad x_2]^T$ and $\boldsymbol{c} = [-3.0 \quad -5.0]^T$. Notice that all constraints, except for the non-negativity constraints, are cast in the less than or equal to form. This is the form adopted throughout this text. In matrix format, (2.4) can be written in the form (2.2) with:

$$\boldsymbol{A} = \begin{bmatrix} 1 & 0 \\ -1 & 0 \\ 0 & 1 \\ 0 & -1 \\ -1 & -1 \end{bmatrix}, \quad \boldsymbol{b} = \begin{bmatrix} 150 \\ -75 \\ 100 \\ -50 \\ -200 \end{bmatrix} \tag{2.5}$$

2.2.3 Power generation example

Two thermal power plants A and B use two different types of coal C_1 and C_2. The minimum power to be generated by each plant is 40.0 and 80.0 MWh, respectively. The quantities of each grade of coal to generate 1.0 MWh and the cost per ton for both plants are given by the following table:

	A (tons)	B (tons)	A (Cost/ton)	B (Cost/ton)
C_1	2.5	1.5	20	18
C_2	1.0	2.0	25	28

Our target is to determine the amount of each grade of coal to be used in generating power so as to minimize the total production cost.

We denote by x_1 and x_2 the quantities in tons of the first type of coal used in both stations. We also denote by x_3 and x_4 the quantities of the second type of coal used in both stations. It follows that the cost of operation is given by $f = 20x_1 + 18x_2 + 25x_3 + 28x_4$. The constraints imply minimum power production in both stations. These constraints are thus formulated as:

$$\begin{aligned} x_1/2.5 + x_3/1.0 &\geq 40 \\ x_2/1.5 + x_4/2.0 &\geq 80 \\ \boldsymbol{x} &\geq 0 \end{aligned} \tag{2.6}$$

The problem can be then written in the following form:

$$\begin{aligned} \text{minimize} \quad & 20x_1 + 18x_2 + 25x_3 + 28x_4 \\ \text{subject to} \quad & -x_1/2.5 - x_3/1.0 \leq -40 \\ & -x_2/1.5 - x_4/2.0 \leq -80 \\ & \boldsymbol{x} \geq 0 \end{aligned} \tag{2.7}$$

Notice that all constraints are put in the less than or equal to form by multiplying both sides by a negative sign. Also, all quantities are forced to be positive through the non-negativity constraint to ensure a physical solution. This system has the form (2.2) with the parameters $\boldsymbol{x} = [x_1 \quad x_2 \quad x_3 \quad x_4]^T$, $\boldsymbol{b} = [-40 \quad -80]^T$, $\boldsymbol{c} = [20 \quad 18 \quad 25 \quad 28]^T$, and:

$$\boldsymbol{A} = \begin{bmatrix} -0.4 & 0 & -1 & 0 \\ 0 & -0.6667 & 0 & -0.5 \end{bmatrix}$$

2.2.4 Wireless communication example

Consider the wireless communication system shown in Figure 2.1. There are n users connected to a base station. The ith user transmits from his cell phone with a power of p_i. The power transmitted by the ith user faces attenuation h_i by the time it reaches the base station. When the station is receiving from the ith user, the sum of powers received from all other users is considered as interference. For reliable communication, the signal to interference ratio (SNR) must exceed a certain threshold γ_i. Our target is to minimize the powers sent by all users while maintaining a reliable communication.

The total power transmitted by all users is $f = p_1 + p_2 + \cdots + p_n$. The SNR for the ith user is $\text{SNR}_i = h_i p_i / \sum_{j, j \neq i} h_j p_j, \forall i$. The condition on the SNR ratio implies that $\text{SNR}_i \geq \gamma_i$ or that $h_i p_i - \gamma_i \sum_{j, j \neq i} h_j p_j \geq 0,\ i = 1, 2, \ldots, n$. The LP problem to solve this problem can thus be cast in the form:

$$\begin{aligned} \text{minimize} \quad & \boldsymbol{c}^T \boldsymbol{x} \\ & \boldsymbol{A}\boldsymbol{x} \leq \boldsymbol{b} \\ & \boldsymbol{x} \geq \boldsymbol{0} \end{aligned}$$

where $\boldsymbol{x} = [p_1 \quad p_2 \quad \ldots \quad p_n]^T$, $\boldsymbol{c} = [1 \quad 1 \quad \ldots \quad 1]^T$, $\boldsymbol{A} = \begin{bmatrix} -h_1 & \gamma_1 h_2 & \ldots & \gamma_1 h_n \\ \gamma_2 h_1 & -h_2 & \ldots & \gamma_2 h_n \\ \vdots & \vdots & \vdots & \vdots \\ \gamma_n h_1 & \gamma_n h_2 & \ldots & -h_n \end{bmatrix}$,

$\boldsymbol{b} = \boldsymbol{0}^T$.

2.2.5 A battery charging example

The electric circuit shown in Figure 2.2 uses a source of 30 V to charge three batteries of values 5.0, 10.0, and 20.0 V. The currents in the branches have the following physical constraints $I_1 \leq 4.0$ A, $I_2 \leq 3.0$ A, $I_3 \leq 3.0$ A, $I_4 \leq 2.0$ A, and $I_5 \leq 2.0$ A. All currents $I_i, \forall i$ must be positive to prevent batteries from discharging. We aim at formulating an LP that determines the values of the currents maximizing the total power transferred to the charged batteries.

The power transferred to each battery is equal to the product of its current (in the charging direction) by the voltage of the battery. The total power is thus given by $f = 5I_2 + 10I_4 + 20I_5$. In addition, circuit current laws impose the constraint that

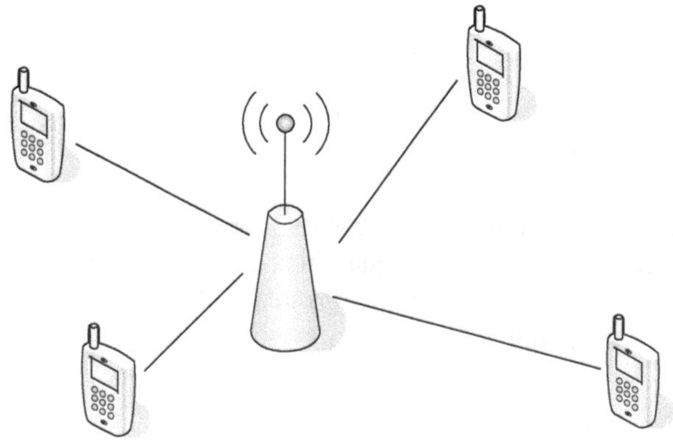

Figure 2.1 A wireless communication example

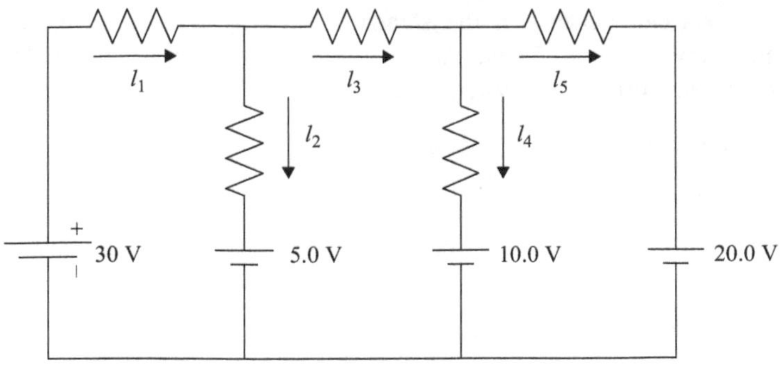

Figure 2.2 A battery charging example

the current flowing into every node must be equal to the currents flowing out from that node. It follows that the charging problem can be cast in the following LP:

$$\begin{aligned} &\text{maximize} \quad 5I_2 + 10I_4 + 20I_5 \\ &\text{subject to} \quad I_1 = I_2 + I_3 \\ &\qquad I_3 = I_4 + I_5 \\ &\qquad I_1 \leq 4.0, \quad I_2 \leq 3.0, \quad I_3 \leq 3.0 \\ &\qquad I_4 \leq 2.0, \quad I_5 \leq 2.0, \quad \text{and} \\ &\qquad \boldsymbol{I} \geq 0 \end{aligned} \tag{2.8}$$

where $\boldsymbol{I}$ is a vector whose components are the currents. This problem can be cast in the compact form:

$$\begin{aligned} &\text{maximize} \quad \boldsymbol{c}^T\boldsymbol{x} \\ &\text{subject to} \quad \boldsymbol{A}\boldsymbol{x} = \boldsymbol{b} \\ &\qquad \boldsymbol{x} \leq \boldsymbol{x}_u \\ &\qquad \boldsymbol{x} \geq \boldsymbol{0} \end{aligned} \tag{2.9}$$

where $\boldsymbol{x} = [I_1 \quad I_2 \quad I_3 \quad I_4 \quad I_5]^T$, $\boldsymbol{c} = [0 \quad 5 \quad 0 \quad 10 \quad 20]^T$,

$\boldsymbol{A} = \begin{bmatrix} 1 & -1 & -1 & 0 & 0 \\ 0 & 0 & 1 & -1 & -1 \end{bmatrix}$, $\boldsymbol{b} = [0 \quad 0]^T$, and $\boldsymbol{x}_u = [4 \quad 3 \quad 3 \quad 2 \quad 2]^T$.

2.3 Standard form of an LP

LP problems are usually cast in the standard form:

$$\begin{aligned} &\min_x \boldsymbol{c}^T\boldsymbol{x} \\ &\boldsymbol{A}\boldsymbol{x} = \boldsymbol{b} \\ &\boldsymbol{x} \geq \boldsymbol{0} \end{aligned} \tag{2.10}$$

Here $\boldsymbol{A} \in \Re^{m \times n}$, where m is the number of equations and n is the number of variables with $m < n$. Other LP variations can also be cast in the standard form (2.10). For example, an LP may have only inequality constraints of the form:

$$\begin{aligned} &\boldsymbol{A}\boldsymbol{x} \leq \boldsymbol{b} \\ &\boldsymbol{x} \geq \boldsymbol{0} \end{aligned} \tag{2.11}$$

These equations can be cast in the standard form by adding an extra "slack" variable to every equation. These variables are positive and they "balance" the equations into equalities. The system of equations in (2.11) then becomes:

$$\begin{aligned} \boldsymbol{Ax} + \boldsymbol{Ix}_s &= \boldsymbol{b} \\ \boldsymbol{x}, \boldsymbol{x}_s &\geq \boldsymbol{0} \end{aligned} \tag{2.12}$$

Notice that in (2.12), one positive slack variable is added to each equation with a unity coefficient. We can thus define a new extended system matrix $\bar{\boldsymbol{A}} = [\boldsymbol{A} \quad \boldsymbol{I}]$ and a new extended vector of unknowns $\bar{\boldsymbol{x}} = \begin{bmatrix} \boldsymbol{x}^T & \boldsymbol{x}_s^T \end{bmatrix}^T$ to convert (2.12) into the standard form:

$$\begin{aligned} &\min_{x} \bar{\boldsymbol{c}}^T \bar{\boldsymbol{x}} \\ &\bar{\boldsymbol{A}}\bar{\boldsymbol{x}} = \boldsymbol{b} \\ &\bar{\boldsymbol{x}} \geq \boldsymbol{0} \end{aligned} \tag{2.13}$$

where $\bar{\boldsymbol{c}}^T = [\boldsymbol{c}^T \quad \boldsymbol{0}^T]$. The original problem (2.11) has been converted into the standard form (2.10) at the expense of increasing the dimensionality of the problem by extra m variables with zero cost coefficients.

Similarly, problems with "greater than or equal to" constraints of the form:

$$\begin{aligned} \boldsymbol{Ax} &\geq \boldsymbol{b} \\ \boldsymbol{x} &\geq \boldsymbol{0} \end{aligned} \tag{2.14}$$

can be converted into the standard form by subtracting a positive slack variable from every equation to get the equality equations:

$$\begin{aligned} \boldsymbol{Ax} - \boldsymbol{Ix}_s &= \boldsymbol{b} \\ \boldsymbol{x}, \boldsymbol{x}_s &\geq \boldsymbol{0} \end{aligned} \tag{2.15}$$

An augmented system matrix can then be defined as in (2.13). If the variables of the problem are not constrained to be positive, these variables can be written as the difference between two positive variables. For example, consider an LP with constraints of the form:

$$\boldsymbol{Ax} = \boldsymbol{b} \tag{2.16}$$

Note that the variables in (2.16) do not have non-negativity constraints. Constraints of the form (2.16) can be converted into the standard form by defining $\boldsymbol{x} = \boldsymbol{x}_p - \boldsymbol{x}_n$, where $\boldsymbol{x}_p$ and $\boldsymbol{x}_n$ are vectors of non-negative variables. In other words, the unconstrained original problem can be cast in the standard form by adding n more variables. The new system to be solved has the form:

$$\begin{aligned} &[\boldsymbol{A} \quad -\boldsymbol{A}] \begin{bmatrix} \boldsymbol{x}_p \\ \boldsymbol{x}_n \end{bmatrix} = \boldsymbol{b} \\ &x_p \geq 0, \quad x_n \geq 0 \end{aligned} \tag{2.17}$$

It follows from (2.11)–(2.17) that any LP can be converted to the standard form (2.10). We will thus focus on the solution of the standard problem (2.10) in the rest of this chapter.

The following examples illustrate the graphical solutions of LP for two-dimensional problems.

Example 2.1: Solve the LP:

$$\begin{aligned} &\text{minimize} && f(x_1, x_2) = -3x_1 + 2x_2 \\ &\text{subject to} && 0 \le x_1 \le 4 \\ &&& 1 \le x_2 \le 6 \\ &&& x_1 + x_2 \le 5 \end{aligned} \tag{2.18}$$

Obtain the solution graphically. Put the LP in the standard form.

Solution: Figure 2.3 shows the feasible region (in gray) satisfying all the constraints. By drawing different constant value contours (lines) of the objective function, we see that the minimum value of the objective function is achieved at $\boldsymbol{x}^* = [4.0 \quad 1.0]^T$ with an objective function value of $f^* = -10.0$. All other points in the feasible region have higher values of the objective function.

To put the problem (2.18) in the standard form, we put all the constraints in the less than or equal to form to obtain:

$$\begin{aligned} &\text{Solve minimize} && f = -3x_1 + 2x_2 \\ &\text{subject to} && x_1 \le 4 \\ &&& x_2 \le 6 \\ &&& -x_2 \le -1 \\ &&& x_1 + x_2 \le 5 \\ &&& \boldsymbol{x} \ge \boldsymbol{0} \end{aligned} \tag{2.19}$$

Then we add slack variables x_3, x_4, x_5, x_6 to balance the equations to get the standard problem:

$$\begin{aligned} &\text{Solve minimize} && f = -3x_1 + 2x_2 \\ &\text{subject to} && x_1 + x_3 = 4 \\ &&& x_2 + x_4 = 6 \\ &&& -x_2 + x_5 = -1 \\ &&& x_1 + x_2 + x_6 = 5 \\ &&& \boldsymbol{x} \ge \boldsymbol{0} \end{aligned} \tag{2.20}$$

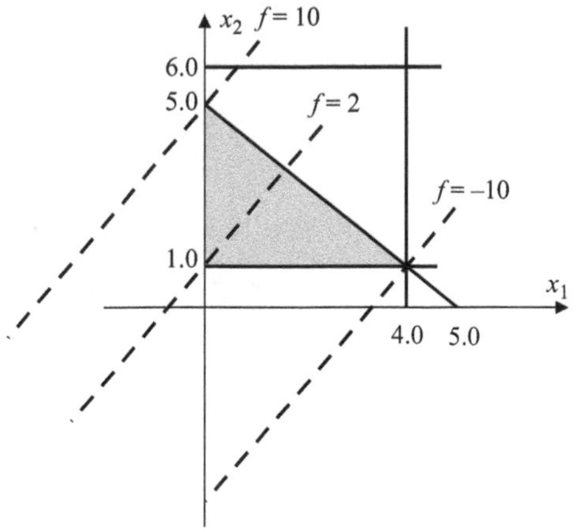

Figure 2.3 A graphical solution for Example 2.1; the objective function assumes its lowest value over the feasible region (shown in gray) at the point $\boldsymbol{x}^* = [4.0 \quad 1.0]^T$

The standard problem (2.20) has six unknown parameters $\boldsymbol{x} = [x_1 \quad x_2 \quad x_3 \quad x_4 \quad x_5 \quad x_6]^T$. Notice that the equivalent standard problem is solved in a higher dimensional space with parameters:

$$\boldsymbol{c} = [-3 \quad 2 \quad 0 \quad 0 \quad 0 \quad 0]^T, \ \boldsymbol{A} = \begin{bmatrix} 1 & 0 & 1 & 0 & 0 & 0 \\ 0 & 1 & 0 & 1 & 0 & 0 \\ 0 & -1 & 0 & 0 & 1 & 0 \\ 1 & 1 & 0 & 0 & 0 & 1 \end{bmatrix}, \text{ and}$$

$$\boldsymbol{b} = [4 \quad 6 \quad -1 \quad 5]^T$$

Example 2.2: Put the problem

$$\begin{aligned} &\text{minimize} && -2x_1 + 3x_2 \\ &\text{subject to} && x_1 + x_2 \leq 5 \\ & && \boldsymbol{x} \geq 0 \end{aligned} \tag{2.21}$$

in the standard form. Obtain a graphical solution for the original problem and the standard problem.

Solution: The original problem is a two-dimensional problem with one constraint in addition to the non-negativity constraints. The feasible region of this problem is illustrated in Figure 2.4. From this figure, it is obvious that the minimum value of the objective function over the feasible region is $f^* = -10$ and is achieved at the point $\boldsymbol{x}^* = [5 \quad 0]^T$. All other points within the feasible region have higher values of the objective function.

To put the problem in the standard form, we add the slack variable x_3 to have the standard problem:

$$\begin{aligned} &\text{minimize} && -2x_1 + 3x_2 \\ &\text{subject to} && x_1 + x_2 + x_3 = 5 \\ & && \boldsymbol{x} \geq 0 \end{aligned} \tag{2.22}$$

This problem is now a three-dimenssional problem as shown in Figure 2.5. The feasible region is the part of the plane $x_1 + x_2 + x_3 = 5$ lying in the positive octant.

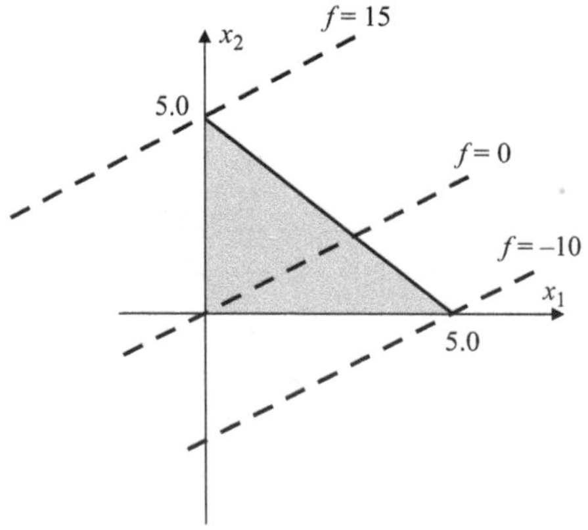

Figure 2.4 The feasible region for Example 2.2; the minimum of the objective function is achieved at the point $\boldsymbol{x}^* = [5 \quad 0]^T$

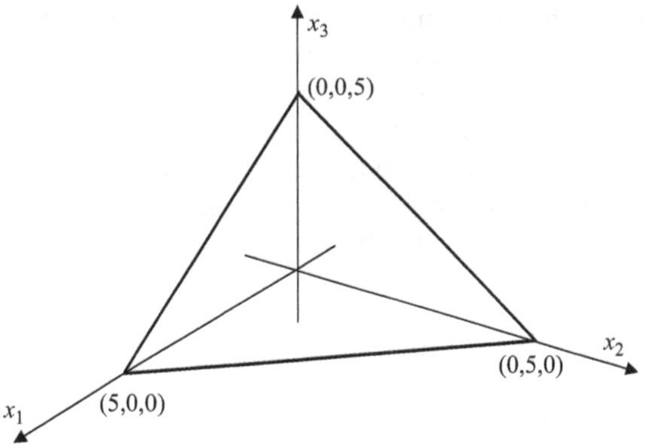

Figure 2.5 The feasible region of the standard form of Example 2.2; the feasible region is the part of the plane $x_1 + x_2 + x_3 = 5$ in the first octant. The projection of this region on the x_1–x_2 plane gives the feasible region of the original two-dimensional problem

The constant objective function surfaces are now planes parallel to the x_3 axis. Notice that the feasible region of the original problem is the projection of the feasible region of the standard problem on the x_1x_2 plane. The feasible region of the standard problem has only three extreme points represented by the vertices $\boldsymbol{x}_{b1} = [5 \quad 0 \quad 0]^T$, $\boldsymbol{x}_{b2} = [0 \quad 5 \quad 0]^T$, and $\boldsymbol{x}_{b3} = [0 \quad 0 \quad 5]^T$. The minimum of the objective function at all these vertices is achieved at $\boldsymbol{x}^* = [5 \quad 0 \quad 0]^T$. This is the same optimal solution of the original two-dimensional problem as the slack variable x_3 is set to zero.

This example illustrates an important point. In most problems, we solve the original problem in a higher dimensional space. In every vertex of the feasible region in this higher dimensional space, some of the parameters are zeros. Also, the optimal solution of the original problem must be achieved at one of these vertices. The theorem discussed in the following section states why this is the case.

2.4 Optimality conditions

The standard problem (2.10) has an infinite number of solutions for the case $n > m$ as mentioned in Section 1.4. The set of all possible solutions of (2.10) is a convex set. This can be seen by considering two solutions $\boldsymbol{x}_1$ and $\boldsymbol{x}_2$ both satisfying $\boldsymbol{Ax}_i = \boldsymbol{b}$. Any convex combination of these two solutions satisfies:

$$\begin{aligned} \boldsymbol{A}(\lambda \boldsymbol{x}_1 + (1-\lambda)\boldsymbol{x}_2) &= \lambda \boldsymbol{A}\boldsymbol{x}_1 + (1-\lambda)\boldsymbol{A}\boldsymbol{x}_2 \\ &= \lambda \boldsymbol{b} + (1-\lambda)\boldsymbol{b} = \boldsymbol{b} \end{aligned} \tag{2.23}$$

Thus it is also a solution. The non-negativity constraint is automatically satisfied for convex combinations. The set of all solutions of (2.10) is called a polytope. The vertices of this polytope cannot be written as a convex combination of other solution as illustrated in Example 2.2. They are extreme points of the set. Every vertex of the polytope is obtained by solving for only m variables while setting all other variables to zero in (2.10). The following theorem states the optimality conditions for the solution of (2.10).

Theorem 2.1: *The solution of (2.10) must be an extreme point of the feasible set defined by the constraints. For a nondegenerate solution, the optimal solution will have only m nonzero components.*

This theorem tells us that the optimal solution of (2.10) will have only m nonzero components in the nondegenerate case. In the degenerate case, the solution has less than m nonzero components. We call a solution with m positive variables and $(n-m)$ zero variables a basic feasible solution. The nonzero components of this solution are denoted as the basic variables while the zero components are denoted as the non-basic variables. The columns corresponding to the basic variables constitute the basic set of the current solution. The following examples illustrate the concept of basic solutions.

Example 2.3: Obtain a basic feasible solution for the LP problem:

$$\begin{aligned} &\text{maximize} \quad f = 2x_1 + x_2 + 3x_3 \\ &\text{subject to} \quad 4x_1 + 2x_2 + 6x_3 \le 12 \\ &\qquad\qquad\quad \boldsymbol{x} \ge \boldsymbol{0} \end{aligned} \tag{2.24}$$

Solution: We first put this problem into the standard form by converting it to a minimization problem and adding a slack variable x_4 to get:

$$\begin{aligned} &\text{minimize} \quad f = -2x_1 - x_2 - 3x_3 \\ &\text{subject to} \quad 4x_1 + x_2 + 6x_3 + x_4 = 12 \\ &\qquad\qquad\quad \boldsymbol{x} \ge \boldsymbol{0} \end{aligned} \tag{2.25}$$

This problem has one equation ($m=1$) and four variables ($n=4$). To obtain a basic solution, we set $(n-m)=3$ of the variables to zero and solve for the remaining $(m=1)$ variable. Setting $x_2=0$, $x_3=0$, and $x_4=0$, we get $x_1=3$, with a corresponding objective function value of $f_1=-6$. This basic solution is a feasible one as it satisfies the non-negativity constraints as well. Alternatively, we may choose $x_1=0$, $x_3=0$, and $x_4=0$ to get $x_2=12$ with a corresponding objective function $f_2=-12$. It is then obvious that this problem has four basic solutions. In general, for a problem with m constraints and n variables, the number of possible basic solutions is:

$$C_m^n = \frac{n!}{m!(n-m)!}. \tag{2.26}$$

This number of possible basic solutions is obtained by choosing m variables at a time out of the n variables. For this example, the number of basic solutions is $4!/(3!1!)=4$. Note that not all basic solutions have to be feasible, i.e., satisfying the non-negativity constraints. In some cases, the basic solution is not in the positive subset of the solution space. The following example illustrates this case.

Example 2.4: Put the LP problem:

$$\begin{aligned} &\text{minimize} \quad f = x_1 - 4x_2 \\ &\text{subject to} \quad x_1 + x_2 \le 2 \\ &\qquad\qquad\quad x_1 - x_2 \le 6 \\ &\qquad\qquad\quad \boldsymbol{x} \ge \boldsymbol{0} \end{aligned} \tag{2.27}$$

in the standard form. Obtain a basic feasible solution for this problem.

Solution: We put the LP (2.27) in the standard form by adding the slack variables x_3 and x_4 to the two inequality constraints to obtain:

$$\begin{aligned} \text{minimize} \quad & f = x_1 - 4x_2 \\ \text{subject to} \quad & x_1 + x_2 + x_3 = 2 \\ & x_1 - x_2 + x_4 = 6 \\ & \boldsymbol{x} \geq \boldsymbol{0} \end{aligned} \tag{2.28}$$

In this linear program, we have $n = 4$ and $m = 2$. To obtain a basic solution, we set $(n - m) = 2$ variables to zero and then solve for the other two variables. We start first by setting the two slack variables to 0. This indicates that the two inequality constraints in (2.27) are active (satisfied as equalities). Solving the two resulting equations in (2.28) for x_1 and x_2 we get $x_1 = 4$ and $x_2 = -2$. The objective function value at this solution is $f_1 = 12$. The solution $\boldsymbol{x}_1 = [4 \quad -2 \quad 0 \quad 0]^T$ is a basic solution but not a feasible solution because it violates the non-negativity constraints. Alternatively, we may set $x_1 = 0$ and $x_3 = 0$ and solve for the other two variables to get $x_2 = 2$ and $x_4 = 8$. The solution $\boldsymbol{x}_2 = [0 \quad 2 \quad 0 \quad 8]^T$ is a basic feasible solution with an objective function value of $f_2 = -8$. It can be shown graphically that the optimal solution of the original two-dimensional problem (2.27) is achieved at the point $[0 \quad 2]^T$ with an optimal objective function value of -8. This solution corresponds to the point $\boldsymbol{x}_2$ in the standard problem (2.28).

2.5 The matrix form

The matrix form offers a more compact expression for the solution of the standard problem. In this form, we obtain a basic solution by considering only m linearly independent columns of the matrix $\boldsymbol{A}$ and then solve for the corresponding variables while setting all other variables to zero. The system of equations can be written in the form:

$$\begin{aligned} \boldsymbol{A}\boldsymbol{x} = \boldsymbol{b} = [\boldsymbol{B} \quad \boldsymbol{D}] \begin{bmatrix} \boldsymbol{x}_b \\ \boldsymbol{x}_n \end{bmatrix} = \boldsymbol{b} \\ \boldsymbol{B}\boldsymbol{x}_b + \boldsymbol{D}\boldsymbol{x}_n = \boldsymbol{b} \Rightarrow \boldsymbol{x}_b = \boldsymbol{B}^{-1}\boldsymbol{b} - \boldsymbol{B}^{-1}\boldsymbol{D}\boldsymbol{x}_n \end{aligned} \tag{2.29}$$

where the matrix $\boldsymbol{B}$ is an $m \times m$ matrix while the matrix $\boldsymbol{D} \in \Re^{m \times (n-m)}$ contains the rest of the columns of the matrix $\boldsymbol{A}$. $\boldsymbol{x}_b$ is the vector of basic variables and $\boldsymbol{x}_n$ is the vector of non-basic variables. By setting the non-basic variables $\boldsymbol{x}_n$ to zero in (2.29), the basic variables are given by:

$$\boldsymbol{x}_b = \boldsymbol{B}^{-1}\boldsymbol{b} \tag{2.30}$$

Notice that the vector of basic variables $\boldsymbol{x}_b$ may not be feasible as explained earlier. Also, in (2.29), we assume, without loss of generality, that we solve for the first m basic variables. Reshuffling the columns of the matrix shuffles the order of the variables but does not change the system of equations.

As per Theorem 2.1, the solution (2.30) is a possible optimal candidate only if it is a basic feasible solution. The Simplex algorithm, to be discussed in a later section, solves this problem by starting with a basic feasible solution and guaranteeing that the next obtained solution is also a basic feasible solution with an improved objective function value. It should be noted that obtaining the inverse of the matrix $\boldsymbol{B}$ is equivalent to carrying out elementary row operations on the whole matrix $\boldsymbol{A}$ to convert $\boldsymbol{B}$ into unit matrix.

2.6 Canonical augmented form

Many of the algorithms for solving linear programs require that the matrix system is put in the canonical augmented format. The matrix $\boldsymbol{A}$ is first augmented by adding the vector $\boldsymbol{b}$ as the $(n+1)$st column. We say that the augmented matrix is in the canonical form if every one of the m basic variables appears only once with a coefficient of one as shown below:

$$\begin{array}{ccccccc} x_1+ & & & a_{1,m+1}x_{m+1} & a_{1,m+2}x_{m+2} & \cdots & a_{1,n}x_n = b_1 \\ & x_2+ & & \vdots & \ddots & & \vdots \\ & & \ddots & \vdots & & \ddots & \vdots \\ & & x_m+ & a_{m,m+1}x_{m+1} & a_{m,m+2}x_{m+2} & & a_{m,n}x_n = b_m \end{array} \tag{2.31}$$

Notice that in (2.31), we assume that the basic variables have been reshuffled to create a unity matrix in the first m columns. The canonical form can be achieved by using slack variables as explained in Section 2.3. Elementary row operations can also be applied to the matrix $\boldsymbol{A}$ to convert m of the columns to the identity matrix.

We assume that, in general, this canonical form is obtained using elementary row operations. This implies that the entries of the jth column represent the coordinates of the expansion of that column in terms of the first m columns of the original matrix as explained in Section 1.4. In other words, we have:

$$\boldsymbol{a}_j = a_{1j}\boldsymbol{a}_1 + a_{2j}\boldsymbol{a}_2 + \cdots + a_{mj}\boldsymbol{a}_m = \sum_{i=1}^{m} a_{ij}\boldsymbol{a}_i, \quad j = m+1, m+2, \ldots, n \tag{2.32}$$

In particular, the last column of the augmented matrix represents the coordinates of the vector $\boldsymbol{b}$ in terms of the first m columns:

$$\boldsymbol{b} = a_{1,n+1}\boldsymbol{a}_1 + a_{2,n+1}\boldsymbol{a}_2 + \cdots + a_{m,n+1}\boldsymbol{a}_m \tag{2.33}$$

The vector $[a_{1,n+1} \quad a_{2,n+1} \quad \cdots \quad a_{m,n+1}]^T$ is thus the solution of the system of equation $\boldsymbol{Bx} = \boldsymbol{b}$, where $\boldsymbol{B}$ is the first m columns of the augmented matrix before elementary row operations were applied. The following example illustrates the application of elementary row operations to obtain a basic solution.

Example 2.5: Apply basic row operations to the LP problem:

$$\begin{aligned} \text{minimize} \quad & 3x_1 - x_2 - x_3 \\ \text{subject to} \quad & 3x_1 + x_2 + x_4 = 4 \\ & 4x_1 + 3x_2 + x_3 + x_4 = 5 \end{aligned}$$

Solution: This system is already in the standard form. We create the augmented tableau (Tableau 1) by including the RHS vector $\boldsymbol{b}$ as a column to get:

Tableau 1

x_1	x_2	x_3	x_4	RHS
3	1	0	1	4
4	3	1	1	5

The variable x_3 already appears only in the second equation with a unity coefficient. We pivot around the element a_{12} to add x_2 as a basic variable to get Tableau 2.

Tableau 2

x_1	x_2	x_3	x_4	**RHS**
3	1	0	<u>1</u>	4
−5	0	1	−2	−7

This step is equivalent multiplying all columns with the inverse of the matrix $\boldsymbol{B}$ formed by the two columns $\boldsymbol{a}_2$ and $\boldsymbol{a}_3$ in the first tableau. The coefficients in the first and fourth columns in Tableau 2 are the expansion coefficients of columns $\boldsymbol{a}_1$ and $\boldsymbol{a}_4$ in terms of the columns $\boldsymbol{a}_2$ and $\boldsymbol{a}_3$ in the original tableau, i.e.:

$$\begin{bmatrix}3\\4\end{bmatrix} = 3\begin{bmatrix}1\\3\end{bmatrix} - 5\begin{bmatrix}0\\1\end{bmatrix}$$
$$\begin{bmatrix}1\\1\end{bmatrix} = 1\begin{bmatrix}1\\3\end{bmatrix} - 2\begin{bmatrix}0\\1\end{bmatrix} \tag{2.34}$$

The solution obtained using Tableau 2 is $\boldsymbol{x}_b = [0 \quad 4 \quad -7 \quad 0]^T$. This solution is not a feasible solution as the non-negativity constraints are not satisfied.

Another basic solution can be obtained by using other variables as basic variables. If we, for example, apply pivoting to Tableau 2 around the element a_{14} to make x_4 enter the basis instead of x_2, we get:

Tableau 3

x_1	x_2	x_3	x_4	**RHS**
<u>3</u>	1	0	1	4
1	2	1	0	1

In Tableau 3, the basic variables are x_3 and x_4. Their corresponding columns have been transformed to a shuffled identity matrix. If the non-basic variables are given zero values, the corresponding solution is $\boldsymbol{x}_b = [0 \quad 0 \quad 1 \quad 4]^T$ which is a basic feasible solution. Notice that Tableau 3 is equivalent to multiplying Tableau 1 by the inverse of the matrix formed by columns $\boldsymbol{a}_4$ and $\boldsymbol{a}_3$. The identity in Tableau 3 has permuted columns with variable x_4 assigned to the first equation and variable x_3 assigned to the second equation. Similar to the Tableau 2, the coefficients in columns $\boldsymbol{a}_1$ and $\boldsymbol{a}_2$ of Tableau 3 are the coefficients of the expansion of these columns in Tableau 1 in terms of columns $\boldsymbol{a}_4$ and $\boldsymbol{a}_3$ in Tableau 1, i.e.:

$$\begin{bmatrix}3\\4\end{bmatrix} = 3\begin{bmatrix}1\\1\end{bmatrix} + 1\begin{bmatrix}0\\1\end{bmatrix}$$
$$\begin{bmatrix}1\\3\end{bmatrix} = 1\begin{bmatrix}1\\1\end{bmatrix} + 2\begin{bmatrix}0\\1\end{bmatrix} \tag{2.35}$$

Another basic solution can be obtained by pivoting at the element a_{11} to make the variable x_1 enter the basis to replace the variable x_4. This gives Tableau 4.

Tableau 4

x_1	x_2	x_3	x_4	RHS
1	1/3	0	1/3	4/3
0	5/3	1	−1/3	−1/3

We can verify that the coefficients of $\boldsymbol{a}_2$ and $\boldsymbol{a}_4$ in Tableau 4 are the expansion coefficients relating these columns of Tableau 1 in terms of $\boldsymbol{a}_1$ and $\boldsymbol{a}_3$ in Tableau 1 as follows:

$$\begin{bmatrix}1\\3\end{bmatrix}=\frac{1}{3}\begin{bmatrix}3\\4\end{bmatrix}+\frac{5}{3}\begin{bmatrix}0\\1\end{bmatrix}$$
$$\begin{bmatrix}1\\1\end{bmatrix}=\frac{1}{3}\begin{bmatrix}3\\4\end{bmatrix}-\frac{1}{3}\begin{bmatrix}0\\1\end{bmatrix} \tag{2.36}$$

The solution obtained in Tableau 4 by setting the non-basic variables to zero is $\boldsymbol{x}=[4/3 \quad 0 \quad -1/3 \quad 0]^T$ which is not a feasible solution. Guaranteeing that the steps progress from a basic feasible solution to another basic feasible solution with a better objective function is the main concept behind the Simplex technique. In the next sections, we lay the theoretical foundations of this method.

2.7 Moving from one basic feasible solution to another

As shown in Example 2.5, moving from one canonical augmented matrix to another canonical augmented matrix implies that a variable x_p will leave the set of basic variables and another variable x_q will enter this set. In the new corresponding augmented matrix, the qth column replaces the pth column in the identity matrix. The qth column is thus assigned to the same equation previously assigned to the pth variable. As explained in the previous section, the entries of the qth column represent the coordinates of the vector $\boldsymbol{a}_q$ with respect to the basis $(\boldsymbol{a}_1, \boldsymbol{a}_2, \ldots, \boldsymbol{a}_m)$ in the original tableau, i.e.:

$$\boldsymbol{a}_q = a_{1q}\boldsymbol{a}_1 + a_{2q}\boldsymbol{a}_2 + \cdots + a_{mq}\boldsymbol{a}_m = \sum_{\substack{i=1,\\ i\neq p}}^{m} a_{iq}\boldsymbol{a}_i + a_{pq}\boldsymbol{a}_p \tag{2.37}$$

The same expression can be written for any non-basic column $\boldsymbol{a}_j$ with the subscript j replacing q in (2.37). To eliminate the pth column $\boldsymbol{a}_p$ from the basis, we express $\boldsymbol{a}_p$ in terms of $\boldsymbol{a}_q$ and the rest of the original basis columns using (2.37) to get:

$$\boldsymbol{a}_p = \frac{1}{a_{pq}}\boldsymbol{a}_q - \sum_{\substack{i=1,\\ i\neq p}}^{m} \frac{a_{iq}}{a_{pq}}\boldsymbol{a}_i \tag{2.38}$$

As the entries of the jth column in the augmented matrix, $j=1, 2, \ldots, n+1$, represent an expansion in terms of the basis $\{\boldsymbol{a}_1, \boldsymbol{a}_2, \ldots, \boldsymbol{a}_m\}$, similar to (2.37), we can eliminate $\boldsymbol{a}_p$ using (2.38) to get:

$$\boldsymbol{a}_j = \sum_{\substack{i=1,\\ i \neq p}}^{m} \left(a_{ij} - a_{pj} \frac{a_{iq}}{a_{pq}} \right) \boldsymbol{a}_i + \frac{a_{pj}}{a_{pq}} \boldsymbol{a}_q \tag{2.39}$$

It follows that the new augmented matrix has the components:

$$a'_{ij} = \left(a_{ij} - a_{pj} \frac{a_{iq}}{a_{pq}} \right), \quad j \neq q \tag{2.40}$$

This update equation is equivalent to applying basic row operations with element a_{pq} as the pivot to convert the qth column into unity vector with only $a_{pq} = 1.0$. It follows that moving from one basic solution to another basic solution implies carrying out elementary row operations at each step. A similar procedure was adopted in Example 2.5.

Moving from one basic solution to another may not satisfy the non-negativity constraint in (2.10) as explained earlier. To guarantee that we move from one basic feasible solution to another basic feasible solution, not any column can be used to exit the basis to allow the qth column to enter the set of basic columns.

To ensure feasibility of the new solution, the qth variable must enter the basic set at the right value. This implies moving along the edge of the polytope from one vertex to another without violating the non-negativity constraints. This can be explained as follows. Assuming that the initial basic set includes the first m variables, the current solution is thus $\boldsymbol{x} = \begin{bmatrix} x_1 & x_2 & \ldots & x_m & \boldsymbol{0}^T \end{bmatrix}^T$. This solution satisfies:

$$x_1 \boldsymbol{a}_1 + x_2 \boldsymbol{a}_2 + \cdots + x_m \boldsymbol{a}_m = \boldsymbol{b} \tag{2.41}$$

The other $n - m$ variables are set equal to zero. Multiplying (2.37) by a small positive quantity α, and subtracting from (2.41), one gets:

$$(x_1 - \alpha a_{1q}) \boldsymbol{a}_1 + (x_2 - \alpha a_{2q}) \boldsymbol{a}_2 + \cdots + (x_m - \alpha a_{mq}) \boldsymbol{a}_m + \alpha \boldsymbol{a}_q = \boldsymbol{b} \tag{2.42}$$

The coefficient of the qth column must be positive so $\alpha \geq 0$. Also, the coefficient α can be increased in value until the coefficient of one of the other columns becomes zero. This guarantees that no negative component is introduced in the solution. Assuming that the coefficient of the pth column is the first one to become zero in (2.42), it then follows that α has the value:

$$\alpha = \frac{x_p}{a_{pq}} = \min_i \frac{x_i}{a_{iq}} \tag{2.43}$$

We thus evaluate the ratio between the elements of the solution ($(n+1)$st column) to the elements in the qth column, which is to enter the basic set. The component with the smallest positive value determines which variable p will leave the basic set.

The discussion in (2.37)–(2.43) assumes that the set of basic variables is the first m variables. This implies that the jth variable is responsible for the jth equation, $j = 1, 2, \ldots, m$. It also means that $a_{jj} = 1$ and $a_{ij} = 0$, $i \neq j$ for the tableau. However, if the pth column is responsible for the ith equation, the discussion above still holds with the pivoting element being a_{iq}. This implies that a_{iq} becomes 1 after pivoting with $a_{jq} = 0, j \neq i$.

The following example illustrates how new basic variables can enter the set of basic variables while keeping the solution feasible.

Example 2.6: Put the LP problem:

$$\begin{aligned} \text{minimize} \quad & f = x_1 - 4x_2 \\ \text{subject to} \quad & x_1 - x_2 \le 2 \\ & x_1 + x_2 \le 6 \\ & \boldsymbol{x} \ge 0 \end{aligned} \tag{2.44}$$

in the standard form. Carry out three iterations showing how to move from one basic feasible solution to another basic feasible solution.

Solution: We first convert this problem to the standard form by adding two slack variables x_3 and x_4 to each equation to get the standard system:

$$\begin{aligned} \text{minimize} \quad & f = x_1 - 4x_2 \\ \text{subject to} \quad & x_1 - x_2 + x_3 = 2 \\ & x_1 + x_2 + x_4 = 6 \\ & \boldsymbol{x} \ge 0 \end{aligned} \tag{2.45}$$

The two slack variables enter in each equation with a unity value. They thus offer the initial set of basic variables as shown in Tableau 1.

Tableau 1

x_1	x_2	x_3	x_4	RHS	Ratio
<u>1</u>	−1	1	0	2	2
1	1	0	1	6	6

In this initial tableau, the current solution is $\boldsymbol{x} = [0 \quad 0 \quad 2 \quad 6]^T$ which is a feasible basic solution. Notice that in Tableau 1, one more column is added to calculate the ratio given by (2.43). We will first allow x_1 ($q = 1$) to enter the basis. We calculate the ratio of the current RHS column to the first column and store it in the ratio column vector. The minimum positive ratio is achieved in the first row which implies that x_3, the variable responsible for this equation, will leave the basic set. Our pivot is thus the element a_{11}. Basic row operations with this element as the pivot results in Tableau 2.

Tableau 2

x_1	x_2	x_3	x_4	RHS	Ratio
1	−1	1	0	2	−2
0	<u>2</u>	−1	1	4	2

The solution corresponding to this tableau is $\boldsymbol{x} = [2 \quad 0 \quad 0 \quad 4]^T$ which is still a basic feasible solution. Next, we choose to make x_2 ($q = 2$) enter the basis. We thus calculate the ratio between the current RHS column and the second column. The smallest positive ratio is 2 and is achieved in the second row. This implies that variable x_4, the one assigned to row 2, will leave the basic set. The pivot element is thus a_{22}. Pivoting around this element results in Tableau 3.

Tableau 3

x_1	x_2	x_3	x_4	**RHS**	**Ratio**
1	0	1/2	1/2	4	
0	1	−1/2	1/2	2	

The solution corresponding to this tableau is $\boldsymbol{x} = [4 \quad 2 \quad 0 \quad 0]^T$ which is also a feasible basic solution. Notice that in all three pivoting steps, we did not check how the objective function is changing. In moving from one basic feasible solution to another, the objective function must decrease until the optimal basic solution is reached. The next section explains how to guarantee continuous reduction in the objective function when moving from one basic feasible solution to another one.

2.8 Cost reduction

The basic idea of the Simplex algorithm is to move from one basic feasible solution to another basic feasible solution so that the objective function value is reduced. The optimal solution is reached when no more reduction in the objective value is possible. Starting from an augmented canonical form with a basic feasible solution, elementary row operations are used to create another augmented canonical matrix with another basic feasible solution but with a better value for the cost function.

To ensure that the new solution has a better value for the objective function, we have to check the improvement in the objective function value if the qth column enters the feasibility basis. At the original feasible basis, the value of the objective function is given by:

$$f_{\text{o}} = c_1 x_1 + c_1 x_1 + \cdots + c_m x_m = \boldsymbol{c}^T \boldsymbol{x} \tag{2.46}$$

Adding the qth column to the feasible set with a value of α to replace the pth variable results in a change in the objective function. The new solution in this case is $[(x_1 - \alpha a_{1q}) \quad (x_2 - \alpha a_{2q}) \quad \cdots \quad (x_m - \alpha a_{mq}) \quad 0 \quad \cdots \quad \underbrace{\alpha}_{q\text{th component}} \quad \cdots \quad 0]^T$. The new value of the objective function is given by:

$$f = c_1(x_1 - \alpha a_{1q}) + c_2(x_2 - \alpha a_{2q}) + \cdots + c_m(x_m - \alpha a_{mq}) + c_q \alpha \tag{2.47}$$

Collecting all terms multiplying α together, we write the new value of the objective function after the qth column is added in terms of the the original objective function value f_{o} corresponding to solution $\boldsymbol{x} = [x_1 \quad x_2 \quad \ldots \quad x_m]^T$:

$$f = f_{\text{o}} + \alpha \left(c_q - \sum_{i=1, i \neq q}^{n} a_{iq} c_i \right) = f_{\text{o}} + \alpha r_q \tag{2.48}$$

The term r_q is the relative cost of adding the qth column. Because $\alpha \geq 0$, to get an improvement in the objective value, we must have $r_q < 0$ for the selected qth column. If all non-basic variables have $r_q > 0$, this implies that the objective function cannot be improved further and that the optimal solution has been reached. Notice that if all non-basic variables are perturbed from their zero value by perturbations α_i, $i = m+1, m+2, \ldots, n$, the corresponding new value of the objective function is given by:

$$f = f_o + \alpha_{m+1} r_{m+1} + \alpha_{m+2} r_{m+2} + \cdots + \alpha_n r_n \tag{2.49}$$

This equation can be treated as an additional equation with f considered as an independent variable and f_o is the right-hand side of this equation. The relative costs are the coefficients of the non-basic variables in this equation. In other words, f is treated as the $(n+1)$st variable and a new augmented matrix is formed. It should be noted that the coefficients of the basic variables in (2.49) are zeros. This equation is updated using row operations at every iteration to determine the new objective function value and the updated relative costs of the non basic variables.

2.9 The classical Simplex method

The Simplex algorithm utilizes all the elements proposed in Sections 2.6–2.9. Starting with an initial basic feasible solution, the algorithm moves to another basic feasible solution with an improved objective function. This is repeated at every iteration until no further reduction is possible. The steps of the Simplex method can be summarized as follows:

1. Set up the initial augmented canonical matrix with a corresponding basic feasible solution.
2. Determine which qth column will enter the feasible set through evaluating the relative cost coefficients for all q as in (2.48). Pick up the column with the most negative relative cost. If there are no column with $r_q < 0$, optimal solution is reached.
3. Determine which pth column to leave the basis set to maintain feasibility using (2.43).
4. Carry out elementary row operations to make $\boldsymbol{a}_q$ replace $\boldsymbol{a}_p$ in the set of basic columns.
5. Go to step 2.

The following examples illustrate the basic ideas in the Simplex method.

Example 2.7: Solve the LP problem:

$$\begin{aligned} \text{minimize} \quad & x_1 - 3x_2 \\ \text{subject to} \quad & -x_1 + 2x_2 \le 6 \\ & x_1 + x_2 \le 5 \\ & x_1, x_2 \ge 0 \end{aligned} \tag{2.50}$$

using a graphical approach. Confirm your solution using the Simplex method.

Solution: The graphs of the feasible region and the objective function are shown in Figure 2.6. It is clear from the graph that the optimal solution is obtained at $\boldsymbol{x}^* = [4/3 \quad 11/3]^T$. The optimal function at this value is $f^* = -29/3$.

To apply the Simplex method, we add two slack variables to put the system of equations in the standard form:

$$\begin{aligned} \text{minimize} \quad & x_1 - 3x_2 \\ \text{subject to} \quad & -x_1 + 2x_2 + x_3 \quad = 6 \\ & x_1 + x_2 \quad + \quad + x_4 = 5 \\ & \boldsymbol{x} \ge 0 \end{aligned} \tag{2.51}$$

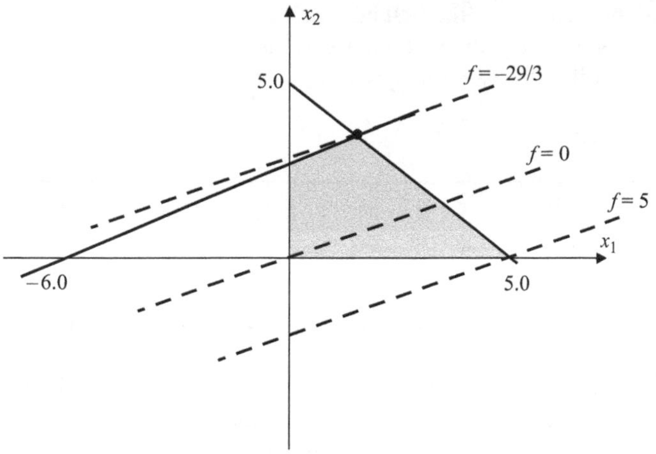

Figure 2.6 The feasible region for Example 2.7; the minimum of the objective function is achieved at the point $\mathbf{x}^* = [4/3 \quad 11/3]^T$ with an objective function value of $f^* = -29/3$

Putting this system in the initial tableau form, we have Tableau 1.

Tableau 1

x_1	x_2	x_3	x_4	$-f$	RHS	Ratio
–1	**2**	1	0	0	6	6/2
1	1	0	1	0	5	5/1
1	**–3**	0	0	–1	0	

Notice that the auxiliary equation (2.49) is added in the last row of the tableau. The value $-f_o = 0$ appears in the RHS column in the last row because the initial solution has zero objective function value. This equation expresses the objective function value as a function of the two non-basic variables (x_1 and x_2). We notice also that x_2 is the parameter with the most negative relative cost (–3). It follows that x_2 ($q = 2$) is to enter the feasible set. Also, the ratio between each element in the current solution (RHS column) and the second column reveals that the lowest positive value is achieved in the first row which is assigned to the parameter x_3. It follows that we pivot at the element a_{12} to allow x_2 to enter the basis and x_3 to leave the basis. Notice that the last equation is also involved in the elimination process to provide the updated relative costs. This gives the Tableau 2:

Tableau 2

x_1	x_2	x_3	x_4	$-f$	RHS	Ratio
–1/2	1	1/2	0	0	3	–6
3/2	0	–1/2	1	0	2	4/3
–1/2	0	3/2	0	–1	9	

The new objective function value is –9. In Tableau 2, x_1 has the most negative updated relative cost (–1/2). It follows that $q = 1$. The ratio between elements of the RHS and the elements of the first column reveal that the lowest positive value is achieved at the second equation which is assigned to x_4. It follows that x_4 is to leave the set of basic variables. We thus pivot at the element a_{21}.

Tableau 3

x_1	x_2	x_3	x_4	$-f$	RHS	Ratio
0	1	2/6	1/3	0	11/3	
1	0	−1/3	2/3	0	4/3	
0	0	8/6	1/3	−1	29/3	

In Tableau 3, all non-basic variables have positive relative cost implying that an optimal solution was reached. The optimal value of the objective function is (−29/3) and is achieved at the optimal values $x^* = [4/3 \quad 11/3 \quad 0 \quad 0]^T$. The two slack variables have been driven out of the basic set in the final solution. This solution confirms the result of the graphical solution.

Example 2.8: Solve the LP problem:

$$\begin{aligned} \text{maximize} \quad & 29x_1 + 45x_2 \\ \text{subject to} \quad & 2x_1 + 8x_2 \le 60 \\ & 4x_1 + 4x_2 \le 60 \\ & x_1, x_2 \ge 0 \end{aligned} \tag{2.52}$$

using the Simplex method.

Solution: The problem is first converted into a minimization problem by multiplying the objective function by a negative sign. Also, two slack variables are added to put it in the standard form. The equivalent standard problem has the form:

$$\begin{aligned} \text{minimize} \quad & -29x_1 - 45x_2 \\ & 2x_1 + 8x_2 + x_3 \quad = 60 \\ \text{subject to} \quad & 4x_1 + 4x_2 + \quad + x_4 = 60 \\ & x \ge 0 \end{aligned} \tag{2.53}$$

The initial tableau is thus as shown in Tableau 1.

Tableau 1

x_1	x_2	x_3	x_4	$-f$	RHS	Ratio
2	**8**	1	0	0	60	60/8
4	4	0	1	0	60	60/4
−29	**−45**	0	0	−1	0	

The non-basic variable with the most negative relative cost is x_2 ($q = 2$). Taking the ratio between the elements of the RHS vector and elements of the second column, we get the lowest possible positive ratio in the first row assigned to x_3. It follows that x_2 enters the basis and x_3 leaves the set. Pivoting is carried out at a_{12} to get Tableau 2.

Tableau 2

x_1	x_2	x_3	x_4	$-f$	RHS	Ratio
0.25	1	0.125	0	0	7.5	30
$\underline{3}$	0	−0.5	1	0	30	10
−17.75	0	5.625	0	−1	337.5	

The non-basic variable with the most negative relative cost is x_1 ($q = 1$). The lowest positive ratio between the RHS and the first column is achieved at the second row assigned to the basic variable x_4. It follows that x_1 enters the basis and x_4 leaves the basis. Pivoting at the element a_{21} we get Tableau 3:

Tableau 3

x_1	x_2	x_3	x_4	$-f$	RHS	Ratio
0	1	0.0833	−0.0833	0	5	30
1	0	−0.1666	0.3333	0	10	1
0	0	1	9.25	−1	515.0	

It is obvious that the non-basic variables x_3 and x_4 have positive relative cost and thus optimality is achieved. The optimal solution is thus $\boldsymbol{x}^* = [10 \quad 5 \quad 0 \quad 0]^T$. The objective function at this point is $f^* = -515.0$.

2.10 Starting the Simplex method

As could be seen from the previous examples, the Simplex method moves from a basic feasible solution to another basic feasible solution with a better objective function value until no more improvements are possible. However, the question arises on how to get the initial basic feasible solution. For problems with inequality constraints, slack variables constitute an initial set of basis variables with a basic feasible solution. However, this approach may not apply to a general problem. We address here three other possible approaches for starting the Simplex method, namely, endless pivoting, the Big M method, and the two-phase Simplex method. The MATLAB listing M2.1 illustrates an implementation of the Simplex method with all inequality constraints. The following example illustrates the application of this listing to solving problems with all inequality constraints.

Example 2.9: Solve the LP:

$$
\begin{aligned}
\text{minimize} \quad & -x_1 - 2x_2 - x_3 \\
\text{subject to} \quad & 2x_1 + x_2 - x_3 \leq 2 \\
& 2x_1 - x_2 + 5x_3 \leq 6 \\
& 4x_1 + x_2 + x_3 \leq 6 \\
& x_1, x_2, x_3 \geq 0
\end{aligned}
$$

Solution: This problem has three unknowns. Three more slack variables are needed to convert this problem to the standard form. These slack variables offer the initial tableau needed for starting the Simplex method. Using the MATLAB listing M2.1, the tableau matrix created by this listing at every iteration is given by:

```
%M2.1
%The Simplex method with all less than or equal to constraints
load Data.mat %load the matrix A, the vector b, the objective function vector c
NumberOfEquations=size(A,1); %this is m
NumberOfParameters=size(A,2); %this is n
%BasicVariableIndexes stores the index of the basic variable of each equation
BasicVariableIndexes=linspace(NumberOfParameters+1, NumberOfParameters+NumberOfEquations,
                    NumberOfEquations); % store indices of the additional m slack variable
Y=zeros(NumberOfEquations+1, NumberOfParameters+NumberOfEquations+1); % tableau storage
Y(1:NumberOfEquations, 1:NumberOfParameters)=A; % copy the matrix A to the tableau
Y(1:NumberOfEquations,NumberOfParameters+1:NumberOfParameters+NumberOfEquations)=
                eye(NumberOfEquations); % copy identity matrix as coefficients for slack variables
Y(1:NumberOfEquations,NumberOfEquations+NumberOfParameters+1)=b; % copy the vector b
Y(NumberOfEquations+1,1:NumberOfParameters)=c'; %store the vector c in the last row
OptimalSolutionFlag=0; %this is the flag of optimality
% find relative costs for initial iteration
[RelativeCost, NewParameterIndex]=min(Y(NumberOfEquations+1,1:NumberOfParameters)); % get index of
                                % nonbasic variable with largest negative relative cost
if(RelativeCost>=0) % no better solution could be obtained
  OptimalSolutionFlag=1;
end
while(OptimalSolutionFlag==0) % start of main loop
 %NewParameterIndex is the index of the nonbasic parameter to enter the basis.
 RatioVector=Y(1:NumberOfEquations,NumberOfEquations+NumberOfParameters+1)./
               Y(1:NumberOfEquations, NewParameterIndex); %build ratio column
 PositiveVector=(RatioVector>=0).*RatioVector; %vector with positive components
 [Value, PivotEquationIndex]= GetPositiveMinimum(RatioVector,PositiveVector); %get minimum
  if(PivotEquationIndex==0)    %no positive ratio found
        OptimalSolutionFlag=1;
        break;
 end;%no variable to enter set with positive value
 BasicVariableIndexes(PivotEquationIndex)=NewParameterIndex; %assign new variable to this equation
 % pivoting element is (PivotEquationIndex, NewParameterIndex)
 Y(PivotEquationIndex,:)=Y(PivotEquationIndex,:)/
                        Y(PivotEquationIndex,NewParameterIndex); %make pivot element equal to 1
 for i=1:(NumberOfEquations+1) %apply pivoting to all rows including the relative cost row
   if(i==PivotEquationIndex) % skip equation of pivot element as it is already unity
     continue;
   end
   Y(i,:)=Y(i,:)-(Y(PivotEquationIndex,:)*Y(i,NewParameterIndex)); %pivoting equation
 end
 %find relative costs for new iteration
 [RelativeCost, NewParameterIndex]=min(Y(NumberOfEquations+1,1:NumberOfParameters)); % get index
                                % of nonbasic variable with largest negative relative cost
 if(RelativeCost>=0) % no better solution could be obtained
   OptimalSolutionFlag=1;
 end
 %we now print the solution at the end of this iteration
 BasicVariableIndexes %print index of basic variables
 Y(1:NumberOfEquations,NumberOfEquations+NumberOfParameters+1)'%tableau
 %print current value of objective function
 -1.0* Y(NumberOfEquations+1,NumberOfParameters+1); % notice that –f is also a variable
end %end of main loop
```

Iteration 0:

Y =

2	<u>1</u>	−1	1	0	0	2
2	−1	5	0	1	0	6
4	1	1	0	0	1	6
1	−2	−1	0	0	0	0

f = 0

Comment: The last row represents the relative cost of each variable. The last column is the value of all the parameters including $(-f)$. The ratio column is not stored in the tableau in this code.

Iteration 1:

Y =

2	1	–1	1	0	0	2
4	0	<u>4</u>	1	1	0	8
2	0	2	–1	0	1	4
3	0	–3	2	0	0	4

f=–4

Iteration 2:

Y=

3.0000	1.0000	0	1.2500	0.2500	0	4.0000
1.0000	0	1.0000	0.2500	0.2500	0	2.0000
0	0	0	–1.5000	–0.5000	1.0000	0
6.0000	0	0	2.7500	0.7500	0	10.0000

f=–10

In iteration 2, it is obvious that the relative cost row has only positive components corresponding to the non-basic variables. It follows that the optimal solution is $\boldsymbol{x}^* = [0 \quad 4.00 \quad 2.00 \quad 0 \quad 0 \quad 0]^T$ which corresponds to an objective function value of -10. This solution is degenerate as one of the basic variables has a zero value.

2.10.1 Endless pivoting

This is a brute force approach to starting the Simplex method. A set of m distinct columns $S = \{\boldsymbol{a}_{j_1}, \boldsymbol{a}_{j_2}, \ldots, \boldsymbol{a}_{j_m}\}$ of the matrix $\boldsymbol{A}$ in (2.10) is selected. Pivoting is carried out at these columns to convert the matrix $\boldsymbol{B}$ formed by these columns into the identity matrix. This is equivalent to multiplying both sides in (2.10) by the matrix $\boldsymbol{B}^{-1}$ as in (2.29). If the obtained basic solution is feasible (with all non-negative components), the Simplex method can be started with the initial basis S. Otherwise, another set of random columns is selected and pivoting is repeated to obtain another solution. This process is repeated until a basic feasible solution is achieved. This approach can be costly for problems with a large number of constraints and parameters.

2.10.2 The big M approach

This approach aims at putting the standard form (2.10) into a canonical form by adding m auxiliary variables. The original objective function is modified by adding large coefficients corresponding to the slack variables. The modified problem has the form:

$$
\begin{aligned}
&\min_{x, x_s} \boldsymbol{c}^T\boldsymbol{x} + \boldsymbol{M}^T\boldsymbol{x}_s \\
&\boldsymbol{A}\boldsymbol{x} + \boldsymbol{I}\boldsymbol{x}_s = \boldsymbol{b} \\
&\quad \boldsymbol{x}, \boldsymbol{x}_s \geq \boldsymbol{0}
\end{aligned}
\tag{2.54}
$$

where $\boldsymbol{M}$ is a vector or large positive cost coefficients and $\boldsymbol{I}$ is the identity matrix. The equations in (2.54) are in the standard canonical form and the Simplex algorithm can be started. The initial set of basic variables contains only the auxiliary variables. Because of the large coefficients assigned to the auxiliary variables, the Simplex method forces these variables out of the basic set to reduce the objective function which leads to the solution of the original problem.

```
% M2.2
%The Two-Phase Simplex Method
load Data.mat %load the matrix A, the vector b, the objective function vector c
NumberOfEquations=size(A,1);
NumberOfParameters=size(A,2);
%BasicVariableIndex stores the indexes of the basic variables associated with each equation
BasicVariableIndexes=linspace(NumberOfParameters+1, NumberOfParameters+NumberOfEquations,
                              NumberOfEquations); %store indices of the %additional m slack variable
Y=zeros(NumberOfEquations+2, NumberOfParameters+NumberOfEquations+2); %tableau
Y(1:NumberOfEquations, 1:NumberOfParameters)=A; %copy the matrix A to the tableau
Y(1:NumberOfEquations,NumberOfParameters+1:NumberOfParameters+NumberOfEquations)=
                    eye(NumberOfEquations); %coefficients for slack variables
Y(1:NumberOfEquations,NumberOfEquations+NumberOfParameters+1)=b; %copy the vector b
Y(NumberOfEquations+1,1:NumberOfParameters)=c'; %store the vector c in the last row
% put the coefficients of the auxiliary objective function
Y(NumberOfEquations+2,NumberOfParameters+1:NumberOfParameters+NumberOfEquations)=1;
%we first put this system in the canonical form by subtracting every row from the auxiliary objective
% function row to convert the cost coefficients corresponding to the basic variables to zero
for j=1:NumberOfParameters+NumberOfEquations
  Y(NumberOfEquations+2,j)=Y(NumberOfEquations+2,j)-sum(Y(1:NumberOfEquations,j));
end
for Phase=2:-1:1 % repeat for the two phases
 OptimalSolutionFlag=0; %this is the flag of optimality
 [RelativeCost,NewParameterIndex]=min(Y(NumberOfEquations+Phase,
      1:NumberOfParameters)); %largest negative relative cost
 if(RelativeCost>=0) % no better solution could be obtained
   OptimalSolutionFlag=1;
 end
 while OptimalSolutionFlag==0 %start of main loop
  %The NewParameterIndex contains the index of the non basic parameter to
  %enter the basis. We then evaluate the ratio with the RHS vector to
  %determine the basic parameter to exit the basis
  PivotColumn=Y(1:NumberOfEquations, NewParameterIndex);
    RatioVector=Y(1:NumberOfEquations,NumberOfEquations+NumberOfParameters+1)./
                                       PivotColumn;
  [Value, PivotEquationIndex]=GetPositiveMinimum(RatioVector, PivotColumn); %get minimum of only
                                                                          %positive components
  if(PivotEquationIndex==0) OptimalSolutionFlag=1; break; end;%no variable to enter with positive
                                                              %value
  BasicVariableIndexes(PivotEquationIndex)=NewParameterIndex; %make new variable related to this
                                                              % equation
  %pivoting element is (PivotEquationIndex, NewParameterIndex)
  %we now pivot around this element
  Y(PivotEquationIndex,:)=Y(PivotEquationIndex,:)/Y(PivotEquationIndex,NewParameterIndex);
 %make pivot element equal to 1
  for i=1:(NumberOfEquations+Phase) %apply pivoting to all rows including the relative cost row
    if(i==PivotEquationIndex) %skip equation of pivot element
      continue;
    end
    Y(i,:)=Y(i,:)-(Y(PivotEquationIndex,:)*Y(i,NewParameterIndex))
  end
  % get index of nonbasic variable with largest negative relative cost
  [RelativeCost, NewParameterIndex]=min(Y(NumberOfEquations+Phase,1:NumberOfParameters));
  if(RelativeCost>=0) % no better solution could be obtained
    OptimalSolutionFlag=1;
  end
 end
```

2.10.3 The two-phase Simplex

The two-phase Simplex approach is closely related to the big M approach. Instead of solving (2.54) with a modified objective function, we create a completely different objective function that depends only on auxiliary variables. We first solve the initial phase of the Simplex algorithm which has the form:

$$\begin{aligned} &\min_{\boldsymbol{x}_s} \mathbf{1}^T \boldsymbol{x}_s \\ &\boldsymbol{Ax} + \boldsymbol{Ix}_s = \boldsymbol{b} \\ &\boldsymbol{x}, \boldsymbol{x}_s \geq \mathbf{0} \end{aligned} \tag{2.55}$$

where **1** is a vector of ones. The minimum value of the objective function in (2.55) is zero and is achieved when all the auxiliary variables are driven out of the set of basic variables. The final set of basic variables in (2.55) is then used to solve the original problem (2.10) with all the slack variables removed and with the original objective function. One problem that appears in the first phase of this approach is that all non-basic variables have zero relative cost in the initial tableau as they do not appear in the objective function (2.55). This contradicts (2.49) where only non-basic variables should have nonzero relative costs. To overcome this, we force all the relative costs of the basic variables to zero by subtracting the sum of all rows ($j = 1, 2, \ldots, m$) from the last row. This allows a start of this phase with nonzero relative costs only for the non-basic variables as in previous examples.

The MATLAB listing M2.2 shows a possible implementation of the two-phase Simplex algorithm. Notice that one additional row is added to the tableau to represent the coefficients of the auxiliary objective function in (2.55). Also, the variable "Phase" determines which objective function row to use. The two-phase algorithm is illustrated by the following example.

Example 2.10: Solve the LP:

$$\begin{aligned} \text{minimize} \quad & f = -x_1 + 3x_2 + x_3 - 3x_4 \\ \text{subject to} \quad & x_1 + x_2 - x_4 = 4 \\ & 2x_1 - x_2 + 3x_3 - 3x_4 = 5 \\ & x_j \geq 0, \quad j = 1, 2, 3, 4 \end{aligned}$$

Solution: For this problem there are $m=2$ equations and $n=4$ variables. The problem is already in the standard form. The two-phase algorithm adds two more auxiliary variables x_5 and x_6 to the two equations. It minimizes, in the first phase, the auxiliary objective function $f_a = x_5 + x_6$. This objective function has a minimum of zero when the two auxiliary variables are driven out from the basic set. Utilizing the MATLAB listing M2.2 for this problem gives the following output:

Iteration 0:

BasicVariableIndexes =

5 6

Y =

1	1	0	–1	1	0	4
2	–1	3	–3	0	1	5
–1	3	1	–3	0	0	0
–3	0	–3	4	0	0	0

Comment: Notice that the sum of the first two rows was subtracted from the auxiliary objective function row (last row) to make the relative cost of the basic variables in the last equation zero. The non-basic variable with the most negative relative cost is x_1.

Iteration 1:

BasicVariableIndexes =

5 1

Y =

0	**1.5000**	–1.5000	0.5000	1.0000	–0.5000	1.5000
1.0000	–0.5000	1.5000	–1.5000	0	0.5000	2.5000
0	2.5000	2.5000	–4.5000	0	0.5000	2.5000
0	**–1.5000**	1.5000	–0.5000	0	1.5000	7.5000

Comment: The non-basic parameter with the most negative cost is x_2.

Iteration 2:

BasicVariableIndexes =

2 1

Y =

0	1.0000	–1.0000	**0.3333**	0.6667	–0.3333	1.0000
1.0000	0	1.0000	–1.3333	0.3333	0.3333	3.0000
0	0	5.0000	**–5.3333**	–1.6667	1.3333	0
0	0	0	0	1.0000	1.0000	9.0000

At the end of this iteration, auxiliary variables are driven out of the basic set. The algorithm switches to phase 2 and the original objective function (row 3) is considered. x_4 is the parameter with the most negative relative cost.

Iteration 3:

BasicVariableIndexes =

4 1

Y =

0	3.0000	–3.0000	1.0000	2.0000	–1.0000	3.0000
1.0000	4.0000	–3.0000	0	3.0000	–1.0000	7.0000
0	16.0000	**–11.0000**	0	9.0000	–4.0000	16.0000
0	0	0	0	1.0000	1.0000	9.0000

Notice that in this iteration even though the relative cost of the parameter x_3 is negative (-11), this variable cannot enter the set at a positive value. This can be seen by taking the ratio of the components of the RHS vector $[3.00 \quad 7.00]^T$ and the corresponding components in the third column $[-3.00 \quad -3.00]^T$. It follows that optimality is reached with the solution of the original problem is $\boldsymbol{x}^* = [7.00 \quad 0 \quad 0 \quad 3.00]^T$ with a corresponding objective function value of $f^* = -16$.

2.11 Advanced topics

I attempted in this chapter to give an overview of the Simplex method and the theory behind it. The target was to give a smooth introduction so that the reader can read if interested in advanced subjects.

One of the advanced subjects is Degeneracy [8]. Degeneracy appears in some problems when some of the basic variables have zero value. We saw an example of that in Example 2.9. This may result in a sequence of Simplex iterations in which the algorithm rotates between a number of solutions with no change in the objective function. This is usually solved by adding randomness in the way we select basic variables to enter the basis. This randomness can break the infinite loop and allow the algorithm to proceed to a better basic feasible solution with better objective function values.

We have seen in formula (2.26) that the number of basic solutions grows exponentially with the number of parameters n. For large-scale problems, this makes the convergence of the Simplex algorithm extremely slow. This motivated research in developing a number of techniques that can accelerate the convergence to the optimal basic feasible solution.

One possibility is to solve the dual problem of (2.10) which exploits duality theorems. It has a number of unknowns equal to the number of rows of the original problem. Solving this problem can be more efficient than solving the original problem especially for large n.

Another family of problems that aim at efficiently solving large-scale problems are interior point methods (IPMs). These methods navigate inside the feasible region in search for better solutions instead of jumping between a large number of feasible basic solutions. These techniques were shown to have a faster convergence especially for large-scale problems. A good review of IPMs can be found in Reference 3.

A2.1 Minimax optimization

A2.1.1 Minimax problem definition

One of the common problems electrical designers solve is to design circuits or structures whose responses should meet certain upper or lower bounds. For example, the reflection coefficient of a microwave or a photonic structure must be below a certain value over the frequency passbands and higher than a certain value in the stopbands. In designing digital or analog filters, similar upper and lower bound constraints are imposed on the transmission over the frequency bands of interest. The target is to push the response above a lower bound as possible and below an upper bound as possible by changing the circuit/structure/filter parameters.

To explain the minimax approach, we consider the response $R(\boldsymbol{x}, \zeta)$ where $\boldsymbol{x}$ is the vector of optimizable parameters. These may include the values of resistors, capacitors, or inductors of an electric circuit. They may represent the dimensions of different discontinuities and material properties of microwave or photonic structures. They may also represent the coefficients of a digital filter under design. There are a variety of interpretations that depend on the application at hand. The parameter ζ is a non-optimizable parameter that also affects the response. It may represent the frequency of operation for frequency domain responses. It may also represent time if the desired response is a time-domain response.

It is usually the case that we are interested in the response over a set of values of the parameter ζ. This set includes the values $\{\zeta_1, \zeta_2, \ldots, \zeta_K\}$. The target of the minimax design problem is to determine the optimal set of the parameters $\boldsymbol{x}^*$ that makes the response satisfy upper or lower limits imposed on the responses $R(\boldsymbol{x}, \zeta_i)$, $\forall i$. The error for the ith response is defined by:

$$q_i = R(\boldsymbol{x}, \zeta_i) - U_i \tag{2.56}$$

for a response with an upper bound where U_i is the value of the bound. For responses with a lower bound, the error is defined as:

$$q_i = L_i - R(\boldsymbol{x}, \zeta_i) \tag{2.57}$$

where L_i is the value of the lower bound. Notice that in (2.56) and (2.57), when the response satisfies the bound, the corresponding error is non-positive. Actually, the more negative the error the better. To satisfy the design specifications, we would like to make all these errors negative. In other words, we would like to push them all to be as negative as possible. This is equivalent to minimizing the maximum of these errors hence the name "minimax". The optimization problem then becomes:

$$\min_{\boldsymbol{x}} \{\max_{i} q_i\} \tag{2.58}$$

It follows that in the problem (2.58), we would like to minimize the objective function $f = \max_i q_i$. This objective function can be shown to be non-differentiable at the solution.

To illustrate this concept, consider the analog filter shown in Figure 2.7. The parameters of this circuits are $\boldsymbol{x} = [L_1 \quad C_1 \quad L_2 \quad C_2]^T$. This circuit exhibits a low-pass response. At low frequencies, inductors represent short circuit and capacitors represent open circuit. Most of the available energy in that case is transmitted to the load resistor. At high frequencies, inductors represent open circuit while the capacitors represent short circuit, thus shorting out the load resistor. We would like to design this circuit to have most of the energy transmitted to the load in the frequency band 1 kHz $\leq$ *freq* $\leq$ 10 kHz. We would like little transmission to the load over the frequency band 13.0 kHz $\leq$ *freq*. Because there are infinitely many frequencies in the pass band, we enforce the transmission condition over

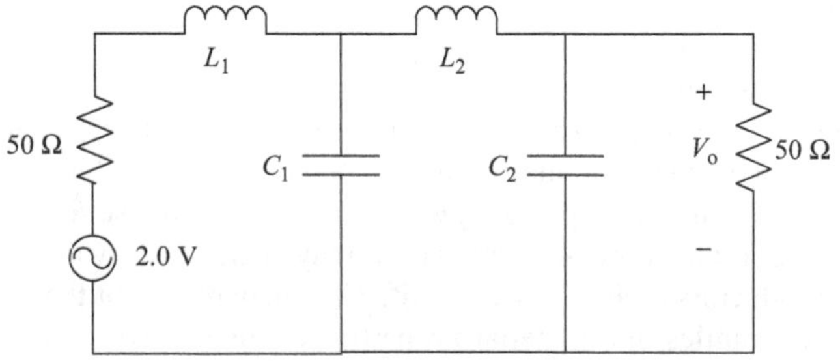

Figure 2.7 A low-pass filter used to illustrate minmax optimization constraints; the modulus of the output voltage should satisfy certain constraints over the pass and stop bands

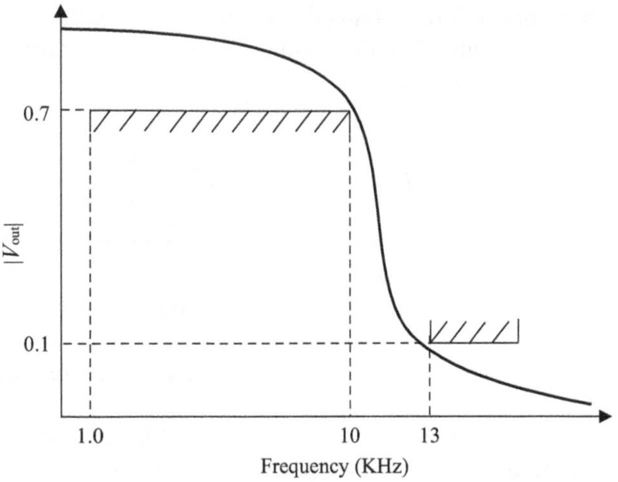

Figure 2.8 An illustration of upper and lower bounds in a minimax problem; for the shown constraints we have $|V_{out}| > 0.7$ for the band (1.0–10.0 KHz) and $|V_{out}| < 0.1$ for frequencies above 13.0 KHz

10 frequencies only {1.0 kHz, 2.0 kHz, ..., 10.0 kHz} to satisfy $|V_{out}| > 0.7$. We also enforce the band stop condition at only one frequency $freq = 13.0$ kHz to satisfy $|V_{out}| < 0.1$. These specifications are shown in Figure 2.8. It follows that we have in this problem 11 errors $\boldsymbol{q} = [q_1 \quad q_2 \quad \cdots \quad q_{11}]^T$ defined by:

$$\begin{aligned}
&q_1 = 0.7 - |V_{out}(\boldsymbol{x}, 1.0 \text{ kHz})|, \quad q_2 = 0.7 - |V_{out}(\boldsymbol{x}, 2.0 \text{ kHz})|,\\
&q_3 = 0.7 - |V_{out}(\boldsymbol{x}, 3.0 \text{ kHz})|, \quad q_4 = 0.7 - |V_{out}(\boldsymbol{x}, 4.0 \text{ kHz})|,\\
&q_5 = 0.7 - |V_{out}(\boldsymbol{x}, 5.0 \text{ kHz})|, \quad q_6 = 0.7 - |V_{out}(\boldsymbol{x}, 6.0 \text{ kHz})|,\\
&q_7 = 0.7 - |V_{out}(\boldsymbol{x}, 7.0 \text{ kHz})|, \quad q_8 = 0.7 - |V_{out}(\boldsymbol{x}, 8.0 \text{ kHz}|,\\
&q_9 = 0.7 - |V_{out}(\boldsymbol{x}, 9.0 \text{ kHz})|, \quad q_{10} = 0.7 - |V_{out}(\boldsymbol{x}, 10.0 \text{ kHz})|,\\
&q_{11} = |V_{out}(\boldsymbol{x}, 13.0 \text{ kHz})| - 0.1
\end{aligned} \tag{2.59}$$

Notice that these errors are all functions of the parameters $\boldsymbol{x}$. By changing the values of the inductors and capacitors we can adjust the values of the errors to become as negative as possible to satisfy the design specifications. It follows that this problem can be cast in the form (2.58). Because the response $|V_{out}|$ is a nonlinear function of the parameters $\boldsymbol{x}$, all the error functions are also nonlinear functions of the parameters.

A2.1.2 Minimax solution using linear programming

There are many techniques in the literature for solving minimax problems. Some of these techniques combine the errors into one objective function that approximates the minimax objective function. The minimizer of this approximate function can then be used to approximate the solution of the minimax problem. The accuracy of the solution can be improved by improving the accuracy of the approximate function.

We discuss an approach for solving the minimax problem that converts it into a sequence of linear programs. The derivation of this problem starts by converting (2.58) to the equivalent problem:

$$\min_{t,x} t$$
$$\text{subject to} \quad q_i(x) \le t, \quad i = 1, 2, \ldots, m \tag{2.60}$$

In (2.60), an auxiliary variable t is introduced. This parameter is forced, through the added constraints, to be greater than all the errors. It follows that by minimizing this variable t, we are actually minimizing the maximum of all the errors. Notice that (2.60) has $n+1$ unknowns $(\boldsymbol{x}^*, t^*)$. Also, (2.60) has a linear objective function but all its m constraints are nonlinear as the errors are, in general, nonlinear functions of the parameters.

The problem (2.60) can be solved as a sequence of linear programs. At the kth iteration, we assume that we have the approximation $\boldsymbol{x}^{(k)}$ to the optimal solution. The ith error function q_i is approximated by its first-order Taylor expansion:

$$L_i = q_i(\boldsymbol{x}^{(k)}) + \nabla q_i(\boldsymbol{x}^{(k)})^T(\boldsymbol{x} - \boldsymbol{x}^{(k)}), \quad i = 1, 2, \ldots, m \tag{2.61}$$

The approximations (2.61) require the values of the errors at the point $\boldsymbol{x}^{(k)}$ and their gradients. These gradients may be estimated through finite differences as discussed in Chapter 1 or through the adjoint sensitivity approaches discussed in Chapter 10. The problem (2.60) can then be cast as the following linear program:

$$\min_{t,x} t$$
$$\text{subject to} \quad (\nabla q_i(\boldsymbol{x}^{(k)})^T \boldsymbol{x} - t) \le (\nabla q_i(\boldsymbol{x}^{(k)})^T \boldsymbol{x}^{(k)} - q_i(\boldsymbol{x}^{(k)})), \quad i = 1, 2, \ldots, m \tag{2.62}$$

In (2.62), all constants were moved to the right-hand side of the equations. The formulation (2.62) is a linear program that can be solved for the unknowns $(\boldsymbol{x}^*, t^*)$. The system (2.62), however, may yield inaccurate solutions because the linearizations (2.61) are only accurate within a small neighborhood of the point $\boldsymbol{x}^{(k)}$. A simple remedy for this limitation is to limit the step taken in (2.62) by adding additional $2n$ constraints on the components of the vector $\boldsymbol{x}$. This limits the step taken to a "trust region" in which the accuracy of the linearizations is acceptable. These constraints are given by:

$$\left|x_j - x_j^{(k)}\right| \le \delta \Rightarrow x_j \le x_j^{(k)} + \delta \quad \text{and} \quad -x_j \le -x_j^{(k)} + \delta, \quad j = 1, 2, \ldots, n \tag{2.63}$$

where δ is the trust region size and x_j is the jth components of the unknown vector $\boldsymbol{x}$. The constraint on the modulus of the perturbation is decomposed in (2.63) into two linear equations. It follows that at the kth iteration, starting from the point $\boldsymbol{x}^{(k)}$, we solve for the better solution $\boldsymbol{x}^{(k+1)}$ by solving the linear program:

$$\min_{t,x} t$$
$$\text{subject to} \quad (\nabla q_i(\boldsymbol{x}^{(k)})^T \boldsymbol{x}^{(k+1)} - t) \le (\nabla q_i(\boldsymbol{x}^{(k)})^T \boldsymbol{x}^{(k)} - q_i(\boldsymbol{x}^{(k)})), \quad i = 1, 2, \ldots, m$$
$$x_j^{(k+1)} \le x_j^{(k)} + \delta \quad \text{and} \quad -x_j \le -x_j^{(k)} + \delta, \quad j = 1, 2, \ldots, n \tag{2.64}$$

The problem (2.64) has the same form as (2.2) with the following matrices and vectors:

$$c = \begin{bmatrix} \mathbf{0}^T & 1 \end{bmatrix}^T$$

$$A = \begin{bmatrix} \nabla q_1(\boldsymbol{x}^{(k)})^T & -1 \\ \nabla q_2(\boldsymbol{x}^{(k)})^T & -1 \\ \vdots & \vdots \\ \nabla q_m(\boldsymbol{x}^{(k)})^T & -1 \\ \boldsymbol{e}_1^T & 0 \\ \boldsymbol{e}_2^T & 0 \\ \vdots & \vdots \\ \boldsymbol{e}_n^T & 0 \\ -\boldsymbol{e}_1^T & 0 \\ -\boldsymbol{e}_2^T & 0 \\ \vdots & \vdots \\ -\boldsymbol{e}_n^T & 0 \end{bmatrix}, \quad \boldsymbol{b} = \begin{bmatrix} (\nabla q_1(\boldsymbol{x}^{(k)})^T \boldsymbol{x}^{(k)} - q_1(\boldsymbol{x}^{(k)})) \\ (\nabla q_2(\boldsymbol{x}^{(k)})^T \boldsymbol{x}^{(k)} - q_2(\boldsymbol{x}^{(k)})) \\ \vdots \\ (\nabla q_m(\boldsymbol{x}^{(k)})^T \boldsymbol{x}^{(k)} - q_m(\boldsymbol{x}^{(k)})) \\ x_1^{(k)} + \delta \\ x_2^{(k)} + \delta \\ \vdots \\ x_n^{(k)} + \delta \\ -x_1^{(k)} + \delta \\ -x_2^{(k)} + \delta \\ \vdots \\ -x_n^{(k)} + \delta \end{bmatrix} \tag{2.65}$$

where $\boldsymbol{e}_i$ is the ith column of the identity matrix. The matrices and vectors in (2.65) can be cast in the more compact matrix form:

$$\boldsymbol{c} = \begin{bmatrix} \mathbf{0}^T & 1 \end{bmatrix}^T, \quad \boldsymbol{A} = \begin{bmatrix} \boldsymbol{J} & -\mathbf{1}_m \\ \boldsymbol{I}_n & \mathbf{0}_n \\ -\boldsymbol{I}_n & \mathbf{0}_n \end{bmatrix}, \quad \boldsymbol{b} = \begin{bmatrix} \boldsymbol{J}\boldsymbol{x}^{(k)} - \boldsymbol{q} \\ \boldsymbol{x}^{(k)} + \delta \mathbf{1}_n \\ -\boldsymbol{x}^{(k)} + \delta \mathbf{1}_n \end{bmatrix} \tag{2.66}$$

where $\boldsymbol{J} \in \Re^{m \times n}$ is the Jacobian matrix of the vector of errors $\boldsymbol{q}$. $\mathbf{1}_m \in \Re^m$ is a column vector of ones and $\mathbf{0}_n \in \Re^n$ is a column vector of zeros.

This system can be solved as in example (2.9) by adding a slack variable for every inequality constraint. Once $\boldsymbol{x}^{(k+1)}$ is determined, the LP (2.64) can be solved again at this new point. The steps are repeated until a convergence criterion is reached. Because the parameter t can be negative, it has to be split into two non-negative parameters $t = t_1 - t_2$.

The trust region size δ is usually dynamically adjusted based on how good the linearized models (2.61) match the actual error functions. If the agreement is very good, the trust region size is increased. If the agreement is poor, the trust region size is decreased. This adjustment of the trust region size is usually done at the end of every iteration.

A2.1.3 A microwave filter example

We illustrate minimax optimization by optimizing the microwave filter shown in Figure 2.9. This filter uses a network of transmission lines and parallel shorted stubs to implement a bandpass filter. The source and load impedances are both 50 Ω. The design specifications on the reflection coefficient, given by the modulus of the scattering parameter S_{11}, are:

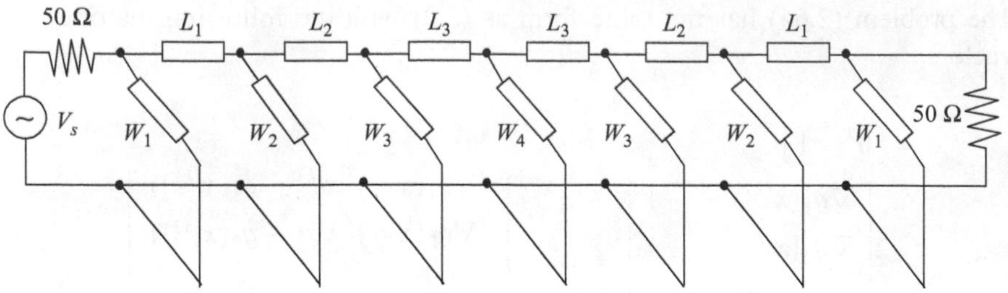

Figure 2.9 The six-section transmission line filter; the parameters are the lengths of the sections L_1, L_2, and L_3 and the lengths of short-circuited stubs W_1, W_2, W_3, and W_4

$$
\begin{aligned}
&|S_{11}| \leq 0.17 \quad \text{for} \quad 5.4\,\text{GHz} \leq freq \leq 9.0\,\text{GHz} \\
&|S_{11}| \geq 0.85 \quad \text{for} \quad freq \leq 5.2\,\text{GHZ} \quad \text{and} \quad |S_{11}| \geq 0.5 \quad \text{for} \quad 9.5\,\text{GHz} \leq freq
\end{aligned}
\tag{2.67}
$$

These constraints impose transmission of most of the energy in the passband 5.4–9.0 GHz. They also impose reflection of most of the energy in the two stopbands $freq \leq 5.2$ GHz and $freq \geq 9.5$ GHz. The designable parameters are the length of the transmission lines and the short circuited stubs $\boldsymbol{x} = [W_1 \ \ W_2 \ \ W_3 \ \ W_4 \ \ L_1 \ \ L_2 \ \ L_3]^T$. Because there are infinitely many frequencies in the passbands and stopbands, we select only few of these frequencies and impose the upper and lower bounds on these points. The errors selected for this problem are thus:

$$
\begin{aligned}
e_1 &= 0.85 - |S_{11}(freq_1)| \ \text{ with } freq_1 = 5.2\,\text{GHz} \\
e_i &= |S_{11}(freq_i)| - 0.17, \ \ i = 2, 3, \ldots, 14 \ \text{ with } freq_i = 5.4\,\text{GHz} + (i-2)^{*}0.3\,\text{GHz} \\
e_{15} &= 0.5 - |S_{11}(freq_{15})| \ \text{ with } freq_{15} = 9.5\,\text{GHz}
\end{aligned}
\tag{2.68}
$$

A spacing of 0.3 GHz is utilized in the passband. A total of 15 error functions are considered in this case. The starting point for this optimization problem is $\boldsymbol{x}^{(0)} = [0.7 \ \ 0.7 \ \ 0.7 \ \ 0.7 \ \ 1.2 \ \ 1.2 \ \ 1.2]^T$ cms. The MATLAB listing M2.3 illustrates the solution of this problem using linear programming. Notice the implementation of the error function and Jacobian estimation are shown in the MATLAB listing M2.4. The MATLAB function **linprog** was used for solving the linear program in the listing M2.3. The function **getFilterResponse** evaluates the modulus of the reflection coefficient of the filter in Figure 2.9 using transmission line theory. The value of the objective function ($f = \max\{q_i\}$) at every iteration is as shown in Figure 2.10. The algorithm takes 36 iterations to converge to a feasible design with all non-positive errors. The convergence could have been accelerated by utilizing second-order derivatives or their approximation as will be illustrated in Chapter 7. The final design is given by $\boldsymbol{x}^* = [0.6759 \ \ 0.4778 \ \ 0.4849 \ \ 1.4115 \ \ 1.4186 \ \ 1.4699 \ \ 1.9635]^T$ cms. The initial and final responses of the filters are shown in Figures 2.11 and 2.12, respectively.

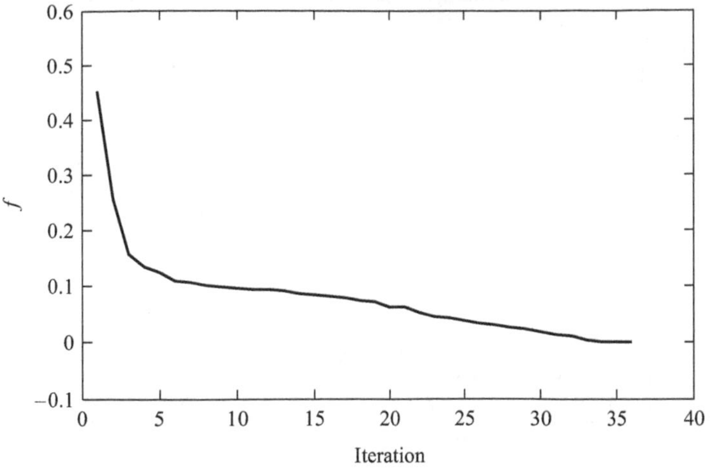

Figure 2.10 The objective function of the microwave filter example at every iteration

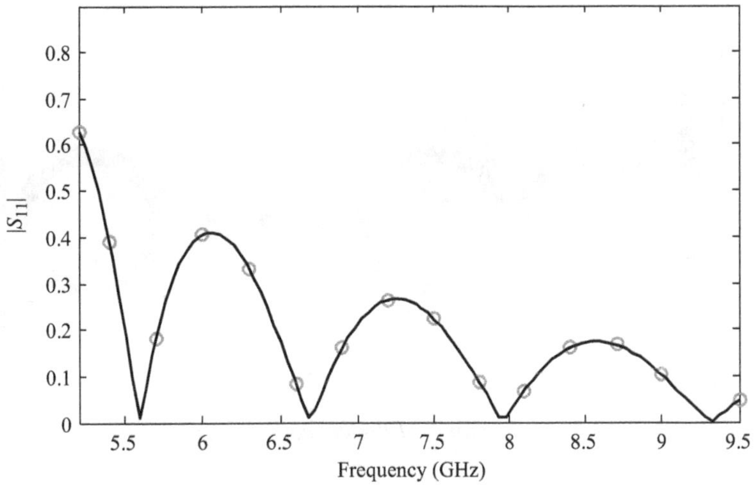

Figure 2.11 The initial response of the microwave filter

A2.1.4 The design of coupled microcavities optical filter

We consider in this application the design of a coupled microcavities filter shown in Figure 2.13. These filters are widely utilized for optical communication systems. In this filter, each ring acts as a resonator that couples light to neighboring rings. The cavity length of the resonator is L_c. By adjusting the coupling from one ring to another, the light can be manipulated differently at different frequencies. The response of the filter is determined the through coupling parameters σ_i, $i = 1, 2, \ldots, N+1$. Assuming weak through couplings, and using the assumption of symmetry:

$$\sigma_i = \sigma_{N+2-i} \tag{2.69}$$

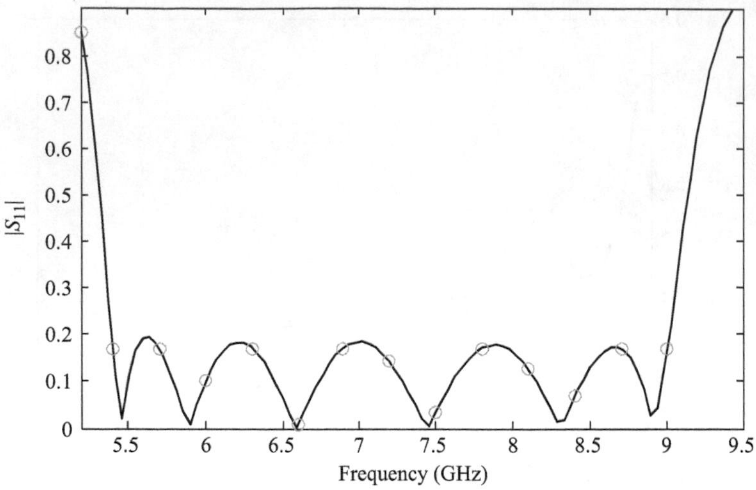

Figure 2.12 The response of the microwave filter at the final design

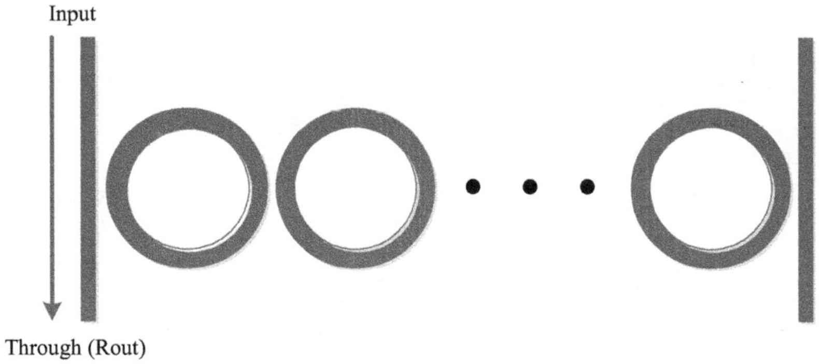

Figure 2.13 The structure of the cascaded microcavities filter

the total through port coupling can be expressed as:

$$\tilde{R}_{out}(\boldsymbol{x},\theta) = e^{-j(N\theta/2)}[2\sigma_1 \cos(N\theta/2) - 2\sigma_2 \cos((N/2-1)\theta) + \cdots + \sigma_{N/2+1}e^{j\pi N/2}] \tag{2.70}$$

where $\boldsymbol{x} = [\sigma_1 \quad \sigma_2 \quad \ldots \quad \sigma_{N/2+1}]^T$ is the vector of optimization parameters. Here, the parameter θ is the normalized angular frequency with $\theta = n_{eff}\omega L_c/c$, n_{eff} is real part of the effective waveguide index, and c is the speed of light. Notice that for an even number of rings (N), the quantity in brackets is real and even function. In this case, the through port transfer function (2.70) can be represented as $\tilde{R}_{out}(\boldsymbol{x},\theta) = e^{-j(N\theta/2)}R'_{out}(\boldsymbol{x},\theta)$. Equation (2.70) is a linear function in the coupling parameters σ_i. Our design problem is to minimize the maximum of the modulus of

```
% M2.3
%This program solves a microwave filter design problem
numberOfParameters=7; %this is n
numberOfErrors=15;   %this is m
oldPoint=[0.7 0.7 0.7 0.7 1.2 1.2 1.2]'; % 1cm is the starting point
trustRegionSize=0.05; %initial trust region size
Epsilon=0.001; %termination condition on trust region
%we then assign memory to the matrix A and vectors b and c
numberOfRows=numberOfErrors+2*numberOfParameters; %number of rows of A
numberOfColumns=numberOfParameters+1; %number of columns of A
A=zeros(numberOfRows, numberOfColumns); %storage for matrix A
b=zeros(numberOfRows,1); %storage for b
c=zeros(numberOfParameters+1,1);
c(numberOfParameters+1,1)=1; %component of c corresponding to t is 1
oldErrors=getErrors(oldPoint); %get vector of errors at x(0)
maxErrors=max(oldErrors)
while(maxErrors>0) %repeat until all errors are negative
 Jacobian=getJacobianMatrix(oldPoint,0.03*min(oldPoint)); %get Jacobian matrix at x(k).
 %now we fill the matrix A and the vector b
 A(1:numberOfErrors, 1:numberOfParameters)=Jacobian; %store Jacobian
 A(1:numberOfErrors,(numberOfParameters+1))=-1; %store -1 in components corresponding to t
 A((numberOfErrors+1):(numberOfErrors+numberOfParameters),1:numberOfParameters)=
                                     eye(numberOfParameters); %store identity
 A((numberOfErrors+numberOfParameters+1): numberOfRows, 1:numberOfParameters)=
                                -1.0*eye(numberOfParameters); %store -1*identity
 b(1:numberOfErrors)=Jacobian*oldPoint-oldErrors; %first m rows of the vector b
 b((numberOfErrors+1):(numberOfErrors+numberOfParameters))=
                              oldPoint+trustRegionSize*ones(numberOfParameters,1);
 b((numberOfErrors+numberOfParameters+1):(numberOfErrors+2*numberOfParameters))=
                           -1.0*oldPoint+trustRegionSize*ones(numberOfParameters,1);
 %we then solve the linear program
 Solution=linprog(c,A,b);
 newPoint=Solution(1:numberOfParameters); %remove value of auxiliary t
 newErrors=getErrors(newPoint); %get new errors
 predictedErrors=oldErrors+Jacobian*(newPoint-oldPoint); %get predicted errors through linearization
 predictedReduction=max(oldErrors)-max(predictedErrors); %This is the predicted reduction
 actualReduction=max(oldErrors)-max(newErrors); %This is the actual  reduction
 if(actualReduction>0) %there is improvement in objective function
   oldPoint=newPoint; %accept new point
   oldErrors=newErrors; %make new error vector current error vector
   maxErrors=max(oldErrors); %get maximum of new errors
 end
 %adjust trust region size according to quality of linearization
 Ratio=actualReduction/predictedReduction; %ratio between actual reduction and predicted reduction
 if(Ratio>0.8) %linearization is goodinearization
   trustRegionSize=trustRegionSize*1.1 %make step bigger
 else
   if(Ratio<0.3) %poor prediction by linearization
     trustRegionSize=trustRegionSize*0.8;
   end
 end
end
```

the function $R'_{out}(\boldsymbol{x}, \theta)$ over the values of interest for θ. A possible optimization problem formulation is thus given by:

$$\begin{aligned} &\min_{\boldsymbol{x}} \quad \max_{\theta\in[\omega_s,\pi]} \left|R'_{out}(\boldsymbol{x}, \theta)\right| \\ &\text{subject to} \quad \left|R'_{out}(\boldsymbol{x}, \theta)\right| \le \xi, \quad \theta \in [0, \omega_p] \\ &\qquad\qquad \delta_1 \le \sigma_i \le \delta_2 \qquad \forall i \end{aligned} \tag{2.71}$$

where ω_s is the normalized stop band angular frequency and ω_p is the normalized passband angular frequency. The parameter ξ is the passband ripple. Physical constraints are also imposed on the coupling parameters. The problem (2.71) is similar to the problem (2.58). The nonlinear functions can be linearized at the

```
% M2.4
%This function returns the Errors at the different frequencies
function Errors=getErrors(Point)
scaleMatrix=0.01*eye(7); % convert all 7 parameters from cms to meters
scaledParameters=scaleMatrix*Point; %get the scaled parameters
Frequencies=1.0e9*[5.2 5.4 5.7 6.0 6.3 6.6 6.9 7.2 7.5 7.8 8.1 8.4 8.7 9.0 9.5]'; % 15 frequencies
Bounds=[0.85 0.17 0.17 0.17 0.17 0.17 0.17 0.17 0.17 0.17 0.17 0.17 0.17 0.17 0.5]'; % bounds
% -ve sign for lower bound, +ve sign for upper bound
boundSigns=[-1.0 1 1 1 1 1 1 1 1 1 1 1 1 1 -1.0]';
numberOfFrequencies=size(Frequencies,1); %this is the number of frequencies
Errors=zeros(numberOfFrequencies,1); %allocate storage for the errors
for i=1:numberOfFrequencies
   currentFrequency=Frequencies(i); %ith frequency
   currentBound=Bounds(i); %bound at ith frequency
   currentBoundSign=boundSigns(i); %sign of bound at ith frequency
   Response=getFilterResponse(scaledParameters, currentFrequency); %get the response |S11|
   Errors(i)=currentBoundSign*(Response-currentBound); %evaluate ith error
end

%This function evaluates the Jacobian of the objective function at a specific point using a given
%perturbation

function Jacobian=getJacobianMatrix(Point, parameterPerturbation)
NumberOfParameters=size(Point,1); % determine the number of parameters
NominalErrors=getErrors(Point); % get the errors
NumberOfErrors=size(NominalErrors,1); % this is the number of responses
Jacobian=zeros(NumberOfErrors, NumberOfParameters); % allocate Jacobian storage
Identity=eye(NumberOfParameters); % get the identity matrix
for i=1:NumberOfParameters % repeat for all parameters
  %get perturbed errors relative to the ith parameter
  PerturbedErrors=getErrors(Point+parameterPerturbation*Identity(:,i));
  %set the ith column of the Jacobian
 Jacobian(:,i)=(PerturbedErrors-NominalErrors)/parameterPerturbation;
end
```

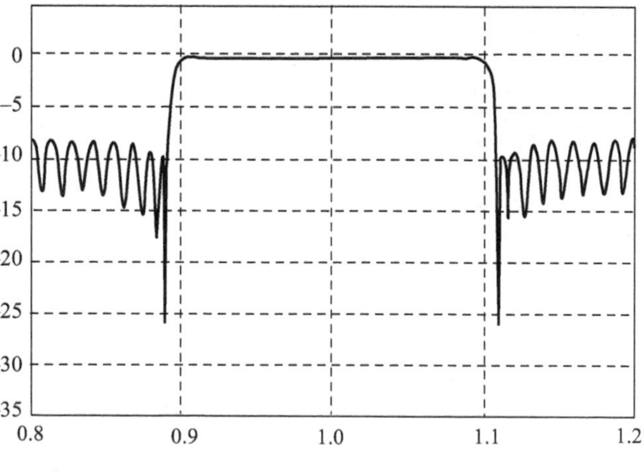

Figure 2.14 The through response of the optimized coupled microcavities filter with 150 microcavities

current solution as explained in the previous example. Figure 2.14 shows the response obtained by solving (2.71) for 150 coupled cavities as reported in References 9 and 10. The free linear programming solver SeDuMi [11] was used in solving this example.

References

1. Singiresu S. Rao, *Engineering Optimization Theory and Practice*, Third Edition, John Wiley & Sons Inc, New York, 1996
2. Erwin Kreyszig, *Advanced Engineering Mathematics*, 7th edn., Wiley
3. Andreas Antoniou and Wu-Sheng Lu, *Practical Optimization Algorithms and Engineering Applications*, Springer, New York, 2007
4. P. Venkataraman, *Applied Optimization with Matlab Programming*, Wiley, 2002
5. Edwin K.P. Chong and Stanislaw H. Zak, *An Introduction to Optimization*, 2nd edn., Wiley, 2001
6. Jorge Nocedal and Stephen J. Wright, *Numerical Optimization*, 2nd edn., Springer, 2006
7. John O. Attia, *Electronics and Circuit Analysis Using Matlab*, CRC Press, 1999
8. Igor Griva, Stephen G. Nash, and Ariela Sofer, *Linear and Nonlinear Optimization*, SIAM, 2009
9. M.A. Swillam, O.S. Ahmed, M.H. Bakr, and X. Li, 'Filter design using multiple coupled microcavities', *IEEE Photonics Technology Letters*, vol. 23, no. 16, pp. 1160–1162, August 2011
10. O.S. Ahmed, M.A. Swillam, M.H. Bakr, and X. Li, 'Efficient design optimization of ring resonator-based optical filters', *IEEE J. Lightwave Tech.*, vol. 29, no. 18, pp. 2812–2817, 2011
11. J.F. Sturm, 'Using SeDuMi 1.02, a MATLAB toolbox for optimization over symmetric cones', *Optimization Methods and Software*, vol. 11–12, pp. 625–653, 1999, retrieved from http://sedumi.mcmaster.ca

Problems

2.1 Consider the optimization problem:

$$\begin{aligned} \text{minimize} \quad & f = x_1 + x_2 \\ \text{subject to} \quad & x_1 + 2x_2 \leq 8.0 \\ & x_2 \leq 2.0 \\ & x_1, x_2 \geq 0 \end{aligned}$$

(a) Draw the feasible region defined by the constraints.
(b) By drawing the linear contours of the objective function, determine graphically the optimal solution.
(c) Put this problem in the standard form and solve it using the Simplex method to verify your answer in (b).

2.2 Repeat problem 2.1 for the following linear program:

$$\begin{aligned} \text{maximize} \quad & f = 2x_1 + x_2 \\ \text{subject to} \quad & 3x_1 + 2x_2 \geq 6.0 \\ & x_2 - x_1 \leq 3.0 \\ & x_2 \leq 6.0 \\ & x_1, x_2 \geq 0 \end{aligned}$$

2.3 Confirm your answers in problems 2.1 and 2.2 using the MATLAB listing M2.1.

2.4 A company produces two types of fridges T_1 and T_2. Type T_1 can be produced through two processes P_1 and P_2 while type T_2 is produced through the two processes P_3 and P_4. The profit per fridge is \$20 for the first type and \$15 for the second type. The machine hours required to produce each type is given by the below table.

	P_1	P_2	P_3	P_4
T_1	3	2		
T_2			3.5	2.0

The maximum machine hours available are 120 per week. Formulate a linear program that maximizes the profit to determine the optimal number of fridges to be produced from each type.

2.5 Modify the MATLAB listing M2.1 to implement the Simplex approach with initialization given by the big M method. Test your code by applying it to the linear program:

$$\begin{aligned} \text{minimize} \quad & f = x_1 + x_2 + x_3 \\ \text{subject to} \quad & x_1 + 2x_2 + x_3 = 8.0 \\ & 2x_1 + x_2 + 2x_3 = 12.0 \\ & x_1, x_2, x_3 \geq 0 \end{aligned}$$

2.6 Repeat problem 2.5 using the two-phase Simplex algorithm.

2.7 Consider the shown Op-Amp bandpass filter in Figure 2.15. It is required to determine the parameters' values such that the following constraints are satisfied:

$$\begin{aligned} |V_o| &\geq 0.75 \quad \text{at } freq = 1.0 \text{ kHz} \\ |V_o| &\leq 0.60 \quad \text{at } freq = 200 \text{ Hz and } 2.0 \text{ kHz} \end{aligned}$$

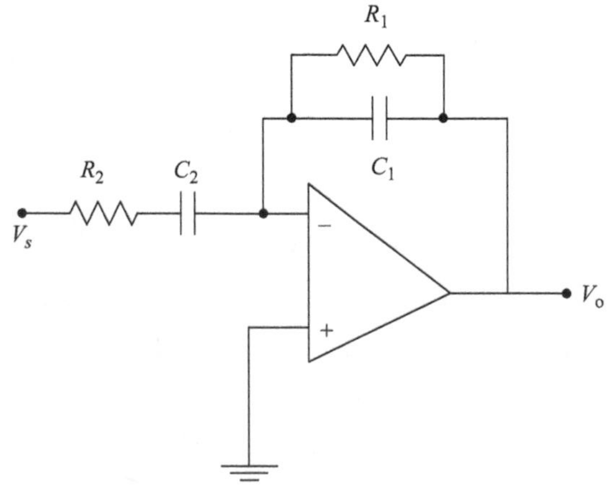

Figure 2.15 The bandpass filter of Problem 2.7

The design parameters of this problem are $[C_1 \quad C_2 \quad R_1 \quad R_2]^T$. The source value is 1.0 V.

(a) Write the ratio between the output voltage V_o and the supply voltage V_s in terms of the circuit parameters.
(b) Formulate the design problem as a minimax design problem.
(c) Utilize the MATLAB codes M2.3 and M2.4, after making the necessary changes, to solve a sequence of linear programs leading to the optimal solution. The starting point is $\boldsymbol{x} = [5.0\ \mu\text{F} \quad 5.0\ \mu\text{F} \quad 2.0\ \Omega \quad 2.0\ \Omega]^T$. Scaling should be applied within this code to avoid ill-conditioning.

Chapter 3
Classical optimization

3.1 Introduction

Before the arrival of the digital computer age, several optimization approaches and theories were utilized to solve simple problems with one or few variables. Powerful tools such as the single-variable Taylor expansion, the multivariable Taylor expansion, and the Karush–Kuhn–Tucker (KKT) conditions were utilized in solving unconstrained and constrained optimization problems. Substitution methods were also used to convert constrained optimization problems to unconstrained optimization ones. Other methods such as the method of constrained variations were also used for solving simple constrained optimization problems.

In this chapter, I focus on explaining some of these classical approaches. The theoretical bases of these techniques are relevant for numerical optimization approaches to be addressed in the following chapters.

3.2 Single-variable Taylor expansion

Taylor's expansion is named after the scientist Brook Taylor who in 1715 was the first one to show that any differentiable function can be approximated through an expansion of its derivatives with known coefficients. For the single-variable case, this expansion is given by:

$$f(x+h) = f(x) + hf'(x) + \frac{h^2}{2!}f''(x) + \cdots + \frac{h^{(n-1)}}{n-1!}\overset{(n-1)}{f}(x) + \frac{h^{(n)}}{n!}\overset{(n)}{f}(x+\theta h), \quad 0 < \theta < 1 \tag{3.1}$$

This expansion estimates the value of a function f at the point $x+h$, which is in the vicinity of the point x, through the function value and its derivatives at the point x. The last term in (3.1) represents the error in the estimation of the function. The derivatives of that last term are estimated at a point somewhere in between x and $x+h$. It should be noted that usually only a limited number of derivatives are known at any point. This expansion is thus valid only for a relatively small h. The more derivatives we know at the point x, the more accurate the estimation is for larger values of h. Actually, if all derivatives are known at a point x, the function value for any other point $x+h$ is known regardless of whether h is small or large. In practical application, we have either the first or the second derivatives available. It follows that we usually utilize either a linear expansion (first-order Taylor expansion):

$$f(x+h) \approx f(x) + hf'(x) \tag{3.2}$$

or a quadratic expansion (second-order Taylor expansion):

$$f(x+h) \approx f(x) + hf'(x) + \frac{h^2}{2!}f''(x) \tag{3.3}$$

The following example illustrates the application of Taylor's first- and second-order derivatives to approximating functions.

Example 3.1: Expand the function $f(x)=x^4$ at the point $x=2$. Use Taylor's expansion to build linear and quadratic approximations of this function. Plot these approximations and compare them to the original function.

Solution: For the function $f(x)=x^4$, the function value, the first-order derivative, and second-order derivative at the point $x=2$ are given by:

$$f(x)|_{x=2} = x^4|_{x=2} = 16,\ f'(x)|_{x=2} = 4x^3|_{x=2} = 32, \quad \text{and} \quad f''(x)|_{x=2} = 12x^2|_{x=2} = 48 \tag{3.4}$$

It follows from (3.2) and (3.3) that the corresponding first- and second-order Taylor approximations around the point $x=2$ are given by:

$$L(x) = 16 + 32(x-2) \quad \text{and} \quad Q(x) = 16 + 32(x-2) + 24(x-2)^2 \tag{3.5}$$

Notice in (3.5) that $h=x-2$ is the deviation from the expansion point $x=2$. Figure 3.1 shows these approximations in comparison with the original function. It is obvious that at the expansion point $x=2$, both models have identical values with the original function. The linear model approximates the original function at a narrow neighborhood around the expansion point. The quadratic model shows a better match than the linear model over a wider neighborhood. It follows that to predict the nonlinear behavior of a function, the more known derivatives at the expansion point the better the accuracy of the expansion is.

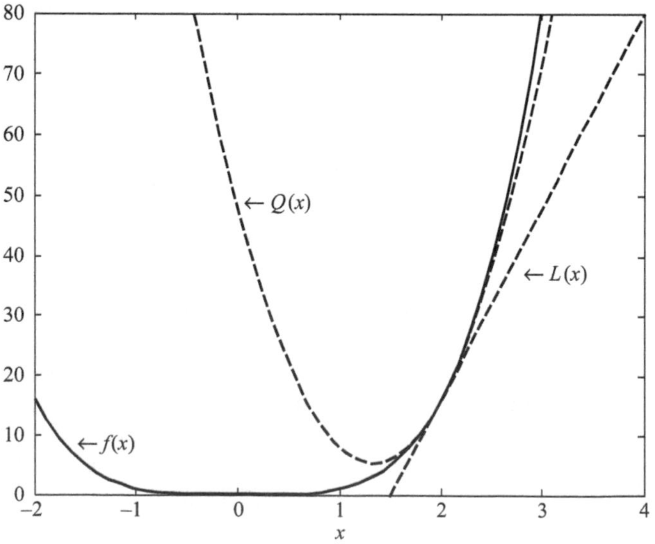

Figure 3.1 A plot of the function $f(x)=x^4$, its linear approximation $L(x)$ at the point $x=2$, and its quadratic approximation $Q(x)$ at the point $x=2$

3.3 Multidimensional Taylor expansion

The one-dimensional Taylor expansion given by (3.1) is extended to the multi-dimensional case through the formula:

$$f(\boldsymbol{x}+\Delta\boldsymbol{x}) = f(\boldsymbol{x}) + df(\boldsymbol{x}) + \frac{1}{2!}d^2 f(\boldsymbol{x}) + \cdots + \frac{1}{(n-1)!}d^{n-1}f(\boldsymbol{x}) + \frac{1}{n!}d^n f(\boldsymbol{x}+\theta\Delta\boldsymbol{x}) \tag{3.6}$$

where $\boldsymbol{x} \in \Re^n$ and $0 \le \theta \le 1$. This expansion estimates the value of a function $f(\boldsymbol{x})$ at a point $\boldsymbol{x}+\Delta\boldsymbol{x}$ in the neighborhood of the point $\boldsymbol{x}$ using the function value and derivatives at the point $\boldsymbol{x}$. The term $d^r f$ is the rth differential of the function. It contains all the possible derivatives of order r and is given by:

$$d^r f(\boldsymbol{x}) = \underbrace{\sum_i \sum_j \cdots \sum_k}_{r \text{ summations}} h_i h_j \cdots h_k \frac{\partial^r f(\boldsymbol{x})}{\partial x_i \partial x_j \cdots \partial x_k}. \tag{3.7}$$

Note that $\boldsymbol{x} = [x_1 \quad x_2 \quad x_3 \ldots x_n]^T$ is a point in an n-dimensional space and $\Delta\boldsymbol{x} = [h_1 \quad h_2 \ldots h_n]^T$. For example, for a function of three variables ($n=3$), the first-order differential ($r=1$) is given by:

$$df = h_1\frac{\partial f}{\partial x_1} + h_2\frac{\partial f}{\partial x_2} + h_3\frac{\partial f}{\partial x_3} = [h_1 \quad h_2 \quad h_3]\begin{bmatrix} \frac{\partial f}{\partial x_1} \\ \frac{\partial f}{\partial x_2} \\ \frac{\partial f}{\partial x_3} \end{bmatrix} = \Delta\boldsymbol{x}^T \nabla f \tag{3.8}$$

Similarly, the second-order differential ($r=2$) for a function of three variables ($n=3$) contains all the possible second-order derivatives of this function. It is given by:

$$d^2 f(x) = h_1^2\frac{\partial^2 f}{\partial x_1^2} + h_2^2\frac{\partial^2 f}{\partial x_2^2} + h_3^2\frac{\partial^2 f}{\partial x_3^2} + 2h_1h_2\frac{\partial^2 f}{\partial x_1\partial x_2} + 2h_1h_3\frac{\partial^2 f}{\partial x_1\partial x_3} + 2h_2h_3\frac{\partial^2 f}{\partial x_2\partial x_3} \tag{3.9}$$

The factor 2 appearing in the mixed derivatives is a result of the multidimensional summations in (3.7) where each mixed derivative appears twice. The second-order differential (3.9) can be cast in the more compact form:

$$d^2 f = [h_1 \quad h_2 \quad h_3]\begin{bmatrix} \frac{\partial^2 f}{\partial x_1^2} & \frac{\partial^2 f}{\partial x_1\partial x_2} & \frac{\partial^2 f}{\partial x_1\partial x_3} \\ \frac{\partial^2 f}{\partial x_2\partial x_1} & \frac{\partial^2 f}{\partial x_2^2} & \frac{\partial^2 f}{\partial x_2\partial x_3} \\ \frac{\partial^2 f}{\partial x_3\partial x_1} & \frac{\partial^2 f}{\partial x_3\partial x_2} & \frac{\partial^2 f}{\partial x_3^2} \end{bmatrix}\begin{bmatrix} h_1 \\ h_2 \\ h_3 \end{bmatrix} = \Delta\boldsymbol{x}^T \boldsymbol{H} \Delta x \tag{3.10}$$

where $\boldsymbol{H}$ is the symmetric Hessian matrix. From (3.6) to (3.9), we can see that the first-order Taylor approximation of an n-dimensional function is given by:

$$f(\boldsymbol{x}+\Delta \boldsymbol{x}) = f(\boldsymbol{x}) + \Delta \boldsymbol{x}^T \nabla f \tag{3.11}$$

which is the equation of an n-dimensional hyper plane. The more accurate second-order Taylor expansion is given by:

$$f(\boldsymbol{x}+\Delta \boldsymbol{x}) = f(\boldsymbol{x}) + \Delta \boldsymbol{x}^T \nabla f + \frac{1}{2}\Delta \boldsymbol{x}^T \boldsymbol{H} \Delta x \tag{3.12}$$

The first- and second-order Taylor expansions are important in optimization theory. These approximations are used to predict the behavior of a function in a certain region of the parameter space. The following example illustrates multidimensional Taylor expansion.

Example 3.2: Obtain the first- and second-order Taylor approximations of the function $f(x_1, x_2) = x_1^2 + x_2^2$ around the point $\boldsymbol{x}_0 = [1.0 \quad 1.0]^T$. Plot both the original function and its approximations.

Solution: The gradient of the function at the expansion point is given by:

$$\nabla f|_{x=[1 \quad 1]^T} = \begin{pmatrix} \dfrac{\partial f}{\partial x_1} \\ \dfrac{\partial f}{\partial x_2} \end{pmatrix}_{[1 \quad 1]^T} = \begin{pmatrix} 2x_1 \\ 2x_2 \end{pmatrix}_{[1 \quad 1]^T} = \begin{pmatrix} 2 \\ 2 \end{pmatrix} \tag{3.13}$$

It follows that the linear approximation is given by:

$$f(\boldsymbol{x}_0+\Delta \boldsymbol{x}) = f(\boldsymbol{x}_0) + \Delta \boldsymbol{x}^T \nabla f|_{\boldsymbol{x}_0} \Rightarrow f(\boldsymbol{x}_0+\Delta \boldsymbol{x}) = 2 + [h_1 \quad h_2]\begin{bmatrix} 2 \\ 2 \end{bmatrix} \tag{3.14}$$

which can be simplified to:

$$f(\boldsymbol{x}_0+\Delta \boldsymbol{x}) = 2 + 2(x_1 - 1) + 2(x_2 - 1) \tag{3.15}$$

In (3.15), we used $h_1 = x_1 - 1$ and $h_2 = x_2 - 1$. Figure 3.2 shows the original function and its linear approximation. It is obvious that the linear approximation has the same value as the function only at the expansion point. The value of the linear approximation is close enough to the function only within a limited neighborhood of the expansion point.

For the second-order approximation, the Hessian matrix is given by:

$$\boldsymbol{H}(\boldsymbol{x}_0) = \begin{bmatrix} \dfrac{\partial^2 f}{\partial x_1^2} & \dfrac{\partial^2 f}{\partial x_1 \partial x_2} \\ \dfrac{\partial^2 f}{\partial x_2 \partial x_1} & \dfrac{\partial^2 f}{\partial x_2^2} \end{bmatrix}\Bigg|_{\boldsymbol{x}_0} = \begin{bmatrix} 2 & 0 \\ 0 & 2 \end{bmatrix} \tag{3.16}$$

This results in the second-order Taylor expansion:

$$\begin{aligned} f(\boldsymbol{x}+\Delta \boldsymbol{x}) &= f(\boldsymbol{x}) + \Delta \boldsymbol{x}^T \nabla f + \frac{1}{2}\Delta \boldsymbol{x}^T \boldsymbol{H} \Delta \boldsymbol{x} \\ &= 2 + [h_1 \quad h_2]\begin{bmatrix} 2 \\ 2 \end{bmatrix} + \frac{1}{2}[h_1 \quad h_2]\begin{bmatrix} 2 & 0 \\ 0 & 2 \end{bmatrix}\begin{bmatrix} h_1 \\ h_2 \end{bmatrix} \end{aligned} \tag{3.17}$$

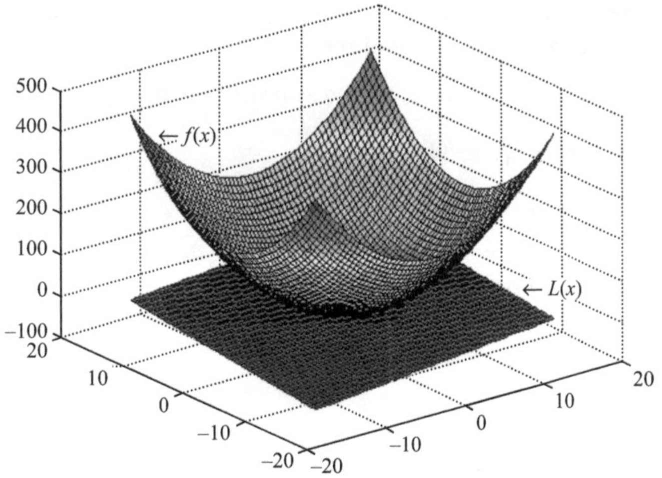

Figure 3.2 A graph of the two-dimensional function $f(\boldsymbol{x}) = x_1^2 + x_2^2$ *and its linear approximation L(x) at the point* $\boldsymbol{x} = [1 \quad 1]^T$

which can be simplified to:

$$f(\boldsymbol{x} + \Delta\boldsymbol{x}) = 2 + 2(x_1 - 1) + 2(x_2 - 1) + (x_1 - 1)^2 + (x_2 - 1)^2 = x_1^2 + x_2^2 \tag{3.18}$$

Notice that the quadratic Taylor approximation is identical to the original function. This is expected as the original function is also quadratic. In this case, a quadratic model is all what is needed to completely predict the response over the whole domain. The assumption that a differentiable function possesses a quadratic behavior over a limited neighborhood is routinely made in optimization algorithms.

3.4 Meaning of the gradient

The Taylor approximation using only first- and second-order differentials is given by (3.11) and (3.12). For a point $\boldsymbol{x} + \Delta\boldsymbol{x}$, which is very close to the point $\boldsymbol{x}$, the second-order term becomes negligible and the first-order term dominates to get (3.11). The value of this linear approximation is a function of $\Delta\boldsymbol{x}$. For every possible perturbation $\Delta\boldsymbol{x}$, the linear approximation will have a different value. If we select only perturbations lying within the neighborhood $\|\Delta\boldsymbol{x}\| \leq \varepsilon$ with ε sufficiently small, it can be shown the maximum value of this function is achieved when:

$$\Delta\boldsymbol{x} = \varepsilon \frac{\nabla f}{\|\nabla f\|} \tag{3.19}$$

In other words, the direction of maximum function increase is the direction of the gradient. The direction of maximum function decrease is then the antiparallel of the gradient. This direction is called the direction of steepest function descent or simply the direction of steepest descent.

It should be noted that (3.11) is valid only for very small perturbation. As we shift away from the expansion point, the effect of the second- and higher-order term becomes dominant. It follows that the steepest descent direction results in maximum function decrease only for sufficiently small perturbations. The following example illustrates the meaning of the gradient.

Example 3.3: Find the gradient of the Rosenbrock function given by:

$$f(\boldsymbol{x}) = 100(x_2 - x_1^2)^2 + (1 - x_1)^2 \tag{3.20}$$

at the point $\boldsymbol{x} = [1 \quad 2]^T$.

Solution: To find the gradient, we start by calculating the derivatives of the function with respect to the two parameters at the given point:

$$\begin{aligned} \frac{\partial f}{\partial x_1} &= 200(x_2 - x_1^2) \times -2x_1 - 2(1 - x_1) \\ \frac{\partial f}{\partial x_2} &= 200(x_2 - x_1^2) \end{aligned} \tag{3.21}$$

At the point $\boldsymbol{x} = [1.0 \quad 2.0]^T$, the gradient is given by:

$$\nabla f|_{[1\ 2]^T} = \left. \begin{bmatrix} \dfrac{\partial f}{\partial x_1} \\ \dfrac{\partial f}{\partial x_2} \end{bmatrix} \right|_{[1\ 2]^T} = \begin{bmatrix} -400 \\ 200 \end{bmatrix} \tag{3.22}$$

Figure 3.3 shows the direction of the gradient at this point and at a number of other points. Using Figure 3.3 and the shown function contours, we can verify that the gradient points in the direction of maximum function increase. We can also see that the direction of the gradient may change from one point to another in the parameter space.

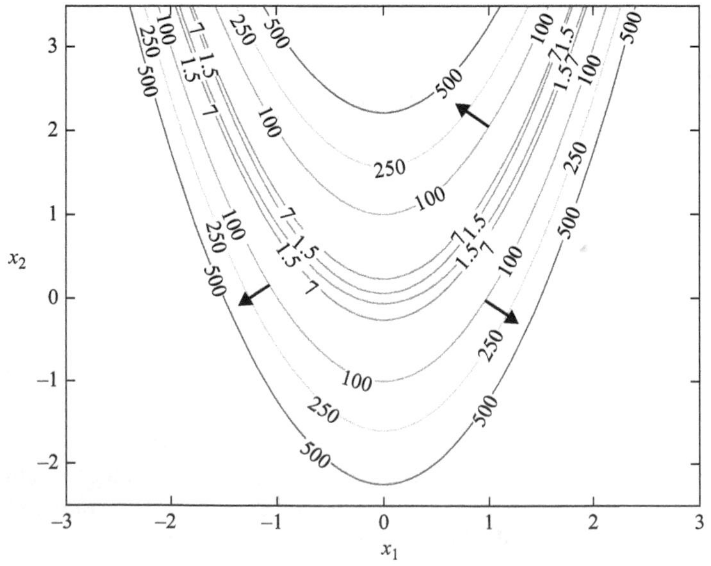

Figure 3.3 The contours of the Rosenbrock function over a subset of the parameter space. The gradient at the point $\boldsymbol{x} = [1.0 \quad 2.0]^T$ is shown along with the gradient at two other points

It can be also seen from Figure 3.3 that the direction of the gradient at a point is always normal to the function contour passing through this point. This is actually a general property of the gradient vector. In general, for a differentiable function of n variables, the gradient is normal to the constant function value surface. When $n = 2$, this surface becomes a contour. For $n = 3$, this surface becomes a three-dimensional surface and the gradient at any point is normal to this surface. The reason for this can be seen by examining the first-order Taylor expansion (3.11). If the gradient ∇f at the expansion point is normal to perturbation $\Delta \boldsymbol{x}$, then the function value does not change ($\nabla f^T \Delta \boldsymbol{x} = 0$). This means that to keep the function value unchanged over a small neighborhood, we have to move from the expansion point in a direction normal to the gradient. This explains why the gradient at any point is normal to the constant function value surface passing through that point.

Another important question to answer is, at any point in the parameter space, what directions result in a reduction in the function value. This is of prime importance since, in optimization, our target is to minimize a given cost (objective) function or maximize some profit function. From the expansion (3.11), we see that if the inner product $\Delta \boldsymbol{x}^T \nabla f$ is negative, the function value will be reduced. This inner product can be expanded as:

$$\Delta \boldsymbol{x}^T \nabla f = \|\Delta \boldsymbol{x}\| \|\nabla f\| \cos \theta \tag{3.23}$$

where θ is the angle between the two vectors ∇f and $\Delta \boldsymbol{x}$. It follows that all directions that make an obtuse angle with the gradient result in a function reduction for a very small perturbation. All directions that make an acute angle with the gradient result in a function increase. Figure 3.4 shows the gradient at a given point and three possible perturbations. The first perturbation results in a function increase

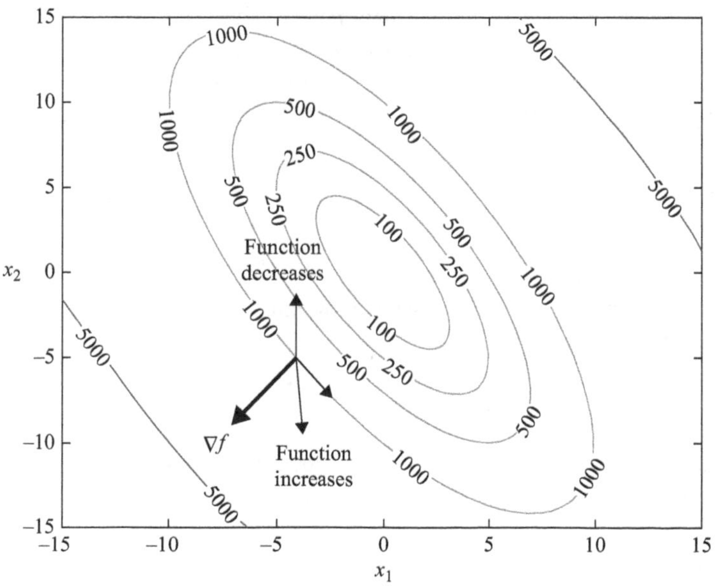

Figure 3.4 The gradient direction of an objective function as compared to three different perturbations where the objective function increases, decreases, or remains unchanged

($\theta < \pi/2$), the second perturbation results in a function decrease ($\pi/2 < \theta < \pi$), while the third one results in no change ($\theta = \pi/2$).

3.5 Optimality conditions

Optimality conditions are an important subject in optimization theory. They help characterize the optimal solutions of a given optimization problem. In some algorithms, these optimality conditions are used as termination conditions of the algorithm. Other possible termination conditions include maximum number of iterations, the size of the step taken in an optimization iteration, etc. Optimality conditions depend on whether the problem is constrained or unconstrained. They can in general be classified into necessary and sufficient conditions. An optimal point must satisfy a necessary condition. However, not all points satisfying necessary conditions are optimal points. Sufficient conditions are stricter. Every point that satisfies sufficient conditions must be an optimal point. No other type of points satisfies sufficient optimality conditions. Sufficient conditions are, however, more difficult to check because they usually require costly information about the function behavior. We discuss in the following sections these cases and illustrate them with examples.

3.6 Unconstrained optimization

In nonlinear unconstrained optimization, our target is to solve the following optimization problem:

$$\boldsymbol{x}^* = \arg \min_{\boldsymbol{x}} f(\boldsymbol{x}) \tag{3.24}$$

where $\boldsymbol{x}^*$ is the desired set of optimal parameters. The vector $\boldsymbol{x}$ can be any point in the whole parameter space with no constraints. The optimization formulation (3.24) states that we seek the optimal point at which the objective function $f(\boldsymbol{x})$ reaches its minimum value. The following theorem states the necessary conditions for the point $\boldsymbol{x}^*$.

Theorem 3.1: If $\boldsymbol{x}^*$ is a local minimum of the function $f(\boldsymbol{x})$, then this point must satisfy $\nabla f(\boldsymbol{x}^*) = \boldsymbol{0}$.

Proof: Utilizing Taylor's expansion at the optimal point $\boldsymbol{x}^*$ with a sufficiently small perturbation $\Delta\boldsymbol{x}$, the linear term dominates to get the relatively accurate approximation:

$$f(\boldsymbol{x}^* + \Delta\boldsymbol{x}) = f(\boldsymbol{x}^*) + \Delta\boldsymbol{x}^T \nabla f(\boldsymbol{x}^*) \tag{3.25}$$

If the gradient at the optimal point $\nabla f(\boldsymbol{x}^*)$ does not vanish, one can always find a direction $\Delta\boldsymbol{x}$ such that $\Delta\boldsymbol{x}^T \nabla f(\boldsymbol{x}^*) < 0$. A possible such direction is $\Delta\boldsymbol{x} = -\varepsilon\nabla f(\boldsymbol{x}^*)$, where ε is a sufficiently small positive scalar. This implies that the function can be further reduced because, from (3.25), $f(\boldsymbol{x}^* + \Delta\boldsymbol{x}) < f(\boldsymbol{x}^*)$. This would imply that $\boldsymbol{x}^*$ is not an optimal point, which leads to a contradiction. It follows that at an optimal point, the gradient must vanish.

It should be noted that points other than a local minimum satisfy the necessary optimality condition. These points include local maxima and inflection points

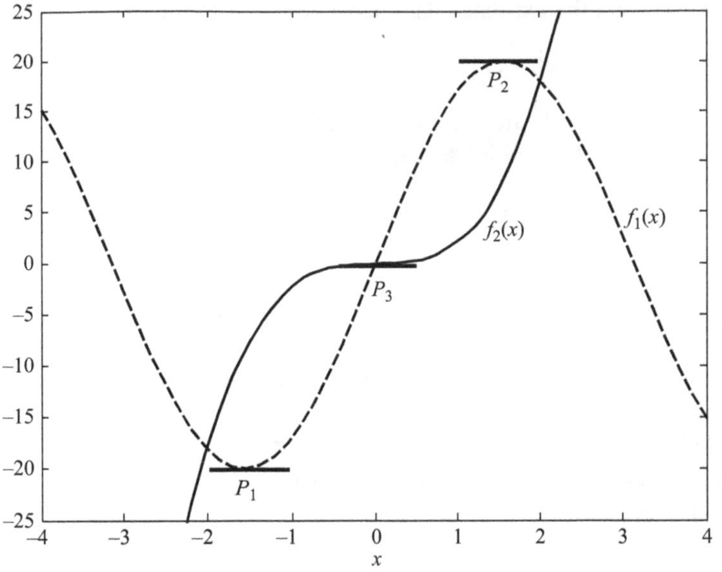

Figure 3.5 An illustration of different types of points at which the gradient vanishes; for the function $f_1(x)$, the gradient vanishes (tangential line is horizontal) at the point P_1 (local minimum) and the point P_2 (local maximum). For the function $f_2(x)$, the gradient vanishes at the point P_3, which is an inflection point

where the "slope" of the function changes sign. Figure 3.5 shows three types of points at which the gradient of the function vanishes for the one-dimensional case.

The sufficient optimality conditions for unconstrained optimization are stricter. They are given by the following theorem.

Theorem 3.2: If $\nabla f(\boldsymbol{x}^*) = 0$, and the Hessian matrix at the optimal point $\boldsymbol{H}(\boldsymbol{x}^*)$ is positive semi-definite, i.e., $\Delta\boldsymbol{x}^T\boldsymbol{H}(\boldsymbol{x}^*)\Delta\boldsymbol{x} \geq 0, \forall\Delta\boldsymbol{x}$, then $\boldsymbol{x}^*$ is a local minimum.

Proof: For a sufficiently small $\Delta\boldsymbol{x}$, the function behavior in the neighborhood of the optimal point $\boldsymbol{x}^*$ is approximated by the second-order Taylor's expansion:

$$f(\boldsymbol{x} + \Delta\boldsymbol{x}^*) = f(\boldsymbol{x}^*) + \Delta\boldsymbol{x}^T\nabla f(\boldsymbol{x}^*) + \frac{1}{2}\Delta\boldsymbol{x}^T\boldsymbol{H}(\boldsymbol{x}^*)\Delta\boldsymbol{x} \tag{3.26}$$

If the gradient $\nabla f(\boldsymbol{x}^*)$ is zero, the first-order term vanishes. The expansion (3.26) is then reduced to:

$$f(\boldsymbol{x} + \Delta\boldsymbol{x}^*) = f(\boldsymbol{x}^*) + \frac{1}{2}\Delta\boldsymbol{x}^T\boldsymbol{H}(\boldsymbol{x}^*)\Delta\boldsymbol{x} \tag{3.27}$$

If the Hessian matrix is positive semi-definite, then for any sufficiently small perturbation, we have $f(\boldsymbol{x} + \Delta\boldsymbol{x}^*) \geq f(\boldsymbol{x}^*)$, which implies that $\boldsymbol{x}^*$ is indeed a local minimum. Notice that as the size of the perturbation $\Delta\boldsymbol{x}$ increases, the third- and higher-order terms become dominant and the theory does not hold. This is why the theory provides optimality conditions for a small neighborhood around the optimal point.

Example 3.4: Find the extreme (stationary) points of the function:

$$f(x_1, x_2) = x_1^3 + x_2^3 + 2x_1^2 + 4x_2^2 + 6$$

Solution: Extreme points are the ones that satisfy the necessary optimality conditions. They can be obtained by setting the components of the gradient to zero:

$$\begin{aligned} \frac{\partial f}{\partial x_1} &= 0 \Rightarrow x_1(3x_1 + 4) = 0 \\ \frac{\partial f}{\partial x_2} &= 0 \Rightarrow x_2(3x_2 + 8) = 0 \end{aligned} \tag{3.28}$$

Setting simultaneously the two gradient components to zero, we get four possible solutions $\boldsymbol{p}_1 = [0 \quad 0]^T$, $\boldsymbol{p}_2 = [0 \quad -8/3]^T$, $\boldsymbol{p}_3 = [-4/3 \quad 0]^T$, and $\boldsymbol{p}_4 = [-4/3 \quad -8/3]^T$. These points may or may not be local minima depending on the Hessian matrix. The Hessian matrix at a general point $\boldsymbol{x}$ is calculated using the second-order derivatives:

$$H(x) = \begin{bmatrix} \dfrac{\partial^2 f}{\partial x_1^2} & \dfrac{\partial^2 f}{\partial x_1 \partial x_2} \\ \dfrac{\partial^2 f}{\partial x_2 \partial x_1} & \dfrac{\partial^2 f}{\partial x_2^2} \end{bmatrix} = \begin{bmatrix} 6x_1 + 4 & 0 \\ 0 & 6x_2 + 8 \end{bmatrix} \tag{3.29}$$

We then check the nature of the curvature of the function (Hessian) at each one of these points to get:

$$\begin{aligned} \boldsymbol{H}(\boldsymbol{p}_1) &= \begin{bmatrix} 4 & 0 \\ 0 & 8 \end{bmatrix} \Rightarrow \text{positive definite} \Rightarrow \boldsymbol{p}_1 \text{ is local minimum} \\ \boldsymbol{H}(\boldsymbol{p}_2) &= \begin{bmatrix} 4 & 0 \\ 0 & -8 \end{bmatrix} \Rightarrow \text{indefinite} \Rightarrow \boldsymbol{p}_1 \text{ is a saddle point} \\ \boldsymbol{H}(\boldsymbol{p}_3) &= \begin{bmatrix} -4 & 0 \\ 0 & 8 \end{bmatrix} \Rightarrow \text{indefinite} \Rightarrow \boldsymbol{p}_3 \text{ is a saddle point} \\ \boldsymbol{H}(\boldsymbol{p}_4) &= \begin{bmatrix} -4 & 0 \\ 0 & -8 \end{bmatrix} \Rightarrow \text{negative definite} \Rightarrow \boldsymbol{p}_4 \text{ is a local maximum} \end{aligned} \tag{3.30}$$

The analytical results obtained in (3.30) can be confirmed by examining the contour plot as shown in Figure 3.6.

3.7 Optimization with equality constraints

The second type of optimality conditions is the one concerning problems with equality constraints. These problems are not though as common as problems with inequality constraints. An optimization problem with such constraints can be cast in the form:

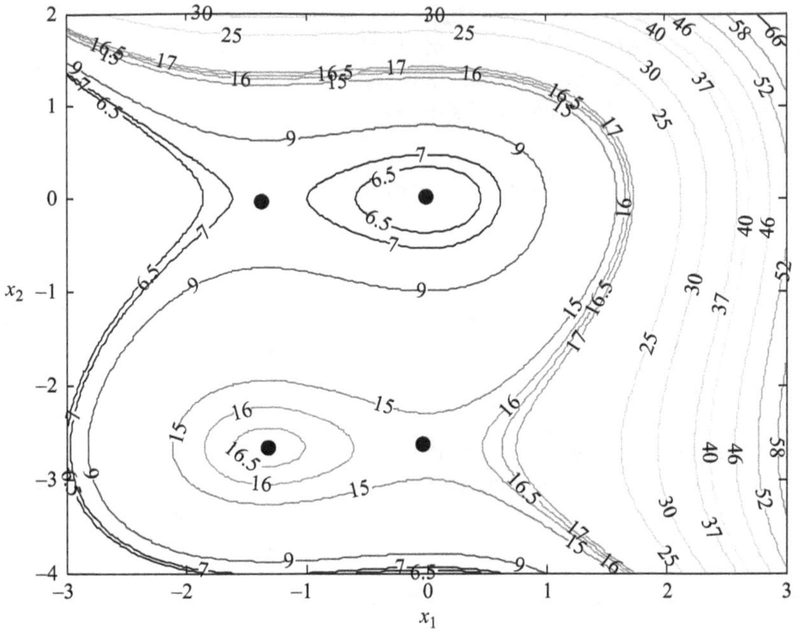

Figure 3.6 The contour plots of the function in Example 3.4; the local maximum, the local minimum, and the two saddle points are indicated by black dots on the plot

$$
\begin{aligned}
& \boldsymbol{x}^* = \arg\min_{\boldsymbol{x}} f(\boldsymbol{x}) \\
\text{subject to} \quad & g_j(\boldsymbol{x}) = 0, \ \ j = 1, 2, \ldots, m \\
& \boldsymbol{x} \in \Re^n
\end{aligned}
\tag{3.31}
$$

One of the classical approaches for solving such a problem is to convert the problem (3.31) into an unconstrained problem by eliminating the constraints. This method is usually referred to as direct substitution, which is first addressed.

3.7.1 Method of direct substitution

One possible classical approach for solving such problems is to use each constraint to eliminate one variable. This converts the problem into an unconstrained optimization problem with fewer parameters. This approach can be illustrated through the following example.

Example 3.5: Solve the optimization problem:

$$
\begin{aligned}
\text{minimize} \quad & f(\boldsymbol{x}) = x_1^2 + (x_2 - 1)^2 \\
\text{subject to} \quad & -2x_1^2 + x_2 = 4
\end{aligned}
\tag{3.32}
$$

Solution: We use the constraint to eliminate the variable x_2 by substituting $x_2 = 4 + 2x_1^2$ in the objective function to get the single-variable new objective function:

$$
\text{minimize} \quad f(x_1) = x_1^2 + (2x_1^2 + 3)^2 \tag{3.33}
$$

The problem (3.33) is an unconstrained optimization problem in a single variable only. The substitution reduces the number of unknowns to only 1. We then apply the necessary unconstrained optimization conditions to the new objective function by setting the first-order sensitivities to zero to get:

$$\frac{df}{dx_1} = 2x_1 + 2(2x_1^2 + 3) \times 4x_1 = 8x_1^3 + 26x_1 = 0 \tag{3.34}$$

Solving this equation, we obtain only one real solution $x_1 = 0$. Substituting into the constraint of the original problem (3.32), we get $x_2 = 4$. It can be confirmed that this point is indeed a local minimum by checking the sign of the Hessian matrix of the function (3.33). Notice that for a single-variable objective function, the Hessian matrix becomes a scalar. In this case, the positive definiteness condition is reduced to the non-negativity condition on the second-order derivative.

As illustrated in this example, the direct substitution approach converts a constrained problem into an unconstrained problem. It should, however, be used with caution; otherwise, wrong and incorrect results may be obtained. The following example illustrates such a case.

Example 3.6: Solve the constrained optimization problem:

$$\begin{aligned} &\text{minimize} \quad f(\boldsymbol{x}) = x_1^2 + x_2^2 \\ &\text{subject to} \quad (x_1 - 1)^3 = x_2^2 \end{aligned} \tag{3.35}$$

Solution: First, I give a graphical solution of this problem. The contours of the objection function are circles with the center at the origin. It follows that we are looking for the contour with the least possible value that touches the constraint. The point at which the contour touches the constraint is the optimal point. This solution is achieved at the point $\boldsymbol{x}^* = [1.0 \quad 0]^T$ as shown in Figure 3.7. Notice that in this example the equality constraint is not differentiable at the optimal point.

It can be seen from Figure 3.7 that the solution is achieved at a point at which the contour of the objective function touches the constraint. This property will be used later for developing another optimization approach for problems with equality constraints.

Using direct substitution, we use $x_2^2 = (x_1 - 1)^3$; the new single-variable optimization problem is given by:

$$\text{minimize} \quad f(x_1) = x_1^2 + (x_1 - 1)^3 \tag{3.36}$$

This new optimization problem is unlimited because as $x_1 \to -\infty, f(x_1) \to -\infty$. This means that the new constrained problem does not have a minimum even though the original problem, as evident from Figure 3.7, has a local minimum. So, what went wrong? Things can be explained as follows. The constraint in (3.35) has the implicit constraint that $x_1 \geq 1$ because $x_2^2 \geq 0$. When we substituted for x_2^2 into the original objective function, we effectively removed this implicit constraint resulting in a completely different optimization problem. The new problem does not have a solution as the original one does.

3.7.2 *Method of constrained variation*

The method of constrained variation can be illustrated through Figure 3.8. This figure shows the contour of the objective function at which the minimum of the

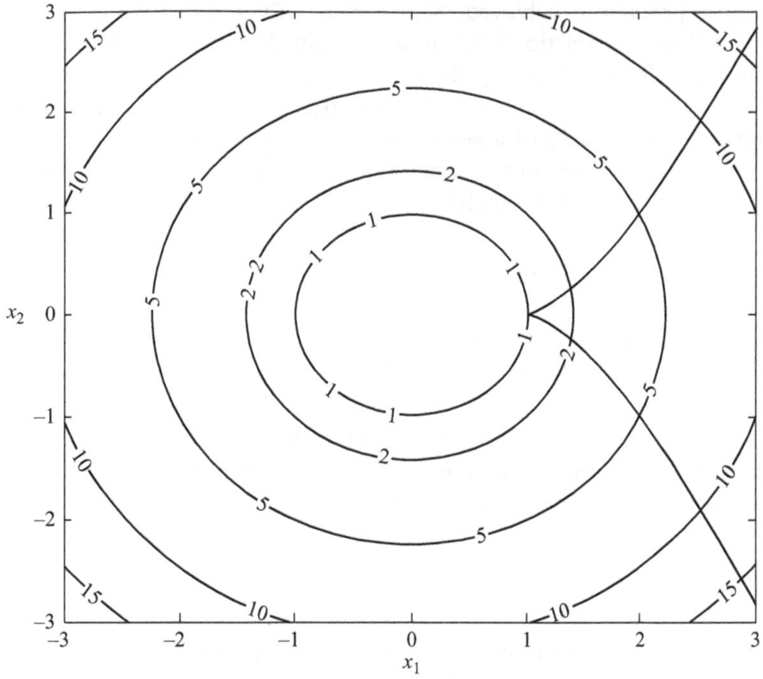

Figure 3.7 *The contours of the function $f(\mathbf{x}) = x_1^2 + x_2^2$ and the constraint $(x_1 - 1)^3 = x_2^2$. Notice that the constraint function has two branches. The point on the constraint with the minimum function value is $\mathbf{x}^* = [1 \quad 0]^T$*

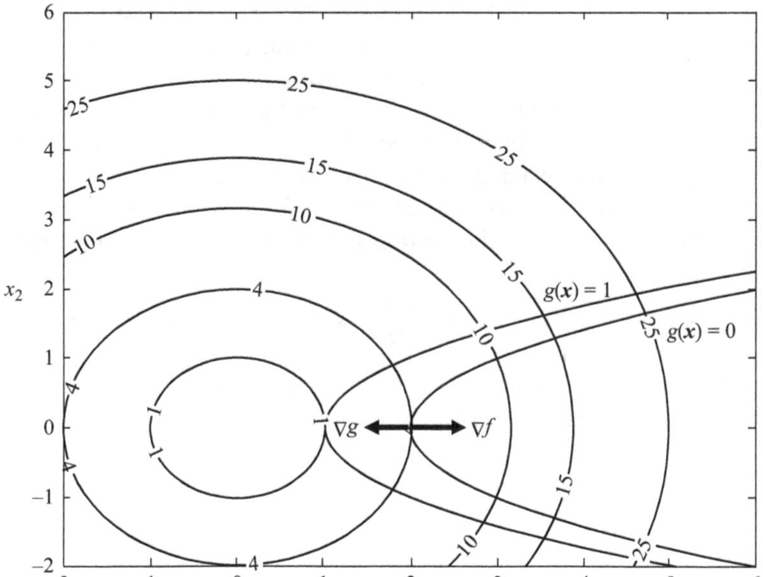

Figure 3.8 *An illustration of an optimization problem with an equality constraint; the differentiable constraint is $g(\mathbf{x}) = x_2^2 - x_1 + 2 = 0$. At the optimal point $\mathbf{x}^* = [2 \quad 0]^T$, the gradient of the constraint and the objective function are antiparallel*

optimization problem is achieved. Notice that at the optimal point, as explained earlier, the objective function contour and the equality constraint are tangential to each other. It follows that their gradients may be parallel or antiparallel at this point depending on the nature of the objective function and the constraint. We illustrate this approach for the case of a single equality constraint problem with two variables. Any infinitesimally small variation $\boldsymbol{dx}$ from the optimal point must also lie on the constraint and it thus satisfies:

$$dg = \frac{\partial g}{\partial x_1} dx_1 + \frac{\partial g}{\partial x_2} dx_2 = 0 \tag{3.37}$$

Equation (3.37) simply states that the constraint is still active at the new point, i.e., $g(\boldsymbol{x} + \boldsymbol{dx}) = 0$. It follows that its first-order change vanishes for the perturbation. This simply implies that the perturbation is normal to the gradient of the constraint at the optimal point. Because the gradient of the objective function and the gradient of the constraint are either parallel or antiparallel at the solution, this perturbation is also normal to the gradient of the objective function:

$$df = \frac{\partial f}{\partial x_1} dx_1 + \frac{\partial f}{\partial x_2} dx_2 = 0 \tag{3.38}$$

Using (3.37), we can express dx_2 in terms of dx_1 to get $dx_2 = (-(\partial g/\partial x_1)/(\partial g/\partial x_2))dx_1$. Substituting into (3.38), we get:

$$\left(\frac{\partial f}{\partial x_1} + \frac{-(\partial g/\partial x_1)}{\partial g/\partial x_2}\frac{\partial f}{\partial x_2}\right) dx_1 = 0 \tag{3.39}$$

Equation (3.39) must be valid for an arbitrary dx_1. It follows that the term multiplying dx_1 is the one that vanishes. This gives one equation that can be solved along with the constraint equation for the two unknowns x_1^* and x_2^*.

The procedure explained above for the case of a two-variable problem can be expanded to the general problem with equality constraint (3.31). At the solution, the constraints must intersect and the objective function contour must be tangential to their curve of intersection. As any perturbation from the optimal point must keep the perturbation feasible, i.e., all constraints are satisfied, we must then have:

$$dg_j = \frac{\partial g_j}{\partial x_1} dx_1 + \frac{\partial g_j}{\partial x_2} dx_2 + \cdots + \frac{\partial g_j}{\partial x_n} dx_n = 0, \qquad j = 1, 2, \ldots, m \tag{3.40}$$

Because the objective function contour passing through the optimal solution is also tangential to the contour of intersection of all the equality constraints, it follows that we have:

$$df = \frac{\partial f}{\partial x_1} dx_1 + \frac{\partial f}{\partial x_2} dx_2 + \cdots + \frac{\partial f}{\partial x_n} dx_n = 0 \tag{3.41}$$

Equations (3.40) and (3.41) represent $m + 1$ equations in n unknowns (the coordinates of $\boldsymbol{x}^*$). We can thus express dx_1, dx_2, ..., dx_m in terms of the remaining $(n - m)$ differential elements to get equations of the form:

$$\alpha_1 dx_{m+1} + \alpha_2 dx_{m+2} + \cdots + \alpha_{n-m} dx_n = 0 \tag{3.42}$$

These equations are then solved with the original m constraints to obtain the unknown n coordinates of the point $\boldsymbol{x}^*$. The following example illustrates the method of constrained variations.

Example 3.7: Solve the following equality constrained problem:

$$\begin{aligned}&\text{minimize} \quad f(x) = \frac{k}{x_1 x_2^2}\\&\text{subject to} \quad x_1^2 + x_2^2 - a^2 = 0, \quad x_1 > 0\end{aligned} \tag{3.43}$$

using the method of constrained variations.

Solution: First, we calculate the gradient of the objective function and the equality constraint to get:

$$\begin{aligned}&\frac{\partial f}{\partial x_1} = -kx_1^{-2}x_2^{-2}, \quad \frac{\partial f}{\partial x_2} = -2kx_1^{-1}x_2^{-3}\\&\frac{\partial g}{\partial x_1} = 2x_1, \quad \frac{\partial g}{\partial x_1} = 2x_2\end{aligned} \tag{3.44}$$

Using (3.39), we get:

$$x_1^{-2}x_2^{-2} = \frac{2x_1}{x_2}x_1^{-1}x_2^{-3} \Rightarrow x_2 = \pm\sqrt{2}x_1 \tag{3.45}$$

Substituting (3.45) into the constraint equation, we obtain:

$$2x_1^2 + x_1^2 = a^2 \Rightarrow x_1 = \frac{a}{\sqrt{3}} \Rightarrow x_2 = \pm\frac{\sqrt{2}a}{\sqrt{3}} \tag{3.46}$$

Negative solutions of x_1 were rejected to accommodate the non-negativity constraint on x_1. It follows that the optimal solutions are given by $\boldsymbol{x}^* = \begin{bmatrix} a/\sqrt{3} & \pm\sqrt{2}a/\sqrt{3} \end{bmatrix}^T$. Figure 3.9 shows the graphical solution of this problem for $k=2$ and $a=3.0$.

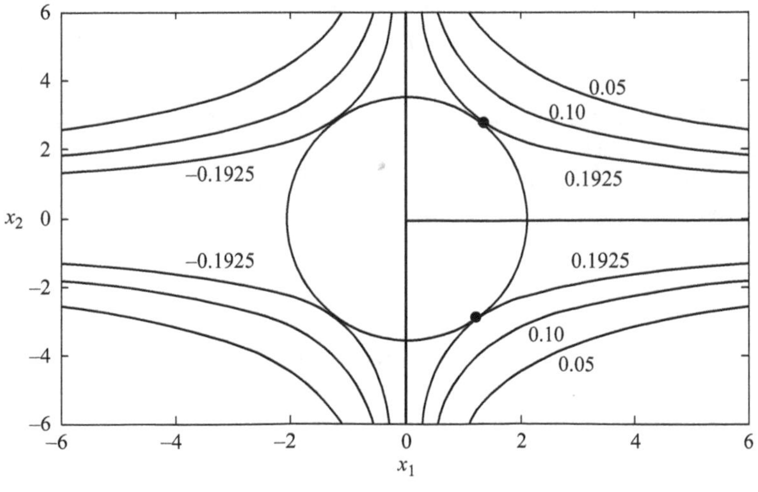

Figure 3.9 The contours of the objective function and the equality constraint of problem 3.7

3.8 Lagrange multipliers

It was mentioned earlier that for the case of an optimization problem with a single equality constraint, the contour of the objective function passing through the optimal point is tangential to the constraint. A possible illustration of this case is shown in Figure 3.8. The gradients of the objective function and the constraint may be parallel or antiparallel. Mathematically, this implies that at the optimal solution we must have:

$$\nabla f(\boldsymbol{x}^*) + \lambda \nabla g(\boldsymbol{x}^*) = 0 \tag{3.47}$$

where λ is a scaling parameter, called the Lagrange multiplier. This parameter may be positive or negative depending on the objective function and the constraint. In Figure 3.8, λ is positive because of the direction of increase in the constraint function.

For the general case, where there is more than one equality constraint, the constant value surface of the objective function passing through the optimal point is tangential to the surface of the intersection of all constraints. It follows that the gradient of the objective function at the optimal point can be expressed as a linear combination of the gradients of all equality constraints at this point. This implies that we have at the optimal point $\boldsymbol{x}^*$:

$$\nabla f(\boldsymbol{x}^*) + \sum_j \lambda_j \nabla g_j(\boldsymbol{x}^*) = 0 \tag{3.48}$$

where λ_j is the Lagrange multiplier of the jth equality constraint. The Lagrange multipliers can be positive or negative depending on the objective function and the constraints. Equation (3.48) represents n equations in $n+m$ unknowns (the coordinates of the optimal point $\boldsymbol{x}^*$ and the values of the m Lagrange multipliers). Combining these equations with the m constraint equations, we obtain the system of equations:

$$\begin{aligned} \nabla f(\boldsymbol{x}^*) + \sum_j \lambda_j \nabla g_j(\boldsymbol{x}^*) &= 0 \\ g_j(\boldsymbol{x}^*) = 0, \quad j &= 1, 2, \ldots, m \end{aligned} \tag{3.49}$$

Figure 3.10 illustrates this case for two equality examples in a two-dimensional problem. The $(n+m)$ equations (3.49) are solved for all unknowns to determine the optimal point $\boldsymbol{x}^*$ and the corresponding Lagrange multipliers. This same target can be achieved by forming the scalar Lagrangian function:

$$L(\boldsymbol{x}, \boldsymbol{\lambda}) = f(\boldsymbol{x}^*) + \sum_j \lambda_j g_j(\boldsymbol{x}^*) \tag{3.50}$$

Differentiating this Lagrangian function relative to $\boldsymbol{x}^*$ and all λ_j the $(n+m)$ equations in (3.49). It follows that we can use the Lagrangian function (3.49) as a new objective function. The following two examples illustrate utilizing the Lagrange multiplier approach for solving problems with equality constraints.

Example 3.8: Find the dimensions of a closed cardboard box with maximum volume and given surface area A.

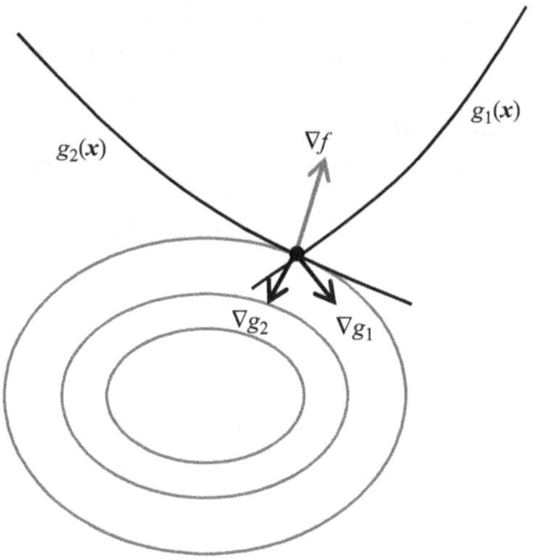

Figure 3.10 An illustration of the optimality conditions for the case of two equality constraints and one objective function. At the optimal point, the gradient of the objective function is a linear combination of the gradients of the two constraints at the optimal point

Solution: The cardboard box has three parameters: the length l, the width w, and the height h. The volume of this box is $V = lwh$, while its surface area is given by $S = 2lh + 2wh + 2wl$. The optimization problem is thus given by:

$$\begin{array}{ll} \text{maximize} & lwh \\ \text{subject to} & 2lh + 2wh + 2wl = A \end{array} \tag{3.51}$$

According to (3.50), the corresponding Lagrangian function is given by:

$$L = -lwh + \lambda(2lh + 2wh + 2wl - A) \tag{3.52}$$

Maximizing the volume is equivalent to minimizing its negative. This is why a negative sign was added to the original objective function. Differentiating (3.52) with respect to the three parameters and the Lagrange multiplier, we get:

$$\begin{aligned} \frac{\partial L}{\partial l} &= -wh + \lambda(2h + 2w) = 0 \\ \frac{\partial L}{\partial w} &= -lh + \lambda(2h + 2l) = 0 \\ \frac{\partial L}{\partial h} &= -lw + \lambda(2l + 2w) = 0 \\ \frac{\partial L}{\partial \lambda} &= 2lh + 2wh + 2wl - A = 0 \end{aligned} \tag{3.53}$$

Because of the symmetry of these equations, the optimal solution has equal values for the parameters l, w, and h. Solving these four equations, we obtain the solution:

$$\boldsymbol{x}^* = [l^* \quad w^* \quad h^*]^T = \left[\sqrt{\tfrac{A}{6}} \quad \sqrt{\tfrac{A}{6}} \quad \sqrt{\tfrac{A}{6}}\right]^T, \quad \lambda^* = \frac{1}{4}\sqrt{\frac{A}{6}} \tag{3.54}$$

A zero solution for (3.53) is not a feasible solution for this problem as it violates the area constraint. Also, because the multiplier is positive, this means that the gradient of the objective function and that of the constraint are antiparallel at the solution. If the constraint is changed to $2lh + 2wh + 2wl - A_2 = 0$, with $A_2 > A$, then the new equality is a more relaxed constraint and an improved (lower) objective function is reached, which corresponds to a bigger volume.

To understand the meaning of the Lagrange multipliers, we consider a constraint of the form $g(\boldsymbol{x}) = b$. If this constraint is relaxed to $g(\boldsymbol{x}) - b - \delta b = 0$, this results in a change $d\boldsymbol{x}$ in the optimal solution. Using first-order Taylor expansion of the constraint, we can write:

$$\delta b = \frac{\partial g}{\partial x_1} dx_1 + \frac{\partial g}{\partial x_2} dx_2 + \cdots + \frac{\partial g}{\partial x_n} dx_n = \sum_i \frac{\partial g}{\partial x_i} dx_i \tag{3.55}$$

The Lagrangian of the new problem is given by:

$$L(\boldsymbol{x}, \lambda) = f(\boldsymbol{x}) + \lambda(g(\boldsymbol{x}) - b - \delta b) \tag{3.56}$$

Differentiating with respect to parameters, we get:

$$\frac{\partial f}{\partial \boldsymbol{x}} + \lambda \frac{\partial g}{\partial \boldsymbol{x}} = 0 \Rightarrow \frac{\partial f}{\partial x_i} = -\lambda \frac{\partial g}{\partial x_i} \forall i \tag{3.57}$$

Substituting (3.57) into (3.55), we get:

$$\delta b = -\frac{1}{\lambda} \sum_i \frac{\partial f}{\partial x_i} dx_i = -\frac{\delta f}{\lambda} \tag{3.58}$$

This implies that the Lagrange multiplier relates the change in the objective function value at the optimal solution to the change in the constraint value b. If λ is positive, a positive change δb in the constraint results in a negative change in the value of the objective function at the optimal point. The relaxed problem has a lower optimal value of the objective function. If λ is negative, a positive change in the value of the constraint results in a positive change in the value of the objective function at the optimal point. It follows that the Lagrange multiplier of a constraint represents the sensitivity of the objective function at the optimal point with respect to the change in this constraint. Figure 3.11 illustrates this interpretation of the Lagrange multipliers.

3.9 Optimization with inequality constraints

Another important class of optimization problems involves inequality constraints. These problems appear very often in engineering applications. An optimization problem with m inequality constraints is given by:

$$\begin{aligned} &\text{minimize} \quad f(\boldsymbol{x}) \\ &\text{subject to} \quad g_j(\boldsymbol{x}) \leq 0, \quad j = 1, 2, \ldots, m \end{aligned} \tag{3.59}$$

The m inequality constraints define a subset of the parameter space. As explained earlier, this subset is denoted as the feasible region or the constrained region. We say that the jth constraint is active at a point $\boldsymbol{x}$ if it is satisfied as an equality

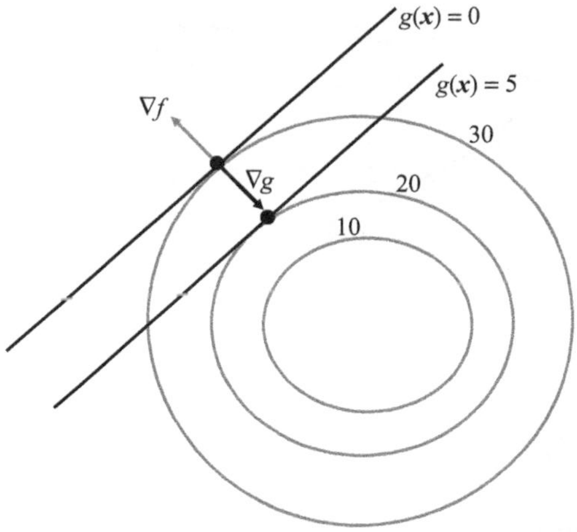

Figure 3.11 Illustration of the meaning of the Lagrange multipliers; at the optimal point the gradients of the constraint and objective function are antiparallel (positive multiplier). A positive relaxation in the constraint ($\delta b = 5$) results in a negative change ($\delta f = -10$) in the value of the objective function at the new optimal point

($g_j(\boldsymbol{x}) = 0$). The constraints in (3.59) can be converted to equality constraints by adding positive slack variables to get:

$$\begin{array}{ll} \text{minimize} & f(\boldsymbol{x}) \\ \text{subject to} & g_j(\boldsymbol{x}) + y_j^2 = 0, \quad j = 1, 2, \ldots, m \end{array} \tag{3.60}$$

The problem (3.60) is an optimization problem with only equality constraints. The transition from (3.59) to (3.60) is achieved at the expense of adding m more positive slack variables. This is very similar to the Simplex method discussed in Chapter 2. A slack variable becomes zero if the corresponding constraint is active, i.e., satisfied as an equality. The Lagrangian of the problem (3.60) has the form:

$$L(\boldsymbol{x}, \boldsymbol{y}, \boldsymbol{\lambda}) = f(\boldsymbol{x}) + \sum_{j=1}^{m} \lambda_j (g_j(\boldsymbol{x}) + y_j^2) \tag{3.61}$$

where $\boldsymbol{y} = [y_1 \quad y_2 \ldots y_m]^T$ and $\boldsymbol{\lambda} = [\lambda_1 \quad \lambda_2 \ldots \lambda_m]^T$. The problem (3.61) has $(n + 2m)$ unknowns $\boldsymbol{x}^*$, $\boldsymbol{y}^*$, and $\boldsymbol{\lambda}^*$. The optimality conditions for this problem are given by:

$$\begin{aligned} &\frac{\partial L}{\partial \boldsymbol{x}} = 0 \Rightarrow \frac{\partial f(\boldsymbol{x})}{\partial \boldsymbol{x}} + \sum_{j=1}^{m} \lambda_j \frac{\partial g_j(\boldsymbol{x})}{\partial \boldsymbol{x}} = 0 \\ &\frac{\partial L}{\partial y_j} = 0 \Rightarrow 2\lambda_j y_j = 0 \Rightarrow y_j = 0 \text{ or } \lambda_j = 0 \\ &\frac{\partial L}{\partial \lambda_j} = 0 \Rightarrow g_j(\boldsymbol{x}) + y_j^2 = 0 \end{aligned} \tag{3.62}$$

These conditions are usually called the Karush–Kuhn–Tucker (KKT) conditions. The first condition implies that at the optimal solution, the gradient of the objective function is a linear combination of the gradients of all constraints with nonzero

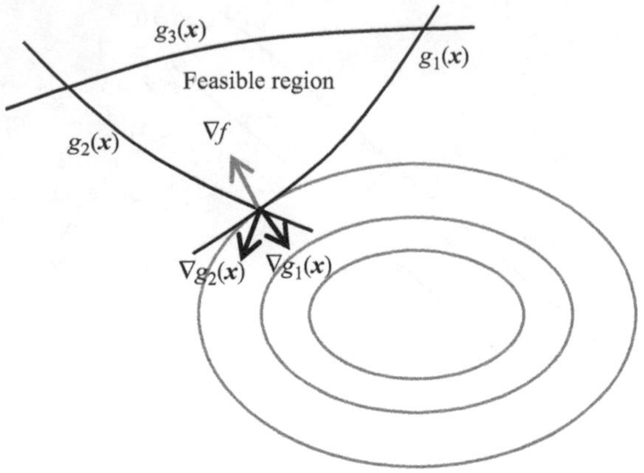

Figure 3.12 An illustration of the optimality conditions for inequality constraints; the feasible region is defined by three constraints and at the optimal point, two of them are active. At this point, the gradient of the objective function is a linear function of the gradients of the active constraints

Lagrange multipliers. The second condition implies that either the Lagrange multiplier is zero, and in that case, the constraint is not active ($y_j \neq 0$), or the Lagrange multiplier is not zero, and in that case the corresponding constraint is active ($y_j = 0$). Combining the first and second sets of constraints, one concludes that at the optimal solution, the gradient of the objective function is a linear combination of the gradients of active constraints. The last set of equations implies that all inequality constraints must be satisfied either with a zero slack variable or with a nonzero slack variable. Figure 3.12 illustrates the optimality conditions for the case $m = 3$ and $n = 2$.

The necessary KKT condition for inequality constraints can thus be cast in the standard form:

$$\begin{aligned} \frac{\partial f}{\partial x_i} + \sum_{j=1}^{m} \lambda_j \frac{\partial g_j}{\partial x_i} &= 0, \quad i = 1, 2, \ldots, n \\ \lambda_j g_j &= 0, \quad j = 1, 2, \ldots, m \\ g_j &\leq 0, \quad j = 1, 2, \ldots, m \\ \lambda_j &\geq 0, \quad j = 1, 2, \ldots, m \end{aligned} \tag{3.63}$$

In (3.63), we imposed the additional condition that the Lagrange multiplier corresponding to an active constraint must be positive. This characteristic distinguishes problems with inequality constraints from problems with equality constraints. The reason for this can be seen in Figure 3.13. This figure shows two possible points on the boundary of the feasible region defined by the inequality constraints. Point $\boldsymbol{p}_1$ is a point at which the gradient of the objective function can be written as a linear combination of the gradients of the two active constraints with negative coefficients (positive Lagrange multipliers). All the feasible points in the neighborhood of $\boldsymbol{p}_1$ have higher value for the objective function implying that this is a local minimum of the problem. The other point ($\boldsymbol{p}_2$) is not a local minimum as the

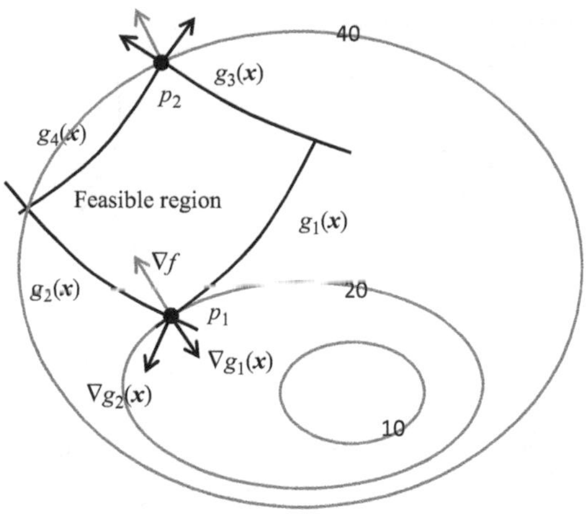

Figure 3.13 An illustration of the sign of the Lagrange multipliers; at optimal point $\boldsymbol{p}_1$, the gradient of the objective function is a linear combination of the gradients of the active constraints with negative coefficients (positive Lagrange multipliers). At point $\boldsymbol{p}_2$, at least one of the coefficients is positive (negative multiplier). There are feasible points in the neighborhood of $\boldsymbol{p}_2$ that have a lower value of the objective function

necessary optimality conditions are violated. It can be shown that the gradient of the objective function at this point can be expanded as a linear combination of the gradients of the active constraints at this point with at least one positive expansion coefficient (negative Lagrange multiplier). There are feasible points in the neighborhood of this point that have a lower value of the objective function. It follows that this point is not a local minimum.

It should be noted that we do not know beforehand which inequality constraints are active at the optimal solution. The objective function may have an unconstrained minimum inside the feasible region. In that case, no constraint is active at the optimal point and all constraints have zero Lagrange multipliers. The gradient of the objective function must vanish at this point, which is consistent with (3.63). Figure 3.14 shows the case where only one constraint is active at the optimal point. The contour of the objective function touches the active constraint at the optimal point. The gradient of the objective function is antiparallel to the gradient of the constraint at this point, which is also consistent with (3.63). Alternatively, one or more constraint may be active at the optimal point. As explained earlier, in this case, the gradient of the objective function at this point is a negative combination of the gradients of the active constraints (all Lagrange multipliers are positive). The following example illustrates the application of the optimality conditions for inequality constraints.

Example 3.9: Solve the constrained optimization problem:

$$\begin{aligned} & f(\boldsymbol{x}) = x_1^2 + x_2^2 - 14x_1 - 6x_2 \\ \text{subject to} \quad & g_1(\boldsymbol{x}) = x_1 + x_2 - 2 \leq 0 \\ & g_2(\boldsymbol{x}) = x_1 + 2x_2 - 3 \leq 0 \end{aligned} \tag{3.64}$$

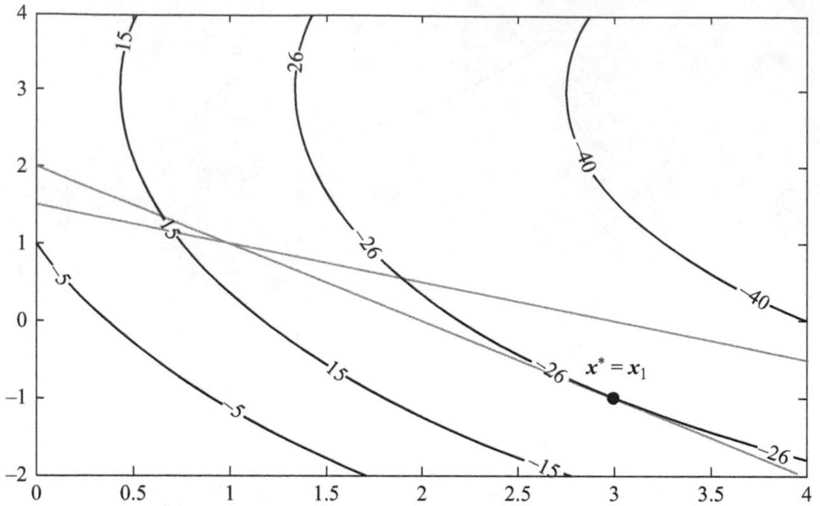

Figure 3.14 An illustration of the objective function contours and constraints of Example 3.9

Solution: The possible solutions of the problem (3.64) are found by solving the necessary optimality conditions (3.63) for the optimal point. The gradients of the objective function and the two constraints are given by:

$$\begin{aligned} \nabla f(\boldsymbol{x}) &= [(2x_1 - 14)\ (2x_2 - 6)]^T \\ \nabla g_1 &= [1 \quad 1]^T \\ \nabla g_2 &= [1 \quad 2]^T \end{aligned} \tag{3.65}$$

The necessary optimality conditions for this problem can then be stated as:

$$\begin{aligned} &\begin{bmatrix} 2x_1 - 14 \\ 2x_2 - 6 \end{bmatrix} + \lambda_1 \begin{bmatrix} 1 \\ 1 \end{bmatrix} + \lambda_2 \begin{bmatrix} 1 \\ 2 \end{bmatrix} = \begin{bmatrix} 0 \\ 0 \end{bmatrix} \\ &x_1 + x_2 - 2 \le 0 \\ &x_1 + 2x_2 - 3 \le 0 \end{aligned} \tag{3.66}$$

We have four equations in four unknowns x_1, x_2, λ_1, and λ_2. As stated earlier, we do not know beforehand which constraints are active at the optimal solution. There are four possibilities for this solution: none of the constraints is active, only the first one is active, only the second one is active, or both constraints are active. If none of the constraints is active, the optimality condition becomes:

$$\begin{bmatrix} 2x_1 - 14 \\ 2x_2 - 6 \end{bmatrix} = \begin{bmatrix} 0 \\ 0 \end{bmatrix} \tag{3.67}$$

Solving, we get $\boldsymbol{x}_0 = [7.0 \quad 3.0]^T$. This point, however, violates both constraints as $g_1(\boldsymbol{x}_0) = 8 > 0$ and $g_2(\boldsymbol{x}_0) = 10 > 0$. Assuming that only the first constraint is active, the optimality conditions become:

$$\begin{bmatrix} 2x_1 - 14 \\ 2x_2 - 6 \end{bmatrix} + \lambda_1 \begin{bmatrix} 1 \\ 1 \end{bmatrix} = \begin{bmatrix} 0 \\ 0 \end{bmatrix}$$
$$x_1 + x_2 - 2 = 0 \tag{3.68}$$

By expressing x_1 and x_2 in terms of λ_1 and substituting in the first constraint, we get:

$$\left(\frac{14 - \lambda_1}{2}\right) + \left(\frac{6 - \lambda_1}{2}\right) = 2 \Rightarrow \lambda_1 = 8 \Rightarrow \boldsymbol{x}_1 = \begin{bmatrix} 3 \\ -1 \end{bmatrix} \tag{3.69}$$

This point has a positive Lagrange multiplier and it also satisfies the second constraint as $g_2(\boldsymbol{x}_1) = -2.0 < 0$. It follows that this point satisfies the necessary KKT conditions. If we assume that only the second constraint is active, we get the following necessary conditions:

$$\begin{bmatrix} 2x_1 - 14 \\ 2x_2 - 6 \end{bmatrix} + \lambda_2 \begin{bmatrix} 1 \\ 2 \end{bmatrix} = \begin{bmatrix} 0 \\ 0 \end{bmatrix}$$
$$x_1 + 2x_2 - 3 = 0 \tag{3.70}$$

Equation (3.70) represents three equations in three unknowns. Expressing x_1 and x_2 in terms of λ and substituting in the second constraint, we get:

$$\left(\frac{14 - \lambda_2}{2}\right) + 2(3 - \lambda_2) = 3 \Rightarrow \lambda_2 = 4 \Rightarrow \boldsymbol{x}_2 = \begin{bmatrix} 5 \\ -1 \end{bmatrix} \tag{3.71}$$

The point $\boldsymbol{x}_2$, however, violates the first constraint as $g_1(\boldsymbol{x}_2) = 2.0 > 0$ and thus does not satisfy the optimality conditions. The last possibility is to assume that both constraints are active and thus both their corresponding multipliers are not zero. In this case, the optimality conditions become:

$$\begin{bmatrix} 2x_1 - 14 \\ 2x_2 - 6 \end{bmatrix} + \lambda_1 \begin{bmatrix} 1 \\ 1 \end{bmatrix} + \lambda_2 \begin{bmatrix} 1 \\ 2 \end{bmatrix} = \begin{bmatrix} 0 \\ 0 \end{bmatrix}$$
$$x_1 + x_2 - 2 = 0 \tag{3.72}$$
$$x_1 + 2x_2 - 3 = 0$$

It follows that we have four equations in four unknowns. Solving the two equations of the constraints, we get $\boldsymbol{x}_3 = [1.0 \quad 1.0]^T$. Substituting into the two other equations involving the Lagrange multipliers, we get:

$$\begin{bmatrix} -12 \\ -4 \end{bmatrix} + \lambda_1 \begin{bmatrix} 1 \\ 1 \end{bmatrix} + \lambda_2 \begin{bmatrix} 1 \\ 2 \end{bmatrix} = \begin{bmatrix} 0 \\ 0 \end{bmatrix} \tag{3.73}$$

resulting in $\lambda_1 = 20$ and $\lambda_2 = -8$. As one of the Lagrange multipliers is not positive, this point is not an optimal point. Figure 3.14 gives a graphical solution of this problem, which confirms the analytical approach.

3.10 Optimization with mixed constraints

A general optimization problem can have both equality and inequality constraints. It can be formulated as:

$$\begin{aligned} &\text{minimize} \quad f(\boldsymbol{x}) \\ &\text{subject to} \quad g_j(\boldsymbol{x}) \le 0, \quad j = 1, 2, \ldots, m \\ &\qquad\qquad\quad h_k(\boldsymbol{x}) \le 0, \quad k = 1, 2, \ldots, p \end{aligned} \tag{3.74}$$

In this case, the optimality conditions are given by:

$$\begin{aligned} \frac{\partial f}{\partial x_i} + \sum_{j=1}^{m} \lambda_j \frac{\partial g_j}{\partial x_i} + \sum_{j=1}^{p} \beta_k \frac{\partial h_k}{\partial x_i} &= 0, \quad i = 1, 2, \ldots, n \\ \lambda_j g_j &= 0, \quad j = 1, 2, \ldots, m \\ g_j &\le 0, \quad j = 1, 2, \ldots, m \\ \lambda_j &\ge 0, \quad j = 1, 2, \ldots, m \\ h_k(\boldsymbol{x}) &= 0, \quad k = 1, 2, \ldots, p \end{aligned} \tag{3.75}$$

Note that the Lagrange multipliers of the inequality constraints must be non-negative, while those for the equality constraints are not sign-constrained.

A3.1 Quadratic programming

Engineering problems with inequality constraints are cast in the form (3.59). The objective function and the constraints are nonlinear functions. In some applications, the objective function is quadratic or can be sufficiently approximated by a quadratic function and the constraints are linear. Such a problem is called a quadratic programming problem. It has the form:

$$\begin{aligned} &\text{minimize} \quad \frac{1}{2}\boldsymbol{x}^T \boldsymbol{H} \boldsymbol{x} + \boldsymbol{b}^T \boldsymbol{x} + c \\ &\text{subject to} \quad \boldsymbol{A}\boldsymbol{x} \le \boldsymbol{d} \end{aligned} \tag{3.76}$$

where $\boldsymbol{H} \in \Re^{n \times n}$ is the Hessian matrix of the objective function, $\boldsymbol{b} \in \Re^n$, and $c \in \Re^1$. The matrix $\boldsymbol{A} \in \Re^{m \times n}$ is the matrix of coefficients of the m constraints. The jth constraint is thus given by:

$$a_{j1}x_1 + a_{j2}x_2 + \cdots + a_{jn}x_n \le d_j \tag{3.77}$$

The gradient of this constraint is the transpose of the jth row of the matrix $\boldsymbol{A}$. Applying the optimality conditions (3.63), we get:

$$\begin{aligned} \boldsymbol{H}\boldsymbol{x} + \boldsymbol{b} + \boldsymbol{A}_a^T \boldsymbol{\lambda} = 0, \quad \boldsymbol{\lambda} \ge 0 \\ \boldsymbol{A}_a \boldsymbol{x} = \boldsymbol{d}_a \end{aligned} \tag{3.78}$$

where $\boldsymbol{A}_a$ is a matrix containing a subset of the rows of the matrix $\boldsymbol{A}$ corresponding to the active constraints and $\boldsymbol{\lambda}$ is the vector of corresponding Lagrange multipliers. The formulation (3.78) results in multiplying every row of the matrix $\boldsymbol{A}_a$ by the corresponding Lagrange multiplier and summing them. $\boldsymbol{d}_a$ is the vector containing the corresponding components of the vector $\boldsymbol{d}$. Because of the special properties of this problem, the system (3.78) can be cast in the compact system of linear equations:

$$\begin{bmatrix} \boldsymbol{H} & \boldsymbol{A}_a^T \\ \boldsymbol{A}_a & \boldsymbol{0} \end{bmatrix} \begin{bmatrix} \boldsymbol{x} \\ \boldsymbol{\lambda} \end{bmatrix} = \begin{bmatrix} -\boldsymbol{b} \\ \boldsymbol{d}_a \end{bmatrix} \tag{3.79}$$

The system (3.79) solves simultaneously for the parameters and the multipliers. The signs of the multipliers are then checked. If they are all non-negative, then (3.79) is indeed a local minimum. Otherwise, the set of active constraints should be changed resulting in a different matrix $\boldsymbol{A}_a$ and a different vector $\boldsymbol{d}_a$.

As explained earlier, the main difficulty in applying (3.79) is that we do not know beforehand which constraints are active at the optimal point and which constraints are not active. I will discuss techniques for iteratively solving this problem in the following chapters. For now, we will illustrate a "brute force" approach to the solution of (3.76). For the given m constraints, there are 2^m possible combinations of these constraints. Each combination is represented by a binary word in which every constraint is represented by a digit. If the value of this digit is "1", then the constraint is active. If the value of the digit is "0", the constraint is not active. We search all these combinations and solve (3.79) for the corresponding set of active constraints until a solution is found that is feasible and has non-negative multipliers. The main drawback of this approach, similar to the Simplex method discussed in Chapter 2, is the exponential number of possible solutions that must be checked. The following example illustrates the solution of a quadratic program by checking the optimality conditions for all possible solutions.

Example 3.10: Solve the quadratic problem:

$$\begin{aligned} &\text{minimize} \quad f(\boldsymbol{x}) = x_1^2 + x_2^2 + 0.5x_3^2 + x_1x_2 + x_1x_3 - 4x_1 - 3x_2 - 2x_3 \\ &\text{subject to} \quad -x_1 - x_2 - x_3 \geq -3 \\ &\qquad\qquad\quad x_1 \geq 0,\ x_2 \geq 0,\ x_3 \geq 0 \end{aligned} \tag{3.80}$$

by using the formulation (3.79) and checking the optimality conditions for all possible solutions.

Solution: First, we cast the problem (3.80) in the standard form (3.76) by writing:

$$\begin{aligned} &\text{minimize} \quad f(\boldsymbol{x}) = \frac{1}{2}\begin{bmatrix} x_1 & x_2 & x_3 \end{bmatrix} \begin{bmatrix} 2 & 1 & 1 \\ 1 & 2 & 0 \\ 1 & 0 & 1 \end{bmatrix} \begin{bmatrix} x_1 \\ x_2 \\ x_3 \end{bmatrix} \\ &\qquad\qquad + \begin{bmatrix} -4 & -3 & -2 \end{bmatrix} \begin{bmatrix} x_1 \\ x_2 \\ x_3 \end{bmatrix} \\ &\text{subject to} \quad \begin{bmatrix} 1 & 1 & 1 \\ -1 & 0 & 0 \\ 0 & -1 & 0 \\ 0 & 0 & -1 \end{bmatrix} \begin{bmatrix} x_1 \\ x_2 \\ x_3 \end{bmatrix} \leq \begin{bmatrix} 3 \\ 0 \\ 0 \\ 0 \end{bmatrix} \end{aligned} \tag{3.81}$$

```
%M3.1
%This program solve a quadratic program by applying optimality conditions to all possible points
numberOfParameters=3; % n
numberOfConstraints=4; % m
H=[2 1 1; 1 2 0; 1 0 1]; %Hessian matrix
b=[-4 -3 -2]'; %vector b
c=0; %constant of the model
A=[1 1 1; -1 0 0; 0 -1 0; 0 0 -1];
d=[2 0 0 0]';
numberOfPossibleSolutions=2^numberOfConstraints; %total number of possible solutions
solutionFound=false; %solution found flag
Counter=0; %counter for solutions
AActive=zeros(numberOfConstraints,numberOfParameters); %matrix to store active constraints
dActive=zeros(numberOfConstraints,1); %vector for store coefficients of active constraints
%the loop repeats until all solutions are exhausted or an optimal solution is found
while((Counter<numberOfPossibleSolutions)&(solutionFound==false))
  binaryWord=dec2bin(Counter,numberOfConstraints); %get binary word corresponding to counter
  numberOfActiveConstraints=0; %initialize number of active constraint
  for k=1:numberOfConstraints %check which constraints are active (binary "1")
    if(binaryWord(k)=='1')
      numberOfActiveConstraints=numberOfActiveConstraints+1; %one more constraint is active
      AActive(numberOfActiveConstraints,:)=A(k,:); %copy the kth constraint
      dActive(numberOfActiveConstraints,1)=d(k,1); %copy the kth coefficient
    end
  end
  %now we build the complete system matrix
  systemSize=numberOfParameters+numberOfActiveConstraints;
  systemMatrix=zeros(systemSize,systemSize); %system matrix
  systemRHS=zeros(systemSize,1);
  %copy the block matrices
  systemMatrix(1:numberOfParameters,1:numberOfParameters)=H; % Hessian of objective function
  %store matrices of active constraints gradients
  systemMatrix(1:numberOfParameters,numberOfParameters+1:systemSize)=
                                             AActive(1:numberOfActiveConstraints,:)';
  systemMatrix(numberOfParameters+1:systemSize,1:numberOfParameters)=
                                             AActive(1:numberOfActiveConstraints,:);
  systemRHS(1:numberOfParameters)=-b; %store -b in the RHS vector
  systemRHS(numberOfParameters+1:systemSize)=dActive(1:numberOfActiveConstraints); %store the
                                                               % active components of d
  solution=inv(systemMatrix)*systemRHS; %solve the system of equations
  Parameters=solution(1:numberOfParameters); %get x
  Lambda=solution(numberOfParameters+1:systemSize);%get the Lagrange multilpliers
  if((A*Parameters<=d))%Is it solution feasible? Check all constraints
    if((size(Lambda,1)>0) &(Lambda>0)) %there at least one active constraint with positive multiplier
      solutionFound=true; %set flag to true
    end
    if((size(Lambda,1))==0) %no active constraint then if this is the unconstrained minimum then
      solutionFound=true; %feasible unconstrained minimum
    end
  end
  Counter=Counter+1; %move to the next solution
end
Parameters
```

In this problem, we have four constraints ($m = 4$). It follows that up to 16 solutions have to be checked to find the optimal solution. The above MATLAB® code M3.1 receives an arbitrary quadratic function and arbitrary number of linear inequality constraints and searches for the optimal solution that satisfies the necessary optimality conditions.

The output for this problem for the given matrices and vectors is:

```
Parameters =
    1.0000
    1.0000
    1.0000
```

This solution $\boldsymbol{x}^* = [1.0 \quad 1.0 \quad 1.0]^T$ is the unconstrained minimum of the objective function. The first constraint $x_1 + x_2 + x_3 \leq 3$ is active at this point but it has a zero

Lagrange multiplier as $\nabla f = 0$ at this point. If this constraint is changed to $x_1 + x_2 + x_3 \leq 2$, the output of the program becomes:

Parameters =
1.5000
0.5000
0

This solution is different from the unconstrained minimum. Only the first constraint is active at this point with a positive Lagrange multiplier $\lambda_1 = 0.5$. This can be checked at this point by writing:

$$\nabla f(\boldsymbol{x})|_{\boldsymbol{x}^*} = \begin{bmatrix} 2 & 1 & 1 \\ 1 & 2 & 0 \\ 1 & 0 & 1 \end{bmatrix} \begin{bmatrix} 1.5 \\ 0.5 \\ 0 \end{bmatrix} + \begin{bmatrix} -4 \\ -3 \\ -2 \end{bmatrix} = \begin{bmatrix} -0.5 \\ -0.5 \\ -0.5 \end{bmatrix} = -0.5\boldsymbol{a}_1 \tag{3.82}$$

where $\boldsymbol{a}_1 = [1.0 \quad 1.0 \quad 1.0]^T$ is the gradient of the first constraint.

A3.2 Sequential quadratic programming

The discussion covered by (3.76)–(3.79) addresses the case of a quadratic objective function with linear constraints. For a problem with a general objective function and general constraints, quadratic programming can be applied in an iterative way to reach the optimal solution. This is referred to as sequential quadratic programming (SQP). It is one of the most widely used techniques in solving nonlinear optimization problems because of its demonstrated convergence properties.

We discuss briefly SQP and illustrate it through an electrical engineering application. The problem we aim at solving is given by:

$$\begin{aligned} &\min_{\boldsymbol{x}} \quad f(\boldsymbol{x}) \\ &\text{subject to} \quad \boldsymbol{g}(\boldsymbol{x}) = \boldsymbol{0} \end{aligned} \tag{3.83}$$

where $\boldsymbol{g}(\boldsymbol{x}) \in \Re^m$ is a vector of general nonlinear functions. The objective function is not, in general, quadratic and the constraints are not linear. The Lagrangian of the problem (3.83) is given by:

$$L(\boldsymbol{x}, \boldsymbol{\lambda}) = f(\boldsymbol{x}) + \boldsymbol{\lambda}^T \boldsymbol{g}(\boldsymbol{x}) \tag{3.84}$$

where $\boldsymbol{\lambda}$ is the vector of Lagrange multipliers. The expression (3.84) is a more compact way of writing (3.49). The necessary KKT optimality conditions are given by:

$$\boldsymbol{h}(\boldsymbol{x}, \boldsymbol{\lambda}) = \begin{bmatrix} \dfrac{\partial L}{\partial \boldsymbol{x}} \\ \dfrac{\partial L}{\partial \boldsymbol{\lambda}} \end{bmatrix} = \begin{bmatrix} \nabla f + \boldsymbol{J}^T \boldsymbol{\lambda} \\ \boldsymbol{g}(\boldsymbol{x}) \end{bmatrix} = \begin{bmatrix} \boldsymbol{0} \\ \boldsymbol{0} \end{bmatrix} \tag{3.85}$$

The equations (3.85) are $(m+n)$ general nonlinear equations in the $(m+n)$ unknowns $\boldsymbol{u} = [\boldsymbol{x}^T \quad \boldsymbol{\lambda}^T]^T$. The matrix $\boldsymbol{J}$ is the Jacobian of the vector of constraints

$g(x)$ with respect to x. The system of nonlinear equations (3.85) can be solved in an iterative way using the Newton–Raphson method discussed in Chapter 1. This approach evaluates the Jacobian of the vector $h(u)$ with respect to u. The step taken by the Newton–Raphson at the kth iteration is given by:

$$J_h(u^{(k)})\Delta u^{(k)} = -h(u^{(k)}) \tag{3.86}$$

where J_h is the Jacobian of the vector of nonlinear functions $h(u)$ relative to the vector of unknowns u. Differentiating (3.85) with respect to u and substituting into (3.86), we get the system of linear equations:

$$\begin{bmatrix} W^{(k)} & J^{(k)T} \\ J^{(k)} & 0 \end{bmatrix} \Delta u^{(k)} = -\begin{bmatrix} \nabla f^{(k)} + J^{(k)T}\lambda^{(k)} \\ g^{(k)} \end{bmatrix} \tag{3.87}$$

The superscript k indicates that all vectors and matrices are estimated at the kth iteration with values $u^{(k)} = [x^{(k)} \quad \lambda^{(k)}]^T$. The matrix $W^{(k)} \in \Re^{n\times n}$ is a matrix of second-order derivatives of the objective function $f(x)$ and the components of the $g(x)$. It is given by:

$$W^{(k)} = H^{(k)} + \sum_{i=1}^{m} \lambda_i^{(k)} H_{g_i}^{(k)} \tag{3.88}$$

where H is the Hessian of the objective function $f(x)$ relative to the parameters x and H_{gi} is the Hessian of the ith constraint $g_i(x)$ relative to x. By writing $\Delta u^{(k)} = [\Delta x^{(k)} \quad \Delta\lambda^{(k)}]^T$, the system (3.87) can be reorganized in the two sets of equations:

$$\begin{aligned} W^{(k)}\Delta x^{(k)} + \nabla f^{(k)} &= -J^{(k)T}(\lambda^{(k)} + \Delta\lambda^{(k)}) = -J^{(k)T}(\lambda^{(k+1)}) \\ J^{(k)}\Delta x^{(k)} + g^{(k)} &= 0 \end{aligned} \tag{3.89}$$

It is a simple exercise to show that the system (3.89) can be obtained by applying the KKT conditions to the quadratic program:

$$\begin{aligned} &\min_{\Delta x^{(k)}} \ 0.5\ \Delta x^{(k)T} W^{(k)} \Delta x^{(k)} + \nabla f^{(k)T} \Delta x^{(k)} \\ &\text{subject to } J^{(k)}\Delta x^{(k)} + g^{(k)} = 0 \end{aligned} \tag{3.90}$$

This program is a typical quadratic program that could be solved for the step $\Delta x^{(k)}$ as in the previous section. Notice that in (3.90) we have only equality constraints and the Lagrange multipliers do not have to be positive as in (3.76). Also, because we have only equality constraints, the set of active constraints does not change as all constraints must be active at all iterations. A linear system of equations similar to (3.79) can be used for solving this quadratic program. The value $u^{(k)}$ is then updated and all the matrices and the vectors are calculated at the new set of parameters. It can be seen a sequence of quadratic programs is solved until the solution of (3.83) is reached.

As an illustration of this approach, consider the microwave imaging scheme shown in Figure 3.15. A number of antennas are placed at specific points in the space around an object with unknown dielectric properties. These properties include the dielectric constant ε_r and the conductivity σ at every point inside the subject domain D_{subject}. These two properties are usually combined into one complex quantity called the dielectric contrast s. The contrast describes how different every point in this domain is from the surrounding medium, which is usually assumed to be air.

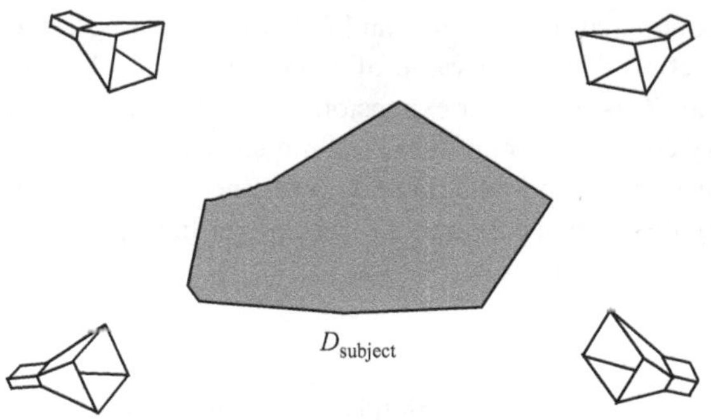

Figure 3.15 An illustration of a microwave imaging system; a number of antennas are used to transmit and receive electromagnetic waves. Based on the measured reflected field $\mathbf{E}_r$ at these antennas, the dielectric profile of the object is determined

The antennas transmit electromagnetic waves toward the object. In the absence of the target, these incident waves result in an incident field $\boldsymbol{E}_i(\boldsymbol{p})$ at every point $\boldsymbol{p}$ in space. This incident field is known either through measurements or through numerical simulations. When the subject domain is radiated, it gives rise to a scattered field $\boldsymbol{E}_r(\boldsymbol{p},\boldsymbol{s})$, where $\boldsymbol{s}$ is the vector of unknown contrasts at every point inside the object. This reflected field is the "foot print" of the subject domain. Every dielectric distribution within the object results in a different reflected field. The total field everywhere in the space is the sum of the incident field and the reflected field at every point in the considered space, i.e., $\boldsymbol{E}_T(\boldsymbol{p},\boldsymbol{s}) = \boldsymbol{E}_i(\boldsymbol{p}) + \boldsymbol{E}_r(\boldsymbol{p},\boldsymbol{s})$. By taking a number of total field measurements with the antennas $\boldsymbol{E}_m^{(j)}, j = 1, 2, \ldots, M$, it is required to recover the dielectric contrast at every point within the object. It should be noted that all fields in this discussion are phasor vectors with three spatial field components.

As stated earlier, the scattered field is a function of the dielectric profile within the object. Their relationship is given by the formula [5]:

$$\boldsymbol{E}_r(\boldsymbol{p}) = \int\limits_{D_{\text{subject}}} s(\boldsymbol{p}')\boldsymbol{G}(\boldsymbol{p},\boldsymbol{p}')\boldsymbol{E}_T(\boldsymbol{p}')\boldsymbol{dp}' \tag{3.91}$$

This equation states that the reflected field anywhere outside the object is obtained by summing contributions from every point within that object. The function $\boldsymbol{G}(\boldsymbol{p},\boldsymbol{p}')$ is the Green's function. It is known either analytically or numerically for every pair of points $\boldsymbol{p}$ and $\boldsymbol{p}'$. Equation (3.91) explains why the solution of this inverse problem is tough. It is a nonlinear equation because the contrast s, which is unknown at every point inside the object, is multiplied by the total field, which is also unknown inside the object. Also, the reflected field appears in both sides of the equation because the total field is the sum of the incident and reflected fields.

The problem (3.91) has infinite number of unknowns as the number of points inside the object is infinite. To make the problem solvable, we divide the subject object into N_1 voxels. We also discretize the domain outside the object. We then assume that dielectric properties and the total fields are constant over each voxel in the entire domain. The dielectric contrast and total field are expanded into:

$$s = \sum_{j=1}^{N_1} s_j P_j \quad \text{and} \quad \boldsymbol{E}_T = \sum_j E_j \boldsymbol{T}_j \tag{3.92}$$

where s_j is the contrast at the jth voxel and E_j is the field amplitude at the jth voxel. P_j is a pulse function that has a value of 1 only over the jth voxel and 0 over all voxels. The vector $\boldsymbol{T}_j$ is the vector expansion over the jth voxel. Expansions similar to (3.92) are very common in numerical techniques used in solving electromagnetic problems such as the method of moments [6] or the finite element method [7].

Substituting this expansion into (3.91) and utilizing a method of moments approach, (3.91) is put in the matrix form [5]:

$$\boldsymbol{Z}(\boldsymbol{s})\boldsymbol{E} = \boldsymbol{V} \tag{3.93}$$

where $\boldsymbol{Z}$ is the impedance matrix, which is a function of the contrast and geometry, and $\boldsymbol{E}$ is the vector of unknown total fields. The problem of solving for the unknown problems aims at finding the dielectric contrasts inside the subject domain that would create fields at the antennas matching the measured fields. A possible objective function that models this problem is given by:

$$f(\boldsymbol{s}) = \sum_{j=1}^{M} \left| \boldsymbol{E}_m^{(j)} - \boldsymbol{E}_T^{(j)}(\boldsymbol{s}) \right|^2 \tag{3.94}$$

The minimization of the objective function (3.94) is subject to the set of nonlinear constraints given by (3.93). This problem may be solved using any constrained optimization approach.

In Reference 5, this problem was solved using an SQP-based algorithm. Figure 3.16 shows the solution obtained through this problem. The actual dielectric

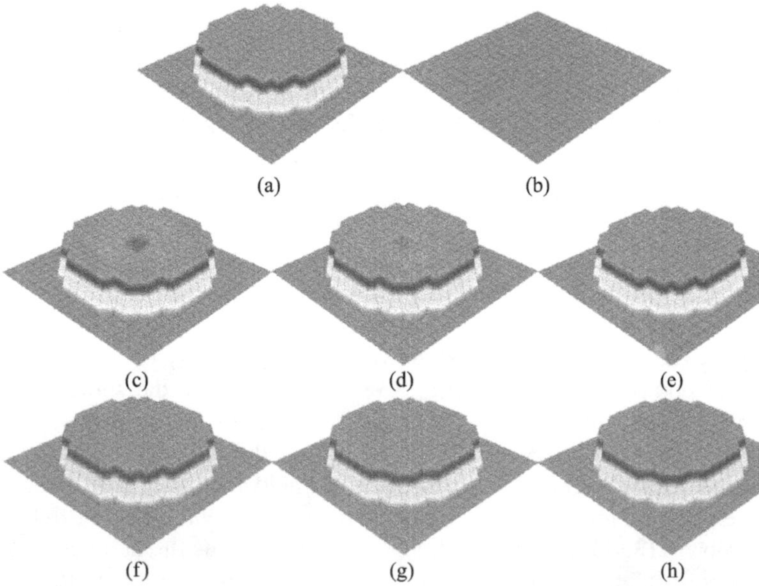

Figure 3.16 An illustration of the SQP method for inverse problems; (a) the actual dielectric profile of the scatterer, (b) the initial guess, and (c)–(h) the reconstructed profile obtained in six iterations

distribution as compared to the background medium is shown in Figure 3.16(a). It is assumed in the beginning that there is no contrast with the background medium as shown in Figure 3.16(b). The SQP algorithm minimizes (3.94) according to the constraints (3.93). It took six iterations to recover the correct contrast as shown in Figure 3.16(c)–(h). Similar approaches are used for microwave imaging of breast and brain cancers inside the human body.

References

1. Singiresu S. Rao, *Engineering Optimization Theory and Practice*, Third Edition, John Wiley & Sons Inc, New York, 1996
2. P. Venkataraman, *Applied Optimization with MATLAB Programming*, Wiley, 2002
3. John W. Bandler and R.M. Biernacki, ECE3KB3 Courseware on Optimization Theory, McMaster University, 1997
4. Edwin K.P. Chong and Stanislaw H. Zak, *An Introduction to Optimization*, 2nd edn., Wiley, 2001
5. Jin-Lin Hu, Zhipeng Wu, Hugh McCann, Lionel Edward Davis, and Cheng-Gang Xie, "Sequential quadratic programming method for solution of electromagnetic inverse problems", *IEEE Transactions on Antennas and Propagation*, vol. 53, no. 8, August 2005
6. Roger F. Harrington, *Field Computations by Moment Methods*, IEEE Press, 1993
7. Jianming Jin, *The Finite Element Method in Electromagnetics*, Wiley, 2002

Problems

3.1 Consider the constrained optimization problem:

$$\begin{aligned} \text{minimize} \quad & f(x_1,x_2) = x_1^2 + 3x_3^2 + 2x_1x_3 + 4x_1 + 6x_2 + 5x_3 \\ \text{subject to} \quad & x_1 + 2x_3 = 3.0 \\ & 4x_1 + 5x_2 = 6.0 \end{aligned}$$

(a) Use the method of direct substitution to solve this problem.
(b) Utilize the quadratic programming MATLAB listing M3.1 to confirm your answer after making the necessary changes to the initialization section.

3.2 Use the method of constrained variations to solve the constrained optimization problem:

$$\begin{aligned} \text{minimize} \quad & f(x_1,x_2) = 2x_1 + x_2^2 \\ \text{subject to} \quad & x_1^2 + x_2^2 = 4.0 \end{aligned}$$

Verify your answer by solving the problem graphically.

3.3 Use the KKT conditions to solve the constrained optimization problem:

$$\begin{aligned} \text{minimize} \quad & f(x_1,x_2) = -x_1x_2 \\ \text{subject to} \quad & x_1^2 + 2x_2^2 = 4.0 \end{aligned}$$

Verify your answer through a graphical solution.

3.4 Consider the linear program:

$$\begin{aligned} \text{minimize} \quad & f(x_1, x_2) = x_1 + 2x_2 + 2.0 \\ \text{subject to} \quad & -x_1 - 2x_2 + 3.0 \leq 0 \\ & -2x_1 - x_2 + 3.0 \leq 0 \end{aligned}$$

Determine the optimal solution of this problem by applying KKT conditions.

3.5 Consider the quadratic program:

$$\begin{aligned} \text{minimize} \quad & f(x_1, x_2) = 2x_1^2 + x_2^2 + 2.0 \\ \text{subject to} \quad & x_2 \geq 2.0 \\ & x_1 \geq x_2 \end{aligned}$$

(a) Using the KKT conditions, obtain the optimal solution of this problem.
(b) Obtain a graphical solution of this problem and compare it with your answer in (a).
(c) Confirm your answers in (a) and (b) using the MATLAB listing M3.1 after making the necessary changes to the initialization section.

3.6 Repeat problem 3.3 for the quadratic program:

$$\begin{aligned} \text{minimize} \quad & f(x_1, x_2) = x_1^2 + x_2^2 + x_1 x_2 - 3x_1 \\ \text{subject to} \quad & x_1, x_2 \geq 0 \end{aligned}$$

Hint: The constant objective value contours are ellipses in this case.

3.7 Find the point on the surface of the sphere $x_1^2 + x_2^2 + x_3^2 = 9.0$, which is closest to the point $\boldsymbol{x}_p = [4.0 \quad -4.0 \quad 4.0]^T$.

Hint: Take the objective function as the square of the distance to the point $\boldsymbol{x}_p$.

3.8 Consider the circuit shown in Figure 3.17. The load resistor $R_s = 30.0\ \Omega$. The target is to minimize the total power dissipated, which is given by:

$$\frac{V_s^2}{R_s + R_L}$$

subject to design constraints. We allow the source voltage and resistance to change to minimize the total power. The optimizable parameters are thus $\boldsymbol{x} = [V_s \quad R_s]^T$. The constraints on the parameters are $V_s \geq 10.0$ V and $10.0\ \Omega \leq R_s \leq 30.0\ \Omega$. Apply the KKT conditions to determine the optimal design $\boldsymbol{x}^*$.

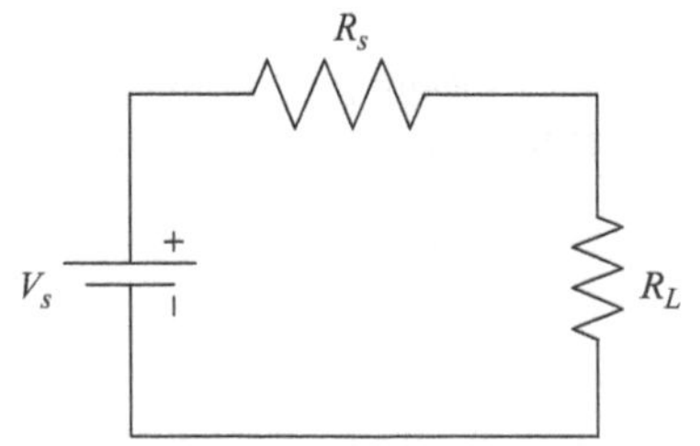

Figure 3.17 The circuit of Problem 3.8

Chapter 4

One-dimensional optimization-line search

4.1 Introduction

Optimization problems with a single variable have special importance in optimization theory. The reason for this is that multidimensional optimization problems are usually solved iteratively through optimization algorithms that utilize single-variable optimization approaches. Typically, at the kth iteration of an algorithm, a promising search direction $\boldsymbol{s}^{(k)} \in \Re^n$ in the n-dimensional parameter space is first determined. Starting from the solution at the kth iteration $\boldsymbol{x}^{(k)}$, a line search is carried out in the direction $\boldsymbol{s}^{(k)}$. A general point along that line is given by $\boldsymbol{x} = \boldsymbol{x}^{(k)} + \alpha \boldsymbol{s}^{(k)}$, where $\alpha > 0$ is the search parameter. The optimal value of the search parameter is the one that minimizes the objective function along the direction $\boldsymbol{s}^{(k)}$, i.e.:

$$\alpha^* = \arg \min_{\alpha} f\left(\boldsymbol{x}^{(k)} + \alpha \boldsymbol{s}^{(k)}\right) \tag{4.1}$$

Notice that because both $\boldsymbol{x}^{(k)}$ and $\boldsymbol{s}^{(k)}$ are constants, the function $f(\alpha) = f\left(\boldsymbol{x}^{(k)} + \alpha \boldsymbol{s}^{(k)}\right)$ is a single-variable function of the parameter α. One-dimensional optimization algorithms may be applied to (4.1). The corresponding solution in the original n-dimensional space is given by $\boldsymbol{x}^{(k+1)} = \boldsymbol{x}^{(k)} + \alpha^* \boldsymbol{s}^{(k)}$. It can be shown that differentiating the function $f(\alpha) = f(\boldsymbol{x}^{(k)} + \alpha \boldsymbol{s}^{(k)})$ relative to α gives $f'(\alpha) = \nabla f(\boldsymbol{x}^{(k)})^T \boldsymbol{s}^{(k)}$. This expression is obtained by first differentiating f relative to its vector argument and then differentiating the vector argument relative to α. The derivative $f'(\alpha)$ is denoted as the directional derivative of the function f in the direction of $\boldsymbol{s}^{(k)}$. It must be negative for a descent direction.

The following example illustrates the concept of one-dimensional line search.

Example 4.1: Consider the function $f(\boldsymbol{x}) = x_1^2 + x_2^2$. Starting from the point $\boldsymbol{x}_s = [1.0 \quad 3.0]^T$, find the minimum along the direction $\boldsymbol{s} = [-2.0 \quad -1.0]^T$.

Solution: Any point along the line passing through $\boldsymbol{x}_s$ pointing in the direction $\boldsymbol{s}$ is given by:

$$\boldsymbol{x} = \begin{bmatrix} 1 \\ 3 \end{bmatrix} + \alpha \begin{bmatrix} -2 \\ -1 \end{bmatrix} = \begin{bmatrix} 1 - 2\alpha \\ 3 - \alpha \end{bmatrix}, \quad \alpha \geq 0 \tag{4.2}$$

The substitution (4.2) converts the two-dimensional objective function into the single-variable function $f(\alpha) = (1 - 2\alpha)^2 + (3 - \alpha)^2$. The value of α at which this function achieves its minimum is obtained by applying the necessary optimality condition:

$$\frac{df}{d\alpha} = 0 \Rightarrow -4(1 - 2\alpha) - 2(3 - \alpha) = 0 \Rightarrow \alpha^* = 1.0 \tag{4.3}$$

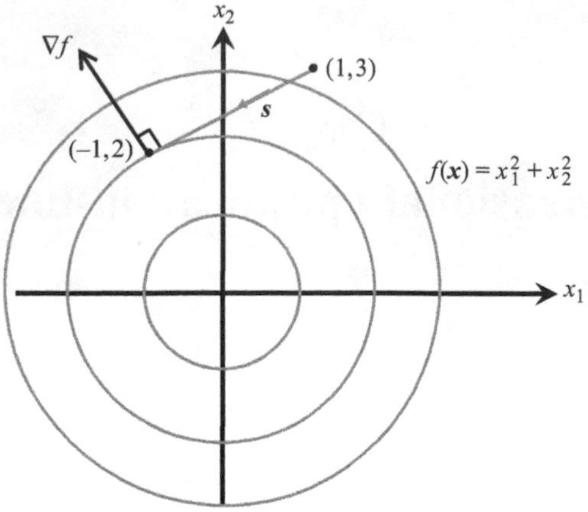

Figure 4.1 An illustration of Example 4.1; starting from the point $[1.0 \quad 3.0]^T$ and searching in the direction $[-2.0 \quad -1.0]^T$, the minimum is reached at the point $[-1.0 \quad -2.0]^T$. At this point, the search direction is tangential to the contour of the objective function passing through that point

It follows that this function is minimized if one moves along this line a distance of 1.0. The corresponding point is obtained from (4.2) as $\boldsymbol{x}^* = [-1.0 \quad 2.0]^T$. Figure 4.1 illustrates the line search process for this example.

It should be noted the gradient of the objective function in the previous example at the optimal point is given by:

$$\nabla f(\boldsymbol{x}^*) = \begin{bmatrix} 2x_1 \\ 2x_2 \end{bmatrix}_{x^*} = \begin{bmatrix} -2 \\ 4 \end{bmatrix} \tag{4.4}$$

This gradient is normal to the direction $\boldsymbol{s}$ as $\nabla f^T \boldsymbol{s} = 0$. This should be expected otherwise that the point $\boldsymbol{x}^*$ would not be the optimal point along that direction. It shows that going any further along the search direction results in an increase in the objective function rather than a decrease. This relationship between the gradient and the search direction is shown in Figure 4.1.

In the following sections, we review a number of techniques used to solving the one-dimensional line search problem (4.1). These techniques vary from bracketing techniques to techniques that obtain an estimate for the search parameter α. This estimate can be obtained through derivative-free approaches, interpolation approaches, or derivative-based approaches.

4.2 Bracketing approaches

In most practical cases, the function value $f(\boldsymbol{x})$ is only available numerically. For every set of parameters $\boldsymbol{x}$, the simulator (modeler) carries out a numerical simulation and supplies the value of the function at this point. We do not usually know the analytical expression of the function and thus a similar approach to that used in Example 4.1 is not applicable. The first step in determining the optimal solution of (4.1) is to determine a range of values (bracket) in which the optimal value α^* lies.

Another approach is then utilized to determine the optimal value of the parameter within this interval.

An important assumption in all line search techniques is the assumption of unimodality. Unimodality means that the function value can be used as an indication of how close a point is to the optimal solution. In other words, the closer a point is to the optimal point, the lower the value of the objective function becomes. Mathematically, this implies that:

$$\begin{aligned} &\text{If } \alpha_1 < \alpha_2 < \alpha^* \quad \text{then } f(\alpha_2) < f(\alpha_1) \\ &\text{and if } \alpha_2 > \alpha_1 > \alpha^* \quad \text{then } f(\alpha_1) < f(\alpha_2) \end{aligned} \tag{4.5}$$

Figure 4.2 illustrates the concept of unimodality for the search interval $0 \leq \alpha \leq 1$. The values of the objective function at a number of values for the parameter α are used to bracket the solution. A multimodal function is a function whose domain can be divided into a number of subdomains within which the function is unimodal. We assume in what follows that within the search interval the objective function is unimodal.

4.2.1 Fixed line search

When carrying out a line search, one starts from a starting point $\boldsymbol{x}_s$ and search in a direction $\boldsymbol{s}$. The point $\boldsymbol{x}_s$ corresponds to $\alpha = 0$. Our target is to determine an interval of α through which the minimum of the function lies. A step size $\Delta\alpha$ may be used along the search direction. Moving in the direction $\boldsymbol{s}$, several approaches can be adopted. In the fixed step size approach, we use a fixed step in the direction $\boldsymbol{s}$ with

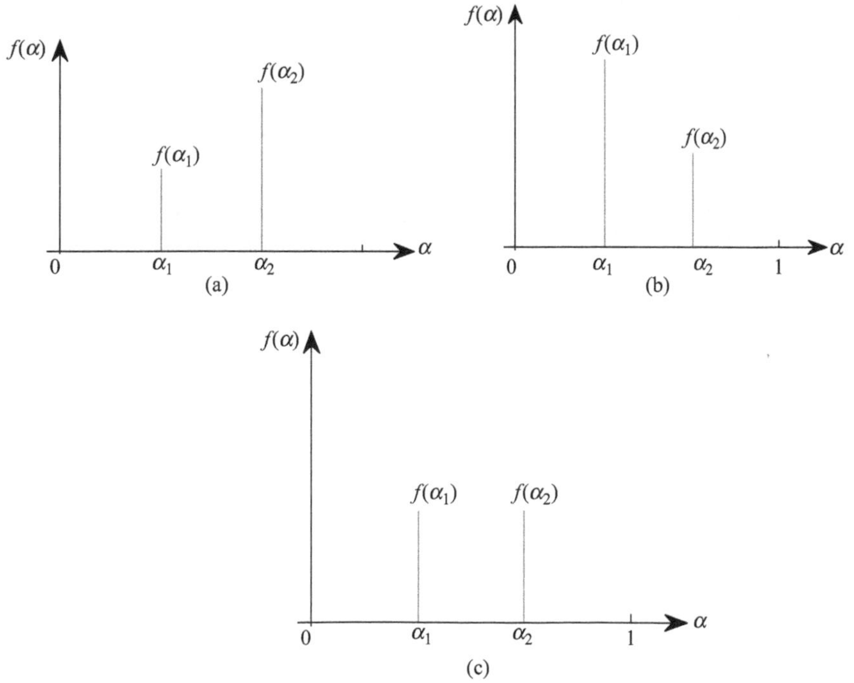

Figure 4.2 Illustrations of the concept of unimodality; (a) the minimum is in the interval $0 \leq \alpha \leq \alpha_1$, (b) the minimum is in the interval $\alpha_2 \leq \alpha \leq 1$, and (c) the minimum is in the interval $\alpha_1 \leq \alpha \leq \alpha_2$

$\alpha_j = j\Delta\alpha$, $j = 1, 2, \ldots$. The values of the objective function $f_j = f(\boldsymbol{x}_s + j\Delta\alpha\boldsymbol{s})$ are then evaluated. Using these values, we can bracket the solution of the problem in a much smaller subdomain. Figure 4.3 illustrates this approach. For the considered function, the function value $f_3 = f(\boldsymbol{x}_s + 3\Delta\alpha\boldsymbol{s})$ is less than f_2 and f_4. Using the assumption of unimodality, the minimum of the objective function along the direction $\boldsymbol{s}$ lies between $2\Delta\alpha$ and $4\Delta\alpha$. Once the solution has been bracketed between two values of α, a more precise technique can be used to determine the value of α^* within the desired accuracy.

The selection of the search parameter $\Delta\alpha$ is an important aspect of this approach. Utilizing a very small step results in carrying out a large number of simulations before the solution has been bracketed. On the other hand, selecting a large value of $\Delta\alpha$ may result in a large final interval. This large interval may reduce the accuracy of the obtained solution.

4.2.2 Accelerated line search

This approach can be explained through Figure 4.4. Instead of using a fixed step size as in Figure 4.3, the step size is doubled in every step to accelerate the

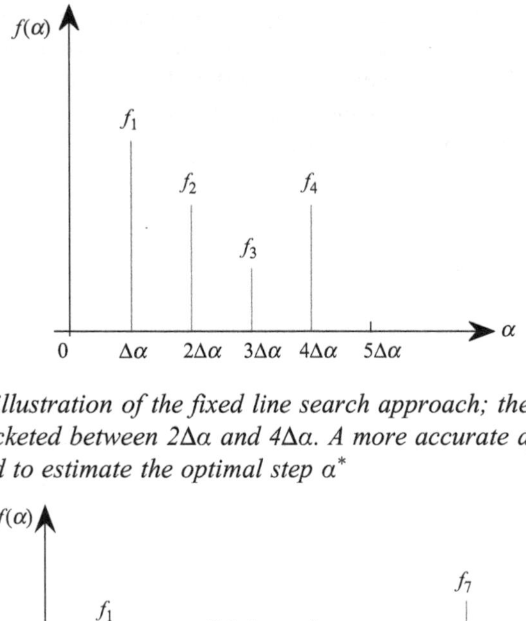

*Figure 4.3 An illustration of the fixed line search approach; the solution is bracketed between 2Δα and 4Δα. A more accurate approach can be used to estimate the optimal step α**

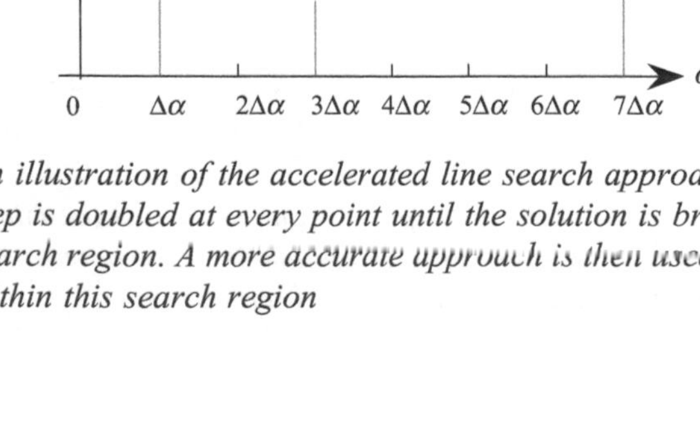

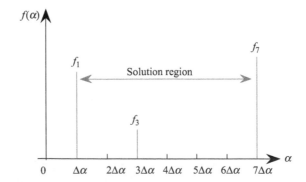

Figure 4.4 An illustration of the accelerated line search approach. The search step is doubled at every point until the solution is bracketed in a search region. A more accurate approach is then used to determine α within this search region*

bracketing approach. As illustrated in Figure 4.4, using unimodality, one can bracket the solution between two values of the parameter α. This approach may be preferred over the constant search step if the minimum is assumed to be far from the starting point $\alpha = 0$. The drawback of this approach is that the interval at which the solution is bracketed may be relatively large. This makes the job of the solution-finding approach more difficult and limits the accuracy of some of these derivative free techniques.

4.3 Derivative-free line search

Once the solution has been bracketed within an interval $[\alpha_s \quad \alpha_f]$, an optimization algorithm is applied to determine the value of the optimal solution within this interval. One possible class of optimization algorithms that can be used for solving the line search problem is derivative-free techniques. These techniques do not use any derivative information. The first-order derivative $f'(\alpha)$ or higher-order derivatives with respect to α are not used. Only the function values $f(\alpha)$ are used to locate the optimal value α^*. As will be seen later, derivative-free optimization is a complete class of optimization approaches that can also be applied to general multi-dimensional problems. We discuss in this section a number of these techniques.

4.3.1 Dichotomous line search

Starting with an initial interval $[\alpha_s \quad \alpha_f]$, the Dichotomous search approach creates two more test values (f_1 and f_2) around the center of the current interval with a spacing of δ as shown in Figure 4.5. Depending on these two function values and using the assumption of unimodality, a part of the interval is eliminated where the solution is not expected to end up with a smaller interval. In Figure 4.5, the solution lies between α_s and α_2. The interval $[\alpha_2 \quad \alpha_f]$ is thus discarded. The next iteration of this approach repeats the same step with α_s unchanged and with α_f updated to α_2 of the previous iteration. These steps are then repeated until the size of the interval has shrunk below a certain termination limit. The MATLAB® code M4.1 gives a possible implementation of the Dichotomous line search.

The MATLAB code M4.1 applies the Dichotomous search to the problem $f(\alpha) = \alpha(\alpha - 1.5)$. Notice that the parameter α is represented in the code by the

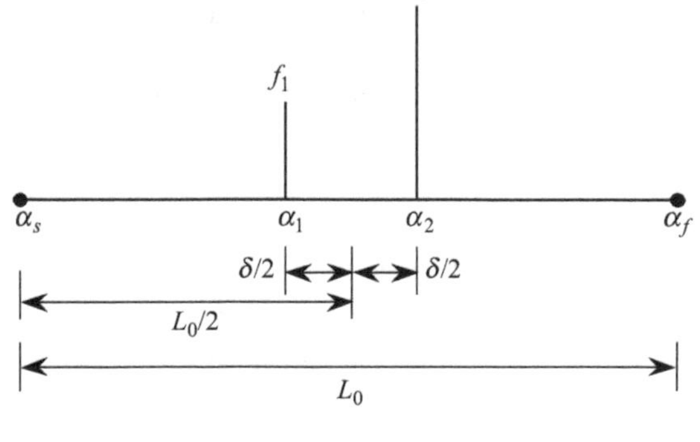

Figure 4.5 The Dichotomous line search; two extra function values are created around the center of the current interval to eliminate a part of the interval at each iteration

```
%M4.1
%an implementation of the Dichotomous search applied to the function x(x-1.5)
xLeft=0; %start of interval
xRight=1.0 %end of interval
delta=1.0e-3; %delta used for perturbation around the centre
xCenter=0.5*(xLeft+xRight) %interval centre
x1=xCenter-0.5*delta; %first test value
x2=xCenter+0.5*delta; %second test value
fLeft=getFunction(xLeft); %function value at start of interval
fRight= getFunction(xRight); %function value at end of interval
f1= getFunction(x1); %function value at first test point
f2= getFunction(x2); %function value at the second test point
L=xRight-xLeft; %current interval length
Epsilon=1.0e-2; %termination limit on interval
while(L>Epsilon) %repeat until interval is small enough
  if(f2>=f1) %if this condition is satisfied, move the right edge of the interval to x2
    xRight=x2; %adjust right edge
    fRight=f2; %adjust right value
    xCenter=0.5*(xLeft+xRight); %get new centre
    x1=xCenter-0.5*delta; %get new test coordinates
    x2=xCenter+0.5*delta; %get second test coordinate
    f1= getFunction(x1);%calculate test values around the centre
    f2= getFunction(x2);
    L=xRight-xLeft;          %adjust interval size
  else              %move the left edge of the interval to x1
    xLeft=x1;               %new left edge
    fLeft=f1;               %adjust new left value
    xCenter=0.5*(xLeft+xRight); %get new centre
    x1=xCenter-0.5*delta;    %get the new test coordinates
    x2=xCenter+0.5*delta;
    f1= getFunction(x1);   %get new test values
    f2= getFunction(x2);
    L=xRight-xLeft;        %adjust interval value
  end
end
xLeft %print final values of interval
xRight
```

symbol “x”. The output of this code is xLeft = 0.7493 and xRight = 0.7581. Applying the necessary optimality condition to this problem, one can see that the optimal solution is $\alpha^* = 0.75$.

One comment about this approach is that we reuse only one of the values created around the center of the problem. In every iteration of this approach, two new test values are evaluated around the center of the interval. The cost of this approach is thus approximately two simulations per iteration. This cost is important if we solve problems where evaluation of the function $f(\alpha)$ requires time-intensive numerical simulations.

4.3.2 The interval-halving method

The interval-halving method aims at reducing the size of the interval in every iteration by one half. The method can be illustrated through Figure 4.6. In the first iteration, three additional function values are evaluated as shown. These points are separated by one quarter of the current interval size. Depending on these values, we remove either the left half of the interval, the right half of the interval, or the two edge quarters of the interval. The assumption of unimodality guarantees that the optimal solution lies within the remaining half. For the following iterations, because we have already evaluated the function value at the center of the interval, only two function values are needed. The MATLAB listing M4.2 shows a possible implementation of the interval-halving approach. The following example illustrates the application of this method.

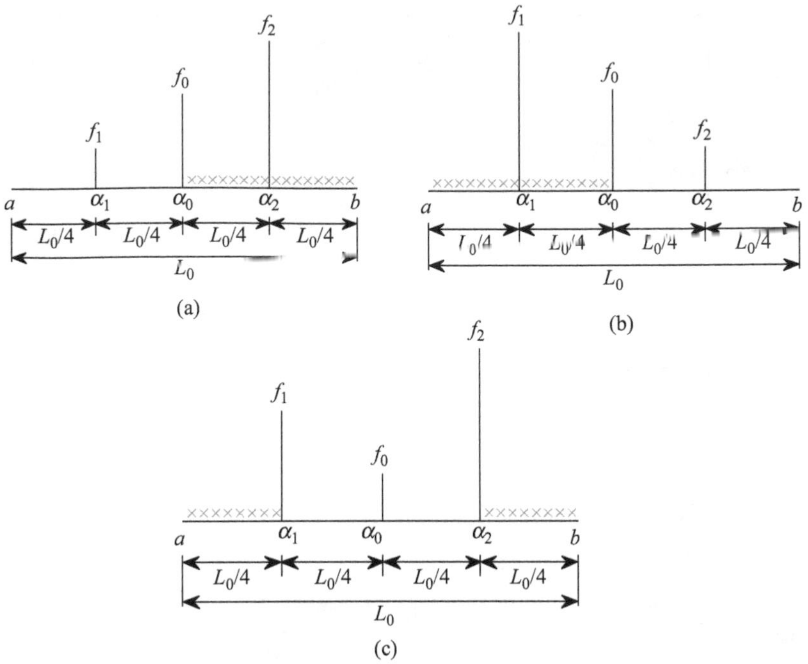

Figure 4.6 An illustration of the interval-halving line search scheme where the interval size is reduced by one half depending on the objective function; (a) the right half of the interval is discarded, (b) the left half of the interval is discarded, and (c) the left and right quarters of the interval are discarded

Example 4.2: Find analytically the minimum of the single-variable function $f(x) = x^5 - 5x^3 - 20x + 5$. Confirm your answer by executing the MATLAB listing M4.2 starting with the interval [0, 5].

Solution: Applying the necessary optimality conditions, we get:

$$\frac{df}{dx} = 0 \Rightarrow 5x^4 - 15x^2 - 20 = 0 \Rightarrow x^4 - 3x^2 - 4 = 0 \Rightarrow (x^2 - 4)(x^2 + 1) = 0 \quad (4.6)$$

Equation (4.6) has the solution $x = 2$ inside the interval. The second-order derivative at this point is $\partial^2 f / \partial x^2|_{x=2} = 100 > 0$. This implies that this point is a local minimum.

The MATLAB listing M4.2 is applied to this problem. The function getFunction returns the value of the function $f(x)$. The output of the code in the first four iterations and the final output are:

Iteration 0: xL = 0 x1 = 1.8750 xc = 2.5000 x2 = 3.1250 xR = 5
Iteration 1: xL = 0 x1 = 1.5625 xc = 1.8750 x2 = 2.1875 xR = 2.5000
Iteration 2: xL = 1.5625 x1 = 1.7969 xc = 1.8750 x2 = 1.9531 xR = 2.1875
Iteration 3: xL = 1.8750 x1 = 1.9141 xc = 1.9531 x2 = 1.9922 xR = 2.1875
Iteration 4: xL = 1.9531 x1 = 1.9629 xc = 1.9922 x2 = 2.0215 xR = 2.1875
Final Answer: xL = 1.9958 x1 = 1.9986 xc = 1.9995 x2 = 2.0004 xR = 2.0032

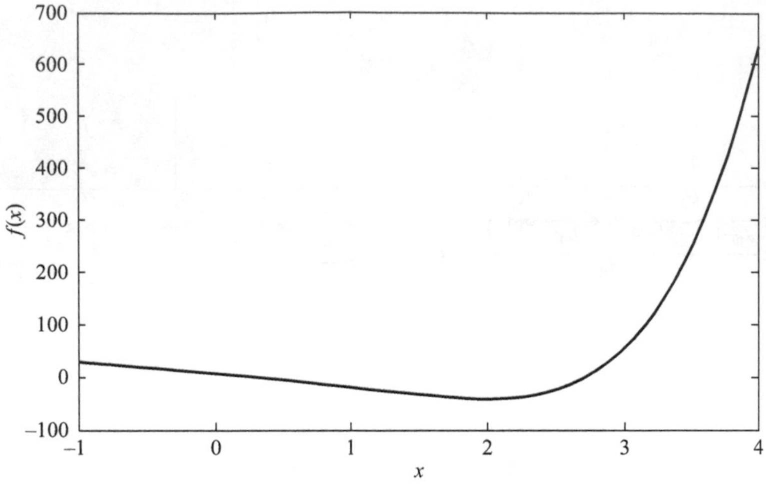

Figure 4.7 The graph of the function of Example 4.2

The answer obtained analytically and numerically can be verified by inspecting the function graph shown in Figure 4.7.

4.3.3 The Fibonacci search

It was shown in the previous section that the interval-halving method reduces the interval size to one half of its size in every iteration. The Fibonacci search, on the other hand, reduces the interval size by a different factor at every iteration. The reduction ratio follows a sequence of numbers satisfying the Fibonacci formula:

$$F_n = F_{n-1} + F_{n-2}, \quad \text{with} \quad F_0 = 1 \quad \text{and} \quad F_1 = 1 \tag{4.7}$$

The sequence of numbers that can be generated using formula (4.7) is 1, 1, 2, 3, 5, 8, 13, 21, 34, 55, 89,.... The Fibonacci search starts by selecting the desired Fibonacci number, say $N=7$ with $F_7=21$. This number, according to the formula (4.7), can be divided into F_6 and F_5. The initial interval of length L_0 is then divided into the two subintervals:

$$L_0 \frac{F_6}{F_7} \quad \text{and} \quad L_0 \frac{F_5}{F_7} \tag{4.8}$$

Two function values are evaluated inside the initial interval at a distance of L_0F_5/F_7 from the two interval edges as shown in Figure 4.8(a). Unimodality assumption is then used to discard the interval part in which the solution is not expected. Notice that if the new interval is L_1 as shown in Figure 4.8, then in the next iteration, we have one available function value at a distance L_0F_5/F_7 from the left edge. Because the new interval $L_1=L_0F_6/F_7$, this point is separated a distance $(F_5/F_7)L_1(F_7/F_6) = (F_5/F_6)L_1$ from the left edge or a distance $(F_4/F_6)L_1$ from the right edge. We then create one more function evaluation separated by $(F_4/F_6)L_1$ from the left edge as shown in Figure 4.8(b). The same steps are repeated for all the following seven steps until the interval has shrunk significantly. Notice how the old left point automatically satisfies the ratio F_{n-1}/F_{n-2} from the right edge of the new interval and vice versa for the right point. The same steps explained for the case $N=7$ can be generalized for any N.

```
%M4.2
%an implementation of the interval halving method
xLeft=0;    %left edge of initial interval
xRight=5.0; %right edge of initial interval
L=xRight-xLeft; %initial interval size
delta=0.25*L; %separation between samples
xCenter=0.5*(xLeft+xRight); %centre of current interval
x1=xCenter-delta; %sample to the left of the centre
x2=xCenter+delta; %sample to the right of the centre
fLeft=getFunction(xLeft); %get all function values
fRight=getFunction(xRight);
f1=getFunction(x1);
f2=getFunction(x2);
fCenter=getFunction(xCenter)
Epsilon=1.0e-2; %termination condition
while(L>Epsilon) %repeat until interval is small enough
  if(fCenter>f1) %if value at centre is greater than left value remove
    xRight=xCenter; %move right interval edge to current centre
    fRight=fCenter;
    xCenter=x1; %left value becomes the new centre
    fCenter=f1;
    L=xRight-xLeft; %new interval length
    delta=0.25*L; %new separation between samples
       x1=xCenter-delta; %the new left point
    x2=xCenter+delta; %create a new right point
    f1=getFunction(x1); %evaluate function at these new points
    f2=getFunction(x2);
  else
    if(f2<fCenter) %if value at centre is greater than right value
     xLeft=xCenter; %move left edge to the current centre
     fLeft=fCenter;
     xCenter=x2; %the right point becomes the new centre
     fCenter=f2;
     L=xRight-xLeft; %the new interval size
     delta=0.25*L; %the new separation between samples
     x1=xCenter-delta; %create a new Left point
     x2=xCenter+delta; %create a new right point
     f1=getFunction(x1); %evaluate function values at these points
     f2=getFunction(x2);
    else        %centre value is less than left and right values
     xLeft=x1;  %we move both left edge and right edge
     fLeft=f1;
     xRight=x2;
     fRight=f2;
     L=xRight-xLeft; %evaluate the new interval size
     delta=0.25*L; %evaluate the new spacing
     x1=xCenter-0.5*delta; %the new Left point
     x2=xCenter+0.5*delta; %the new right point
     f1=getFunction(x1); %get the function values at the new points
     f2=getFunction(x2);
    end
  end
end
```

The Fibonacci search requires two simulations within the interval in the first iteration. In every subsequent iteration, it uses only one extra function evaluation. The interval drops to $(F_k/F_N)L_0$, where L_0 is the initial interval size, at the kth iteration. This approach is illustrated through the following example.

Example 4.3: Minimize $f(x) = 0.65 - (0.75/(1+x^2)) - 0.65\tan^{-1}(1/x)$ in the interval [0, 3] using the Fibonacci method with $N=6$.

Solution: The first two function evaluations are calculated at a distance $d=(F_4/F_6)\times 3=(5/13)\times 3=1.153846$ from the interval edges. It follows that for the first

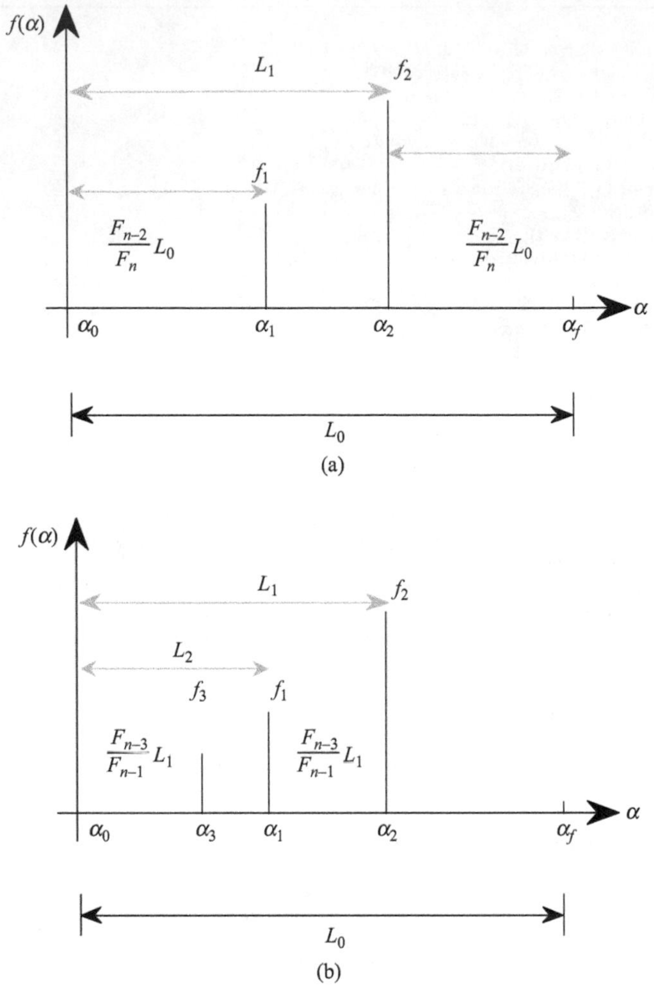

Figure 4.8 An illustration of the Fibonacci line search; (a) in the first iteration, the two points α_1 and α_2 are chosen at a distance $(F_{n-2}/F_n)L_0$ from the edges of the interval. Based on these function values, the interval L_1 is considered in the next iteration and (b) in the second iteration, a new point α_3 is created at a distance from the left edge equal to the distance from α_1 to the new right edge

iteration we create the two points $x_1 = 1.153846$ and $x_2 = 3 - 1.153846 = 1.846154$. The function values at these two points are as shown in Figure 4.9(a).

From the given function values and assuming unimodality, the interval $[x_2, 3.0]$ is eliminated as the minimum does not lie in this range. For the second iteration, the right edge of the interval is at 1.846154 and $x_2 = 1.153846$. Notice that x_2 of the second iteration is x_1 of the first iteration. Instead of using the Fibonacci sequence to create the new x_1, we simply select it at a distance of $1.846154 - 1.153846 = 0.6923$ from the left edge of the interval. This satisfies the interval ratios of the Fibonacci sequence as explained earlier. The second iteration is shown in Figure 4.9(b). The same steps are repeated for four more iterations as shown in Figure 4.9. The final interval is [0.23077, 0.46154]. The ratio between the final interval and the initial interval is $L_6/L_1 = (0.46154 - 0.23077)/3.0 = 0.076923$.

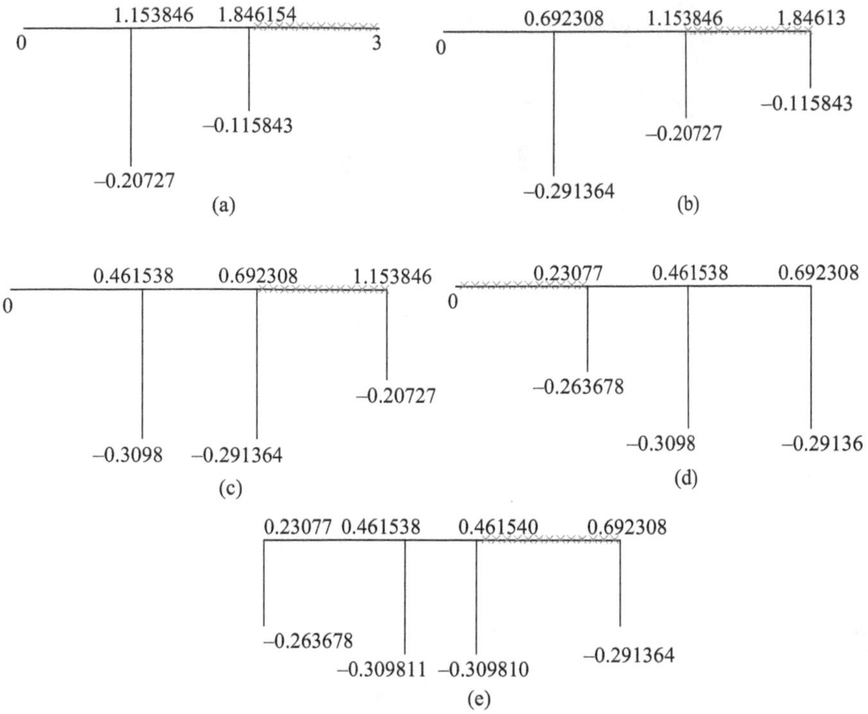

Figure 4.9 An illustration of the first five iterations of Example 4.3 using Fibonacci's line search

4.3.4 The Golden Section method

The Golden Section method is a special case of the Fibonacci method. As the order of the Fibonacci number gets higher and higher, one notices that the ratio between any two consecutive numbers approaches a certain value:

$$\frac{F_8}{F_9} = \frac{34}{55} = 0.618 \quad \text{and} \quad \frac{F_9}{F_{10}} = \frac{55}{89} = 0.6179 \tag{4.9}$$

Actually, as the order of the Fibonacci number becomes too large, this ratio satisfies:

$$\operatorname*{limit}_{N \to \infty} \frac{F_{N-1}}{F_N} = 0.618 \tag{4.10}$$

This ratio fascinated scientists for 2400 years and some scientists still research its properties. This ratio finds applications in mathematics, biology, architecture, etc.

The basic idea of the Golden Section line search is to use the same ratio in all iterations rather than using a variable ratio as in the Fibonacci sequence of numbers. The basic steps for the Golden Section search are otherwise very similar to those of the Fibonacci search. The following example illustrates the application of the Golden Section line search approach.

Example 4.4: Use the Golden Section method to find the minimum of the function $f(x) = x^4 - 14x^3 + 60x^2 - 70x$ in the interval [0, 2]. Locate the value of x^* to within a range of 0.3.

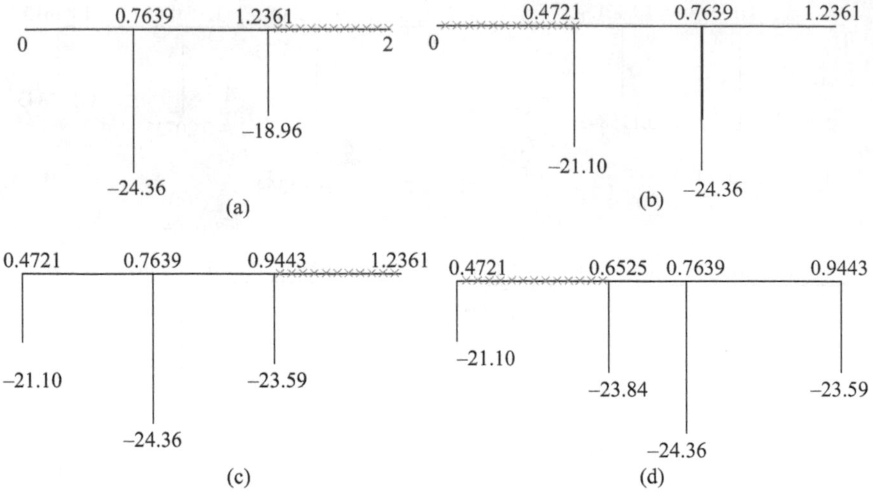

Figure 4.10 An illustration of the Golden Section approach for Example 4.4

Solution: As the reduction in the size of the interval is the same in all iterations, the number of needed Golden Section iterations is given by:

$$2(0.61803)^N < 0.3 \tag{4.11}$$

The smallest number of iterations N that would result in the necessary interval reduction is $N=4$. It follows that four Golden Section iterations are required. Similar to the Fibonacci approach, two points are created inside the interval in the first iteration. The first point is $x_2 = 0.61803 \times 2 = 1.2361$. The point x_1 is selected at the same distance from the left edge at $x_1 = 2.0 - 1.236 = 0.7639$. This is shown in Figure 4.10(a). Based on the function values at these two points and using the assumption of modality, we can eliminate the interval $[x_2, 2]$. In the second iteration, x_1 becomes x_2. Instead of using the Golden Section ratio again to calculate the new x_1, we select it at the same distance from the left edge as x_2 from the right edge. The new x_1 is thus selected at $x_1 = 1.226 - 0.7639 = 0.4721$. This step is shown in Figure 4.10(b). The new function value at x_1 is then calculated and using unimodality, the interval [0, 0.4721] is eliminated. These steps are then repeated as shown in Figure 4.10. The final interval is $L_4 = [0.6525, 0.9443]$, which is within the required accuracy.

4.4 Interpolation approaches

Another important class for solving the line search problem (4.1) is interpolation-based approaches. These approaches aim at approximating the function $f(\alpha)$, which is usually not known analytically, by a polynomial function of α. This polynomial, which usually has a quadratic or cubic order, is then minimized analytically to determine its minimum $\bar{\alpha}$. Either this minimum reasonably approximates the solution α^* or the interpolation model is improved using the available data points.

We will illustrate two different approaches within this class: quadratic modeling and cubic modeling.

4.4.1 Quadratic models

A general quadratic interpolation model is given by:

$$Q(\alpha) = a + b\alpha + c\alpha^2 \tag{4.12}$$

The three parameters of this model (a, b, and c) determine the function value and derivatives for every point α. The minimum of the quadratic function is determined by applying the first-order optimality condition $dQ(\alpha)/d\alpha = 0$, which gives:

$$b + 2\alpha c = 0 \Rightarrow \bar{\alpha} = -\frac{b}{2c} \tag{4.13}$$

The extreme point $\bar{\alpha}$ is a local minimum if the sufficient optimality condition is satisfied, i.e., if $d^2Q(\alpha)/d\alpha^2$ is positive. This implies that $c > 0$ must be satisfied for the point (4.13) to be a local minimum.

Determining the coefficients a, b, and c requires three pieces of data for the approximated function. Assuming that the function pairs $((\alpha_1, f(\alpha_1)), (\alpha_2, f(\alpha_2)), (\alpha_3, f(\alpha_3)))$ are known as shown in Figure 4.11. The three equations used to solve for the coefficients a, b, and c are given by:

$$f_i = f(\alpha_i) = a + b\alpha_i + c\alpha_i^2, \quad i = 1, 2, 3 \tag{4.14}$$

These three equations are then solved numerically for the coefficients by solving the system of equations:

$$\begin{bmatrix} 1 & \alpha_1 & \alpha_1^2 \\ 1 & \alpha_2 & \alpha_2^2 \\ 1 & \alpha_3 & \alpha_3^2 \end{bmatrix} \begin{bmatrix} a \\ b \\ c \end{bmatrix} = \begin{bmatrix} f_1 \\ f_2 \\ f_3 \end{bmatrix} \tag{4.15}$$

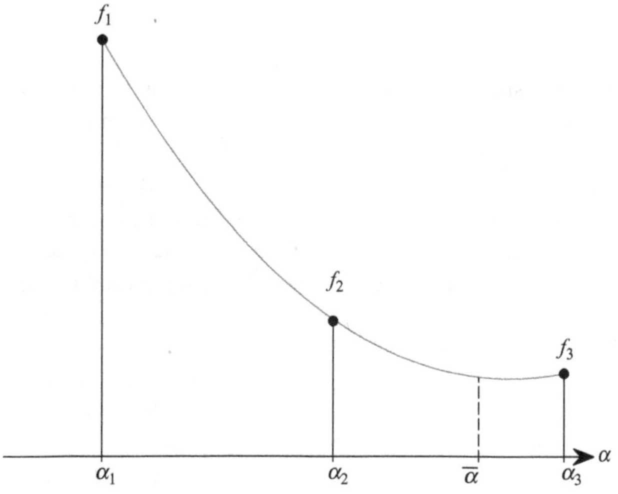

Figure 4.11 An illustration of the quadratic interpolation approach; using the function values at three values of the line search parameter α, a quadratic model is constructed. The minimizer of this model approximates the minimizer of the function α^*

Once the system of equations (4.15) is solved for the coefficients, the minimum of the quadratic approximation is then found using (4.13). A closed form of (4.13) in terms of the available data samples is given by [1]:

$$\alpha^* = \frac{f_1(\alpha_2^2 - \alpha_3^2) + f_2(\alpha_3^2 - \alpha_1^2) + f_3(\alpha_1^2 - \alpha_2^2)}{2[f_1(\alpha_2 - \alpha_3) + f_2(\alpha_3 - \alpha_1) + f_3(\alpha_1 - \alpha_2)]} \tag{4.16}$$

The sufficient condition of a positive second-order derivative can also be expressed in terms of the function's values using:

$$c > 0 \Rightarrow f_1(\alpha_2 - \alpha_3) + f_2(\alpha_3 - \alpha_1) + f_3(\alpha_1 - \alpha_2) < 0 \tag{4.17}$$

If the data points used to create the quadratic approximation are closely spaced, the approximation will reasonably approximate the function $f(\alpha)$. On the other hand, if the data points are widely separated, the quadratic model will not be a good approximation of the original function. It follows that the solution should be first bracketed within a reasonably small interval and then a quadratic model can be created to approximate the function behavior.

Once the quadratic model is created and an approximation of the optimal solution is found using (4.16), the accuracy of the quadratic model $Q(\alpha)$ is checked. We can check whether the function value $f(\bar{\alpha})$ sufficiently matches $Q(\bar{\alpha})$. Alternatively, we can check if the value of the directional derivative at the point $\bar{\alpha}$ is small enough. These two checks are given by:

$$\begin{aligned} &\left|\frac{Q(\bar{\alpha}) - f(\bar{\alpha})}{f(\bar{\alpha})}\right| \le \varepsilon_1 \\ &\left|\frac{f(\bar{\alpha} + \Delta\alpha) - f(\bar{\alpha} - \Delta\alpha)}{2\Delta\alpha}\right| \le \varepsilon_2 \end{aligned} \tag{4.18}$$

where ε_1 and ε_2 are small positive thresholds. If these accuracy checks are not satisfied, the quadratic model is to be "refitted". This means that the model accuracy should be improved to better predict the optimal point. The new available function value $f(\bar{\alpha})$ and the assumption of unimodality are used to achieve this goal. Figure 4.12 illustrates the refitting approach. There are four different refitting scenarios in this figure. If $f(\bar{\alpha}) < f_2$ and $\alpha_2 < \bar{\alpha} < \alpha_3$, the interval $[\alpha_1, \alpha_2]$ is discarded. A new improved quadratic model is constructed using the three data points $((\alpha_2, f(\alpha_2)), (\bar{\alpha}, f(\bar{\alpha})), (\alpha_3, f(\alpha_3)))$. If $f(\bar{\alpha}) < f_2$ and $\alpha_1 < \bar{\alpha} < \alpha_2$, the interval $[\alpha_2, \alpha_3]$ is eliminated. A new improved quadratic model is constructed using the data points $((\alpha_1, f(\alpha_1)), (\bar{\alpha}, f(\bar{\alpha})), (\alpha_2, f(\alpha_2)))$. On the other hand, if $f(\bar{\alpha}) > f_2$ and $\alpha_2 < \bar{\alpha} < \alpha_3$, then the interval $[\bar{\alpha}, \alpha_3]$ is discarded. A new improved quadratic model is constructed using the points $((\alpha_1, f(\alpha_1)), (\alpha_2, f(\alpha_2)), (\bar{\alpha}, f(\bar{\alpha})))$. Finally, if $f(\bar{\alpha}) > f_2$ and $\alpha_1 < \bar{\alpha} < \alpha_2$, we eliminate the interval $[\alpha_1, \bar{\alpha}]$. A new improved quadratic model is constructed using the points $((\bar{\alpha}, f(\bar{\alpha})), (\alpha_2, f(\alpha_2)), (\alpha_3, f(\alpha_3)))$. The listing M4.3 shows a possible MATLAB implementation of this approach.

```
%M4.3
%This program finds the minimum of a function using a quadratic approximation
% The three points in the interval are A, B, and C in order
A=-0.5; %start of interval
C=0.5;  %end of interval
B=0.5*(A+C); %interval middle point
L=C-A; %current interval length
epsilon=1.0e-5; %termination condition
%get all function values at starting points
fA=getFunction(A);
fB=getFunction(B);
fC=getFunction(C);
while (L>epsilon) %repeat until interval is small enough
   Alpha=0.5*(fA*(B*B-C*C)+fB*(C*C-A*A)+fC*(A*A-B*B))/(fA*(B-C)+fB*(C-A)+fC*(A-B));
%find minimizer of quadratic approximation
   fAlpha=getFunction(Alpha); %get function value at new point
   %four refitting scenarios
   %solution in first interval
   if((A<Alpha)&(Alpha<B))
      if(fAlpha<fB) %move right point to current B.
         C=B;
         fC=fB;
         B=Alpha;
         fB=fAlpha;
      else  %move left point to the minimizer
         A=Alpha;
         fA=fAlpha;
      end
   end
   %solution in second interval
   if((B<Alpha)&(Alpha<C))
      if(fAlpha<fB) %move left edge to B
         A=B;
         fA=fB;
         B=Alpha;
         fB=fAlpha;
      else  %move right edge to minimizer
         C=Alpha;
         fC=fAlpha;
      end
   end
   L=C-A; %new interval size
end
```

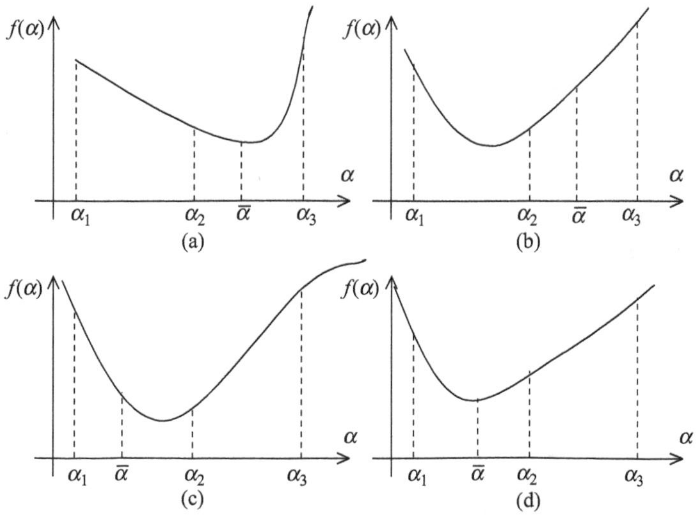

Figure 4.12 Four different scenarios for fitting a quadratic model to improve the result of the line search approach

Example 4.5: The function $f(\alpha) = -5\alpha^5 + 4\alpha^4 - 12\alpha^3 + 11\alpha^2 - 2\alpha + 1$ is unimodal in the interval $[-0.5, 0.5]$. Use a quadratic approximation to find the minimum with a range of uncertainty less than 10^{-5}.

Solution: Using the MATLAB listing M4.3, the three data points in the first three iterations are found to be:

Iteration 0:	$\alpha_1 = -0.5$	$\alpha_2 = 0$	$\alpha_3 = 0.500$
Iteration 1:	$\alpha_1 = 0$	$\alpha_2 = 0.2214$	$\alpha_3 = 0.5$
Iteration 2:	$\alpha_1 = 0$	$\alpha_2 = 0.1316$	$\alpha_3 = 0.2214$
Iteration 3:	$\alpha_1 = 0$	$\alpha_2 = 0.1193$	$\alpha_3 = 0.1316$

In the final iteration, we have $\alpha_1 = 0.1099$, $\alpha_2 = 0.1099$, and $\alpha_3 = 0.1099$. The best estimate of the solution is at $\bar{\alpha} = 0.1099$. Figure 4.13 shows the graph of the function $f(\alpha)$. It is obvious that the solution obtained using quadratic approximation is very close to the actual minimizer $\alpha^* = 0.1099$.

4.4.2 Cubic interpolation

The quadratic interpolation approach discussed in the previous subsection utilizes a second-order polynomial to approximate the behavior of the function and predict the optimal solution. Higher-order polynomials can also be used to obtain even more accurate approximation at the expense of more required data. For example, a cubic approximation model of a general function is given by:

$$C(\alpha) = a + b\alpha + c\alpha^2 + d\alpha^3 \tag{4.19}$$

This model has four unknown coefficients. Four pieces of information are required to determine these coefficients. This information may include the function values at four different points within the interval of interest. Using so many points may make the algorithm cumbersome. An easier approach is to utilize both function values and derivatives at only two points to solve for these coefficients. The first-order derivative of (4.19) is given by:

$$C'(\alpha) = b + 2c\alpha + 3d\alpha^2 \tag{4.20}$$

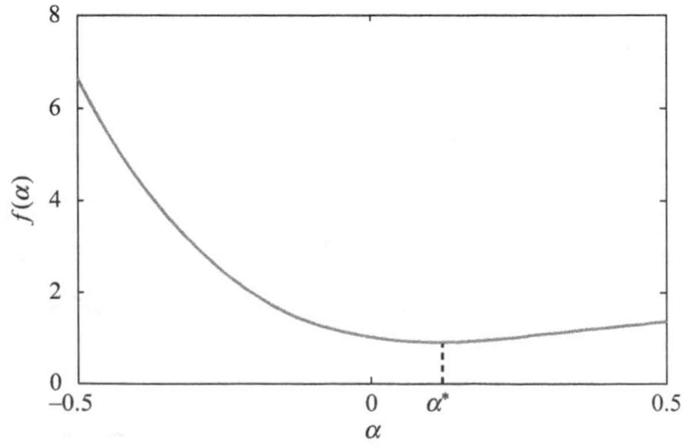

Figure 4.13 The graph of the function of Example 4.5

Knowing the function values and its derivatives at two different points α_1 and α_2 results in the following system of equations:

$$\begin{bmatrix} 1 & \alpha_1 & \alpha_1^2 & \alpha_1^3 \\ 0 & 1 & 2\alpha_1 & 3\alpha_1^2 \\ 1 & \alpha_2 & \alpha_2^2 & \alpha_2^3 \\ 0 & 1 & 2\alpha_2 & 3\alpha_2^2 \end{bmatrix} \begin{bmatrix} a \\ b \\ c \\ d \end{bmatrix} = \begin{bmatrix} f(\alpha_1) \\ f'(\alpha_1) \\ f(\alpha_2) \\ f'(\alpha_2) \end{bmatrix} \tag{4.21}$$

These first-order derivatives at these two points should have opposite signs to ensure that the zero-gradient point (α^*) lies within the interval. The minimum of the cubic model is found by setting (4.20) to zero. Because (4.20) is a quadratic model, two possible solutions exist. These solutions are given by:

$$\bar{\alpha} = \frac{-c \pm \sqrt{c^2 - 3bd}}{3d} \tag{4.22}$$

These solutions may be inside or outside the interval $[\alpha_1, \alpha_2]$. Only the solution inside the desired region should be considered. The sign of the second-order derivatives of the solution inside this interval should be checked for optimality. This check is given by:

$$\frac{d^2C}{d\alpha^2} > 0 \Rightarrow 2c + 6d\bar{\alpha} > 0 \tag{4.23}$$

Similar to quadratic interpolation, the accuracy of the obtained minimum is checked using (4.18) with $Q(\alpha)$ replaced by $C(\alpha)$. If the required accuracy is achieved, the final solution is accepted. Otherwise, the model is refitted with the new obtained point. The point $\bar{\alpha}$ replaces the point with the same derivative sign to keep the solution bracketed between two points of opposite derivatives. The MATLAB listing M4.4 shows an implementation of the cubic interpolation approach.

Example 4.6: Utilize the MATLAB listing M4.4 to find the minimum of the function $f(\alpha) = -3\alpha \sin(0.75\alpha) + e^{-2\alpha}$ in the interval $\alpha \in [0, 2\pi]$.

Solution: Utilizing the MATLAB code M4.4 with the functions getFunction() and getDerivative() implemented to evaluate the value and derivative of the given function, the code output for the first few iterations is given by:

Iteration 0: A = 0   B = 6.2832   Alpha = 0.7208

Iteration 1: A = 0.7208   B = 6.2832   Alpha = 1.4511

Iteration 2: A = 1.4511   B = 6.2832   Alpha = 2.0359

Iteration 3: A = 2.0359   B = 6.2832   Alpha = 2.3862

The final output of the code is $\bar{\alpha} = 2.6952$. The solution obtained by the MATLAB code can be verified by plotting the graph of the function $f(\alpha)$ as shown in Figure 4.14.

```
%M4.4
%This file implements the Cubic Interpolation Method.
A=0; %left point
B=2*pi; %right point
delta=1.0e-3; %delta for sensitivity analysis
fA=getFunction(A); %get function value at A
fB=getFunction(B); %get function value at B
fpA=getDerivative(A, delta); %get derivative through finite difference at A
fpB=getDerivative(B, delta); %get derivative through finite difference at B
L=B-A; %solution interval
Epsilon=1.0e-1; %interval termination condition
fpAlpha=10; %initial gradient at solution
while(abs(fpAlpha)>Epsilon) %repeat until interval is small enough
 %first we construct system of equations
 Matrix=[1  A  (A*A)  (A*A*A); 1  B  (B*B)  (B*B*B);
         0  1  (2*A)  (3*A*A);  0  1  (2*B)  (3*B*B)];
 RHS=[fA fB  fpA  fpB]'; %this is the vector of RHS
 Coeff=inv(Matrix)*RHS; %Get coefficients
 b=Coeff(2,1); %2nd coefficient
 c=Coeff(3,1);  %3rd coefficient
 d=Coeff(4,1); %4th coefficient
 %two solutions exist. We pick the one in interval
 Alpha1=(-c+sqrt(c*c-3*b*d))/(3*d);
 Alpha2=(-c-sqrt(c*c-3*b*d))/(3*d);
 if((Alpha1>A)&(Alpha1<B))
   Alpha=Alpha1;
 else
   Alpha=Alpha2;
 end
 fAlpha=getFunction(Alpha); %get value at new minimum
 fpAlpha=getDerivative(Alpha,delta); %get derivative at new minimum
 %now we narrow down the interval
 % The point Alpha replaces the point with the same derivative sign
 if(fpAlpha*fpA>0) %on same side of minimum
   A=Alpha;  %move left value to Alpha
   fA=fAlpha;
   fpA=fpAlpha;
 else
   B=Alpha; %move right value to Alpha
   fB=fAlpha;
   fpB=fpAlpha;
 end
end
```

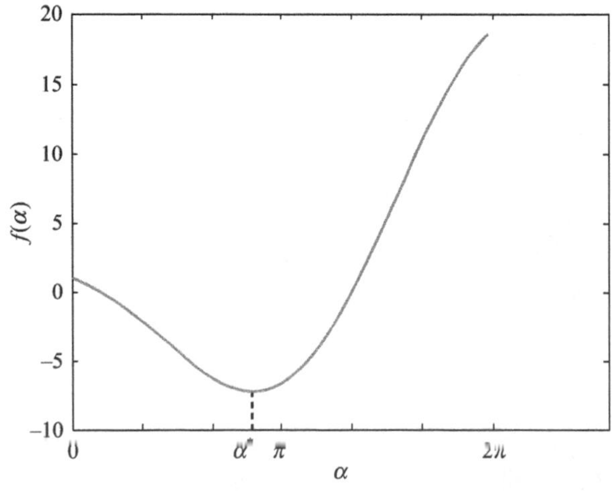

Figure 4.14 The graph of the function of Example 4.6

4.5 Derivative-based approaches

There is a wealth of techniques that utilize first- and second-order derivatives in solving optimization problems including one-dimensional ones. We will delay discussing most of these techniques to Chapters 6 and 7. We will focus in this section only on Newton-like techniques that convert the optimization problem into a root-finding problem.

4.5.1 The classical Newton method

This technique is very powerful in solving optimization problems. It requires both first- and second-order derivatives of the objective function. The technique effectively uses a Taylor expansion of the gradient to find the point at which the first-order derivative vanishes. It converges very fast to the solution if the starting point is close enough to the optimal solution. It may, however, diverge if the starting point is relatively far from the optimal design. In Chapter 7, I will discuss techniques to improve the convergence of this approach regardless of the starting point.

Newton method aims at finding a root for the equation:

$$f'(\alpha) = 0 \tag{4.24}$$

starting from a given initial point $\alpha^{(0)}$. All extreme points of the objective function are possible solutions. Expanding (4.24) using Taylor's expansion around α_i, the ith known iterate, we get:

$$f'(\alpha) = f'(\alpha_i) + f''(\alpha_i)(\alpha - \alpha_i) \tag{4.25}$$

Setting (4.25) equal to zero and solving for α we obtain:

$$\alpha_{i+1} = \alpha_i - \frac{f'(\alpha_i)}{f''(\alpha_i)} \tag{4.26}$$

The iteration (4.26) gives a better estimate to the optimal solution α^* using first- and second-order derivatives at the point α_i. The step (4.26) is repeated at the new point α_{i+1}. The technique terminates when $|f'(\alpha)| \leq \varepsilon$.

The steps of the Newton method can be illustrated through Figure 4.15. This figure shows the function $f'(\alpha)$. Equation (4.25) approximates the function $f'(\alpha)$ at the point α_i with a tangent line. Equation (4.26) effectively intersects this tangent line with the horizontal axis to determine the new point α_{i+1}. A new tangent line is

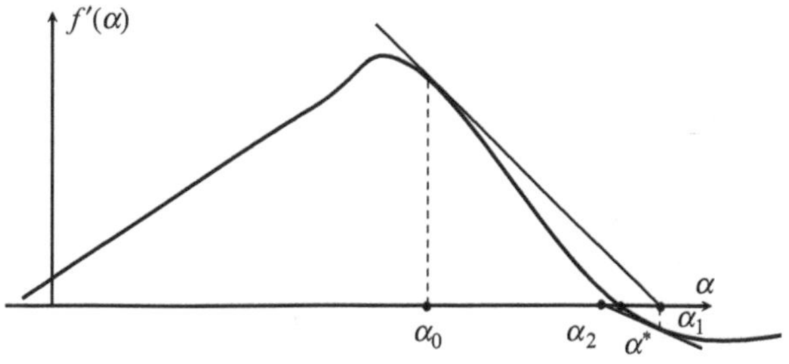

Figure 4.15 An illustration of a converging Newton sequence; starting from the point α_0, the tangent line at this point intersects the horizontal axis at point α_1. The tangent at the point α_1 intersects the horizontal axis at the point α_2. Notice that $|\alpha_2 - \alpha^*| \leq |\alpha_1 - \alpha^*| \leq |\alpha_0 - \alpha^*|$

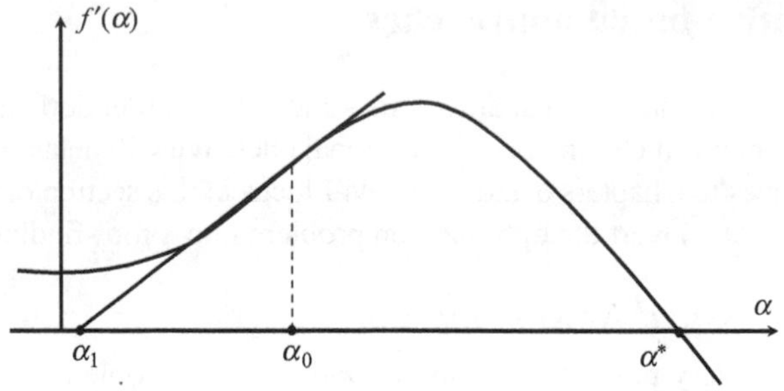

Figure 4.16 One step of a diverging Newton sequence; starting from the point α_0, the tangent line at this point intersects the horizontal axis at point α_1. Notice that $|\alpha_0 - \alpha^*| \leq |\alpha_1 - \alpha^*|$

then drawn at the point α_{i+1} and intersected with the horizontal axis to get a better approximation. If the starting point is close enough to the optimal point, the technique converges to a point at which the first-order derivative vanishes as shown in Figure 4.15.

Figure 4.16 shows a diverging Newton step. In this case, the intersection of the linearization with the horizontal axis gives a new estimate, which is even farther from the optimal solution. Newton's method is illustrated through the following example:

Example 4.7: Using Newton's method, find the minimizer of the function $f(\alpha) = (\alpha^2/2) - \sin(\alpha)$. The initial value is $\alpha_0 = 0.5$. The required accuracy is $\|\alpha_{k+1} - \alpha_k\| < 10^{-5}$.

Solution: Applying Newton's method requires both the first- and second-order derivatives of the function. The first-order derivative of the function is given by $f'(\alpha) = \alpha - \cos(\alpha)$, while the second-order derivative is given by $f''(\alpha) = 1 + \sin(\alpha)$. The iterations of the Newton method are thus given by:

$$\alpha_{i+1} = \alpha_i - \frac{\alpha_i - \cos(\alpha_i)}{1 + \sin(\alpha_i)} \tag{4.27}$$

Starting with $\alpha_0 = 0.5$, we get the following iterates:

$$\alpha_1 = \alpha_0 - \frac{\alpha_0 - \cos(\alpha_0)}{1 + \sin(\alpha_0)} = 0.5 - \frac{0.5 - \cos(0.5)}{1 + \sin(0.5)} = 0.7552$$

$$\alpha_2 = \alpha_1 - \frac{\alpha_1 - \cos(\alpha_1)}{1 + \sin(\alpha_1)} = 0.7552 - \frac{0.7552 - \cos(0.7552)}{1 + \sin(0.7552)}$$

$$= 0.7552 - \frac{0.02710}{1.685} = 0.7391$$

$$\alpha_3 = \alpha_2 - \frac{\alpha_2 - \cos(\alpha_2)}{1 + \sin(\alpha_2)} = 0.7391 - \frac{0.7391 - \cos(0.7391)}{1 + \sin(0.7391)}$$

$$= 0.7391 - \frac{0.0094}{1.673} = 0.7390$$

The final iteration satisfying the termination condition is $\alpha^* = 0.7390$. Notice that this must be a local minimum as the second-order derivative at the solution is positive as evident from the previous iterations.

4.5.2 A quasi-Newton method

The main difficulty in applying Newton's method explained in Section 4.5.1 is that the analytical second-order derivatives are not usually available. In most practical problems, the objective function is only known numerically. In other words, we can obtain $f(\alpha)$ as a number for each value of α. In this case, the first- and second-order derivatives must be estimated through finite differences as explained in Chapter 1. Estimation of the second-order derivatives requires $O(n^2)$ simulations. This cost can be formidable for problems with time-intensive models where one function evaluation may require hours or even days. A whole class of techniques called the quasi-Newton methods is developed to efficiently approximate the second-order derivatives. We will discuss these techniques in detail in Chapter 7. In this section, we discuss only techniques that utilize finite differences to approximate the first- and second-order derivatives.

A possible quasi-Newton approach for solving one-dimensional line search approaches utilizes the central finite difference approximations:

$$\begin{aligned} f'(\alpha_i) &= \frac{f(\alpha_i + \Delta\alpha) - f(\alpha_i - \Delta\alpha)}{2\Delta\alpha} \\ f''(\alpha_i) &= \frac{f(\alpha_i + \Delta\alpha) - 2f(\alpha_i) + f(\alpha_i - \Delta\alpha)}{(\Delta\alpha)^2} \end{aligned} \tag{4.28}$$

where $\Delta\alpha$ is a sufficiently small perturbation. The Newton's iteration (4.26) then becomes:

$$\alpha_{i+1} = \alpha_i - \frac{\Delta\alpha}{2} \frac{f(\alpha_i + \Delta\alpha) - f(\alpha_i - \Delta\alpha)}{f(\alpha_i + \Delta\alpha) - 2f(\alpha_i) + f(\alpha_i - \Delta\alpha)} \tag{4.29}$$

The formula (4.29) does not require analytical first- or second-order derivatives. Only function values are used to approximate the required derivatives. The MATLAB listing M4.5 illustrates a possible implementation of this quasi-Newton method.

```
%M4.5
%This matlab listing implements the Quasi-Newton method
StartingPoint=11; %starting guess
delta=0.001; %perturbation used in derivative analysis
MaxIterationCount=100; %we do not allow more than 100 iterations
Epsilon=1.0e-4; %terminating condition for gradient
gradient=1.0e4; %initial value of gradient
Counter=0; %intialize iteration counter
CurrentPoint=StartingPoint; %initialize current point
while((Counter<MaxIterationCount)&(abs(gradient)>Epsilon))
  f=getFunction(CurrentPoint); %function at point
  fp=getFunction(CurrentPoint+delta); %function with +ve perturbation
  fn=getFunction(CurrentPoint-delta); %function with -ve perturbation
  gradient=(fp-fn)/(2*delta); %get first-order derivative approximation
  Hessian=(fp-2*f+fn)/(delta*delta); %get second-order derivative approximation
  NewPoint=CurrentPoint-(gradient/Hessian); %get the new point using formula (4.29)
  CurrentPoint=NewPoint; % update new point
  Counter=Counter+1 %increment iteration counter
  CurrentPoint
end
```

Example 4.8: Utilize the MATLAB listing M4.5 to find the minimizer of the function $f(\alpha)=\alpha^3-12.2\alpha^2+7.45\alpha+42$ starting with $\alpha_0=11.0$.

Solution: Utilizing the listing M4.5 with the given function we obtain the following output at every iteration:

Counter = 0	CurrentPoint = 11.0
Counter = 1	CurrentPoint = 8.5469
Counter = 2	CurrentPoint = 7.8753
Counter = 3	CurrentPoint = 7.8161
Counter = 4	CurrentPoint = 7.8156
Counter = 5	CurrentPoint = 7.8156

The optimal solution is thus $\alpha^*=7.8156$. This answer can be verified by plotting the function graph as shown in Figure 4.17.

4.5.3 The Secant method

The Secant method offers another variation of the original Newton method. This approach assumes that the solution to the root-finding problem (4.24) has been bracketed in the interval $[\alpha_A, \alpha_B]$ where α_A and α_B have opposite sign of $f'(\alpha)$ with $f'(\alpha_A)f'(\alpha_B) < 0$. The second-order derivative required by the Newton iteration (4.26) is then approximated by:

$$f''(\alpha_i) \approx \frac{f'(\alpha_B) - f'(\alpha_A)}{(\alpha_B - \alpha_A)} \tag{4.30}$$

The Newton iteration then becomes:

$$\alpha_{i+1} = \alpha_A - \frac{f'(\alpha_A)(\alpha_B - \alpha_A)}{f'(\alpha_B) - f'(\alpha_A)} \tag{4.31}$$

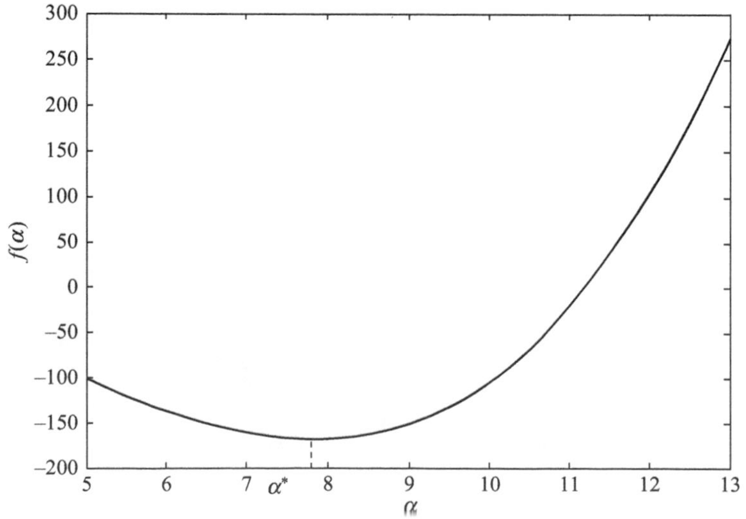

Figure 4.17 The graph of the function in Example 4.8

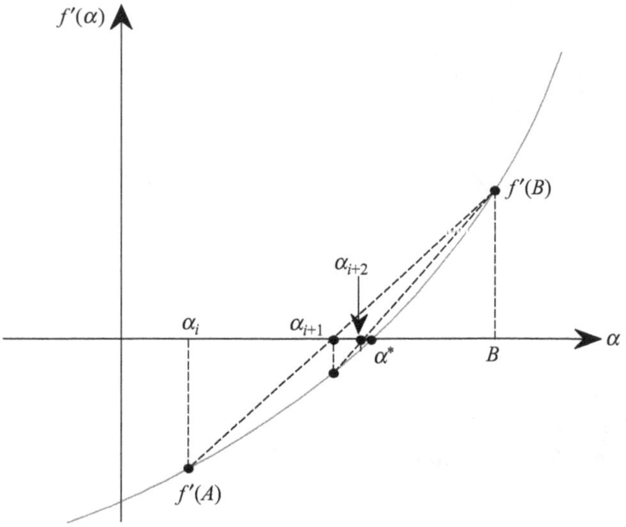

Figure 4.18 *An illustration of the Secant method; the solution is first bracketed between the interval [A, B]. The line connecting the two points [A, f′(A)] and [B, f′(B)] intersects the horizontal axis at the point α_{i+1}. The point α_{i+1} replaces the interval edge with the same derivative sign ($\alpha_i = A$)*

The step (4.31) is equivalent to intersecting the line connecting the two points $(\alpha_A, f'(\alpha_A))$, $(\alpha_B, f'(\alpha_B))$ with the horizontal axis as shown in Figure 4.18. The new point α_{i+1} replaces either the point α_A or α_B depending on the sign of $f'(\alpha_{i+1})$.

4.6 Inexact line search

The techniques discussed so far are all exact line search techniques. They aim at obtaining an accurate solution to the line search problem (4.1). They require repeated objective function evaluations until the solution is obtained. If evaluating the objective function is expensive, one would probably be content to get as close as possible to the optimal solution with the minimum number of function evaluations. Techniques that aim at obtaining an approximate solution with the minimum number of evaluations are called inexact line search techniques. These techniques would terminate the line search if a point that satisfies certain conditions is reached. Possible examples of these conditions include the sufficient reduction condition or Wolf's conditions. The sufficient condition states that the line search can be terminated if the current iteration satisfies:

$$f(\boldsymbol{x}_k + \alpha \boldsymbol{s}_k) \leq l(\alpha) = f(\boldsymbol{x}_k) + c_1 \alpha \nabla f^T(\boldsymbol{x}_k) \boldsymbol{s}_k \tag{4.32}$$

where $c_1 < 1.0$ is a positive quantity. This condition effectively states that at the stopping value of α, there should be a sufficient reduction in the objective function value from that at the starting point. Equation (4.32) is satisfied by all points whose objective function values are below a line with a slope $c_1 \nabla f^T \boldsymbol{s}_k$, which is higher than the slope at the starting point. This condition, however, may terminate with a small value of α, which is undesirable. This approach is illustrated in Figure 4.19.

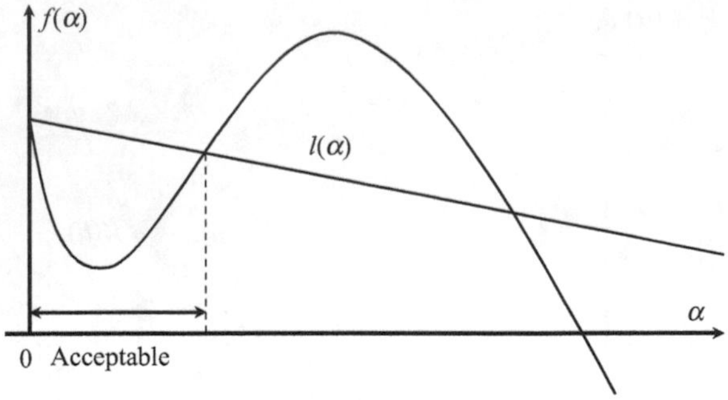

Figure 4.19 An illustration of the sufficient reduction condition; any point whose function value is below the line l(α) qualifies as a solution for the line search problem. Notice that points very close to the starting point α = 0 also qualify

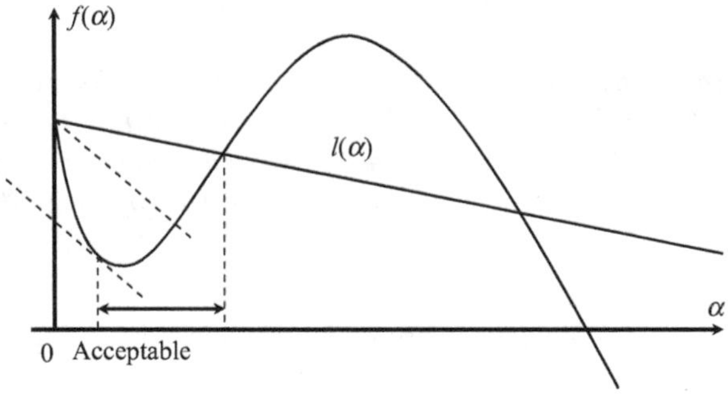

Figure 4.20 An illustration of Wolf's conditions; all points whose function values are below the line l(α) and whose slope has increased beyond the slope at α = 0 according to (4.33) are acceptable

An alternative condition for inexact line search exploits Wolf's condition. This condition states that line search can be terminated early if the value of α satisfies the sufficient reduction condition (4.32) in addition to the sufficient gradient change condition:

$$\nabla f^T(\boldsymbol{x}_k + \alpha \boldsymbol{s}_k)\boldsymbol{s}_k \geq c_2 \nabla f^T(\boldsymbol{x}_k)\boldsymbol{s}_k \tag{4.33}$$

where $c_2 < 1.0$ is a positive number. This condition states that at the terminating value of α, the directional derivative must have changed significantly. The points satisfying (4.33) can be made as close as possible to the actual optimal solution by making c_2 as small as possible. Notice that both directional derivatives in (4.33) are nonpositive. The combined conditions (4.32) and (4.33) guarantee that a significant reduction in the objective function was achieved while moving significantly away from the starting point $\alpha = 0$. Figure 4.20 illustrates Wolf's termination conditions.

A4.1 Tuning of electric circuits

A problem that appears very often in electrical engineering is to change (tune) the response of an electric circuit or structure by changing the value of one parameter. For example, the center frequency of a bandpass filter can be controlled by changing a variable capacitor. The gain of an amplifier can be controlled by a variable resistor. The center frequency of a microwave waveguide filter may be controlled by a tuning screw. The examples of such applications are so many. In some cases, we know exactly the value of the parameter that corresponds to a certain desired response. In most cases though, we can only determine the response corresponding to a parameter value through a numerical simulation. It follows that a one-dimensional optimization problem should be solved to determine the parameter value. We illustrate this approach through three examples.

A4.1.1 Tuning of a current source

Consider the simple MOSFET circuit shown in Figure 4.21. This circuit is a typical current mirror circuit where the current passing through the load resistor R_L is controlled by the value of the resistor R. The circuit has an exact solution for the tuning problem but it is sufficient to explain the basic concepts. If the two transistors are identical, the current flowing through the load resistor R_L is approximately equal to the current passing through R. Our target in this problem is to find the value of R that results in a target current source value of $I_s = 1.0$ mA. The corresponding objective function to be minimized is chosen to be:

$$f(R) = A(I(R) - I_s)^2 \tag{4.34}$$

The current corresponding to each value of R is obtained by solving a quadratic equation and picking the solution that satisfies the saturation condition for the transistor T_1. This equation enforces saturation on T_1 and is given by:

$$I_{DS} = K_n(V_{GS} - V_T)^2 = K_n(V_{DD} - I_{DS}R - V_T)^2 \tag{4.35}$$

The implementation of the objective function calculations is shown in the listing M4.6. The initial guess for the value of R is 2 kΩ. This is the starting point for the optimization problem. We utilize a variation of the MATLAB listing M4.5 in solving this problem as shown in listing M4.7. The quasi-Newton approach is utilized to minimize the objective function (4.34) to zero by finding the value of R that results in the target current. Notice in the listing M4.6 that optimization is done with respect to the scaled variable R/1 kΩ. This is important to guarantee that all the termination conditions are applicable. The scaling is removed every time a call is made to the simulator.

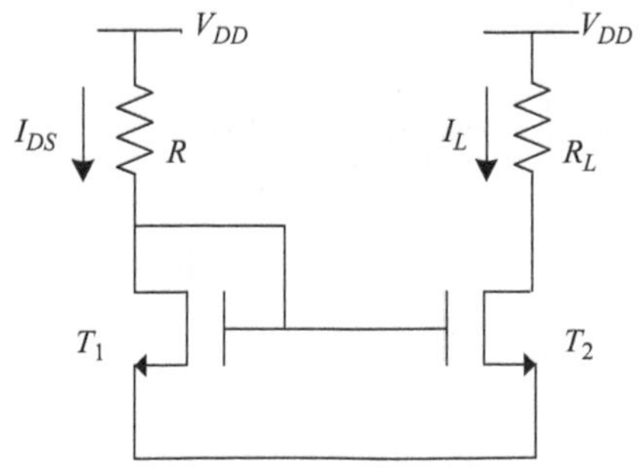

Figure 4.21 The MOSEFET circuit of Section A.4.1

The output of code at every iteration is given by:

Iteration 0	CurrentR = 2	f = 0.1656
Iteration 1	CurrentR = 2.4372	f = 0.0373
Iteration 3	CurrentR = 2.9651	f = 1.0127e-04
Itertaion 4	CurrentR = 2.9997	f = 9.4235e-09
Iteration 5	CurrentR = 3.0007	

The objective function value dropped from 0.1656 at the starting point to 9.42e-9 at the final point. The optimal design of this one-dimensional problem is thus $R = 3.0007$ kΩ.

```
%M4.6
% This function evaluates the error between the actual current and the target current
function errorFunction=getFunction(Kn, VDD, VT, CurrentR, targetIDS);
 IDS=getIDS(Kn, VDD, VT, CurrentR); %get value of IDS
 errorFunction=1.0e6*(IDS-targetIDS)^2;

%this function determines the value of IDS for the transistor mirror
%circuit of application 4.1
function IDS=getIDS(Kn, VDD, VT, R1)
CoefficientA=R1*R1; %first coefficient
CoefficientB=-1.0*(2*R1*(VDD-VT)+(1/Kn)); %second coefficient
CoefficientC=(VDD-VT)^2; %third coefficient
IDS1=(-1*CoefficientB+sqrt(CoefficientB^2-(4*CoefficientA*CoefficientC)))/(2*CoefficientA);
IDS2=(-1*CoefficientB-sqrt(CoefficientB^2-(4*CoefficientA*CoefficientC)))/(2*CoefficientA);
%does solution satisfy saturation conditions?
if(((VDD-IDS1*R1)>VT)&(IDS1>0))
  IDS=IDS1;
else
  if(((VDD-IDS2*R1)>VT)&(IDS2>0))
    IDS=IDS2;
  end
end
```

```
%M4.7
%This matlab listing solves the current mirror circuit
Kn=1.0e-3; %current coefficient
VDD=5.0; %source value
VT=1.0; %threshold value
StartingR=2;%starting guess for R1 in Kilo Ohms
scaleFactor=1.0e3; %multiply by 1K when calling the simulator
delta=0.05; %perturbation used in derivative analysis
MaxIterationCount=100; %we do not allow more than 100 iterations
Epsilon=1.0e-3; %terminating condition for gradient
gradient=1.0e4; %initial value of gradient
Counter=0; %initialize iteration counter
CurrentR=StartingR; %initialize current point
targetIDS=1.0e-3; %the target value for IDS
while((Counter<MaxIterationCount)&(abs(gradient)>Epsilon))
  f=getFunction(Kn, VDD, VT, scaleFactor*CurrentR, targetIDS); %function at point
  fp=getFunction(Kn, VDD, VT, scaleFactor*(CurrentR+delta), targetIDS); %function with +ve
                                                                        % perturbation
  fn=getFunction(Kn, VDD, VT, scaleFactor*(CurrentR-delta), targetIDS); %function with -ve
                                                                        %perturbation
  gradient=(fp-fn)/(2*delta); %get first-order derivative approximation
  Hessian=(fp-2*f+fn)/(delta*delta); %get second-order derivative approximation
  NewR=CurrentR-(gradient/Hessian); %get the new point using formula (4.29)
  CurrentR=NewR; % update new point
  Counter=Counter+1 %increment iteration counter
  CurrentR
end
```

A4.1.2 Coupling of nanowires

Another example of the one-dimensional tuning is the connection of nanowires to a plasmonic power splitter shown in Figure 4.22. The nanowire made of silicon (Si) couples light to the plasmonic wire splitter made of silver (Ag) [5]. When the intrusion (L) of the nanowire into the plasmonic waveguide is changed, the coupling of light varies. A fixed line search with a step of 5.0 nm is carried out to locate the optimal value of L that results in the maximum light coupling. The result of this line sweep is shown in Figure 4.23. As the Si waveguide moves

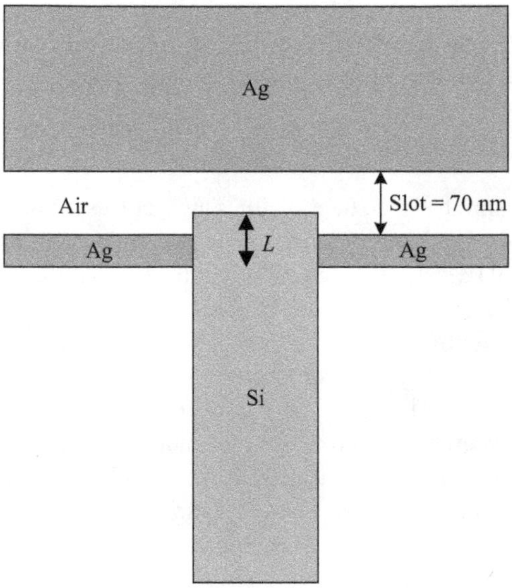

Figure 4.22 The junction of a nanowire with a nanoplasmonic slot waveguide of application

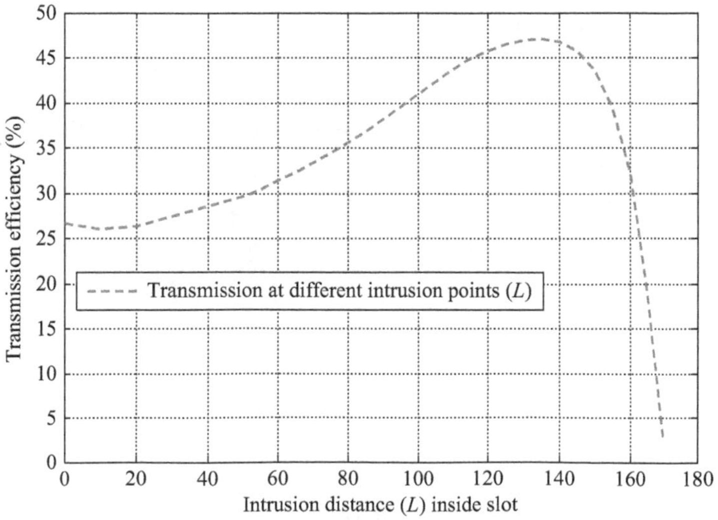

Figure 4.23 The transmission efficiency as a function of the intrusion L of the silicon waveguide inside the slot of the PSW

inside the slot, the transmission efficiency increases. The maximum transmission is achieved when the Si waveguide is 30 nm inside the slot ($L = 130$). The transmission then starts to decrease as the Si waveguide extends further inside the slot. This structure was simulated using the photonics simulator of Lumerical [6].

A4.1.3 Matching of microwave antennas

Another application of one-dimensional search is shown in Figure 4.24. A microwave patch antenna is fed through a microstrip line. It is required to determine the length of the inset y such that the input impedance of the antenna is as close as possible to 50 Ω at 2.4 GHz. This indicates resonance and maximum power radiation at that frequency. A possible objective function to be minimized is given by:

$$f(y) = (\mathrm{Re}(Z_{\mathrm{in}}(y)) - 50)^2 + (\mathrm{im}(Z_{\mathrm{in}}(y)))^2 \tag{4.36}$$

where $Z_{in}(y)$ is the input impedance of the patch corresponding to a certain inset length. The operators Re() and im() denote the real and imaginary parts of the input impedance at 2.4 GHz, respectively. The length and width of the patch are $L =$ 29.0 mm and $W = 29.0$ mm, respectively. The width of the microstrip feed line is 2.28 mm. The substrate of this antenna is FR4 with $\varepsilon_r = 4.4$ and loss tangent $= 0.02$ with a height of 1.2 mm [7]. The structure is simulated using the high frequency structural simulator (HFSS) [8].

Utilizing a line search algorithm, the minimum value of the objective function is found at $y^* = 9.20$ mm. This result can be verified by plotting the real and imaginary parts of the input impedance versus y as shown in Figure 4.25. The reflection coefficient is plotted versus frequency in Figure 4.26. It is obvious that the antenna resonates at 2.4 GHz as desired showing strong transmission.

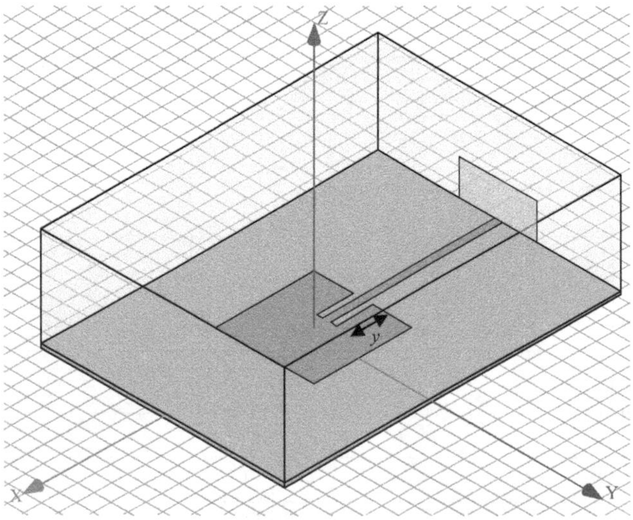

Figure 4.24 A patch antenna with an inset; changing the length of the inset y changes the input impedance of the antenna

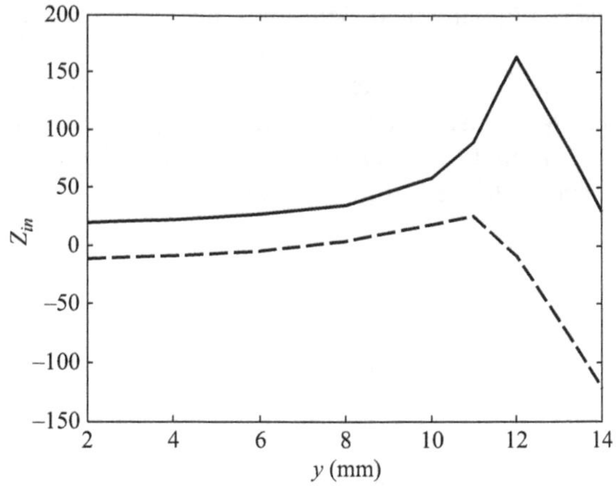

Figure 4.25 *The real part (–) and the imaginary part (– –) of the input impedance at 2.4 GHz as a function of the inset length y*

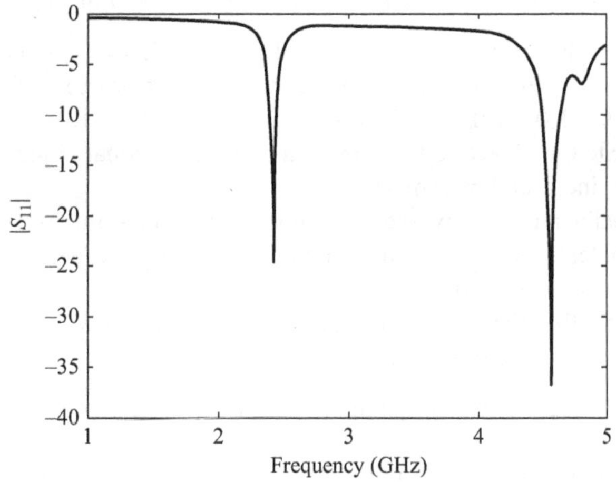

Figure 4.26 *The reflection coefficient of the antenna at the optimal value of the inset length y; the antenna shows resonance as expected at 2.4 GHz*

References

1. Singiresu S. Rao, *Engineering Optimization Theory and Practice*, Third Edition, John Wiley & Sons Inc, New York, 1996
2. Andreas Antoniou and Wu-Sheng Lu, *Practical Optimization Algorithms and Engineering Applications*, Springer, 2007
3. Edwin K.P. Chong and Stanislaw H. Zak, *An Introduction to Optimization*, 2nd edn., Wiley, 2001
4. Jorge Nocedal and Stephen J. Wright, *Numerical Optimization*, 2nd edn., Springer, 2006
5. M.H. El Sherif, O.S. Ahmed, M.H. Bakr, and M.A. Swillam, "Polarization-controlled excitation of multilevel plasmonic nano-circuits using single silicon nanowire", *Optics Express*, vol. 20, no. 11, pp.12473–12486, May 2012

6. F.D.T.D. *Lumerical*, Lumerical Soultions Inc., retreieved from http://www.lumerical.com
7. N. Anantrasirichai, C. Benjangkaprasert, and T. Wakabayashi, "Rectangular patch antenna with inset feed and modified ground-plane for wideband antenna", *SICE Annual Conference*, Tokyo, Japan 2008
8. HFSS, Ansys Inc., retrieved from www.ansys.com

Problems

4.1 Consider the polynomial $-4x^5+3x^4-12x^3+11x^2-2x+1$.
   (a) Utilize the MATLAB listing M4.1 to find the minimum of this function in the interval [−0.5, 0.5]. Make any necessary changes to the code.
   (b) Verify your answer in (a) by plotting that graph of the function.

4.2 The function $e^{3x}+3e^{-2x}$ is known to be unimodal in the interval [−1.0, 1.0].
   (a) Utilize the MATLAB listing M4.2 to find the minimum in this interval.
   (b) Verify your answer in (a) by plotting the graph of this function.

4.3 Consider the three-dimensional function $f(\boldsymbol{x})=(x_1+1)^2+(x_2-2)^2+(x_3+1)^2$. A line search is to be carried out starting from the point $\boldsymbol{x}_0=[1.0 \quad 1.0 \quad 1.0]^T$ in the direction $\boldsymbol{s}=[-1.0 \quad -1.0 \quad -1.0]^T$.
   (a) Write the objective function as a one-dimensional objective function of the line search parameter.
   (b) Obtain analytically the minimum of the one-dimensional line search problem using the function in (a). Obtain the corresponding three-dimensional point.
   (c) Show that the gradient of the objective function at this optimal point is normal to the direction $\boldsymbol{s}$.

4.4 Consider the convex quadratic function $f(\boldsymbol{x})=5x_1^2+3x_2^2+2x_1x_2-3x_1-x_2$.
   (a) Find analytically the minimum of this function along the line starting from $[0 \quad 0 \quad 0]^T$ in the direction $\boldsymbol{s}=[-1.0 \quad -2.0 \quad 0]^T$.
   (b) Verify your answer in (a) using the MATLAB listing M4.3. Select appropriate start and end values of the line search parameter.
   (c) Show that the gradient at the final two-dimensional point is normal to the search direction $\boldsymbol{s}$.

4.5 Consider the function $f(\boldsymbol{x})=10(x_2-x_1^2)^2+(1-x_1)^2$. Starting from the point $\boldsymbol{x}=[-1.0 \quad 1.0]^T$, a line search is carried out in the direction $\boldsymbol{s}=[1 \quad 0]^T$. Find the minimum along that line using the using quasi-Newton code M4.5. Use an initial fixed line search with a step of 0.5 to bracket your solution.

4.6 Given the function $f(x)=\sqrt{1+x^2}-\frac{1}{2\sqrt{1+x^2}}$.
   (a) Minimize this function in the interval [−2.0, 2.0] using the cubic interpolation code M4.4.
   (b) Verify your answer by plotting the graph of the function.

4.7 Change the MATLAB listing M4.5 to implement the Secant method discussed in Section 4.5.3. Apply your code to Problem 4.2.

Chapter 5
Derivative-free unconstrained techniques

5.1 Why unconstrained optimization?

In the previous chapter, we addressed the one-dimensional optimization problem. This problem has a special prominence in optimization theory as line search is a common step in many algorithms. A general optimization problem, unconstrained or constrained, however, can have many variables. Over the years, many algorithms have been developed for solving these problems. We focus in this chapter on solving unconstrained optimization problems using only function values without utilizing any derivative information.

Unconstrained optimization is a class of optimization techniques where the solution of the n-dimensional problem can be any point in the whole parameter space $\boldsymbol{x}^* \in \Re^n$. In practical applications though there are usually several constraints on the optimal design. These constraints limit the solution to a subset of the n-dimensional space called the feasible region. The target of a general optimization problem is to minimize the objective function over all points belonging to the feasible region. In the case of unconstrained optimization, the feasible set is the whole n-dimensional space.

One may wonder why would we then study unconstrained optimization. There are a number of reasons for this. First, the concepts and mathematical approaches developed for unconstrained optimization form the basis for many constrained optimization techniques. Also, in some cases, the optimal solution obtained by ignoring the constraints is the same optimal solution of the constrained optimization problem. We will discuss such optimization problems in this and subsequent chapters.

5.2 Classification of unconstrained optimization techniques

Unconstrained optimization techniques are classified according to the way the solution is obtained. Derivative-free optimization assumes that only function values are available. We assume that no derivative information is available about the optimized function.

Another class of unconstrained optimization problems utilizes the available first-order derivative information. These derivatives may be obtained through finite differences that require repeated evaluations of the objective function as explained in Chapter 1. Evaluating these approximations may be formidable for a large number of parameters or if the objective function is evaluated through a time-intensive numerical simulation. Another approach for estimating the required sensitivities is through the more efficient adjoint sensitivity analysis. A number of adjoint-based approaches will be discussed in more detail in Chapter 10. We will discuss a number of first-order-based optimization techniques in Chapter 6.

Higher-order sensitivities may also be used in the optimization process. These sensitivities are even more computationally expensive than first-order derivatives. Second-order sensitivities may be approximated to accelerate the optimization process. The class of optimization techniques that utilize approximate second-order sensitivities are called quasi-Newton methods. I gave an introduction to these techniques for the one-dimensional case in Chapter 4. We will cover these techniques in more detail in Chapter 7.

We focus in this chapter on derivative-free techniques for solving unconstrained problems. There are many such techniques in the literature. We will focus on the basic ones and illustrate some of them with MATLAB® implementations.

5.3 The random jump technique

This technique aims at generating random points in the interval of interest. The objective function is evaluated at all these points. An approximation to the optimal solution $\boldsymbol{x}^*$ is the random point with the lowest possible objective function.

This approach assumes that the local minimum is expected in the interval:

$$l_i \leq x_i \leq u_i, \quad i = 1, 2, \ldots, n \tag{5.1}$$

where l_i and u_i are lower and upper bounds on the ith parameter. It follows that the subset of the solution is a hyperbox. The technique generates a sequence of random numbers within this interval of the form:

$$\boldsymbol{x} = \begin{bmatrix} x_1 \\ x_2 \\ \vdots \\ x_n \end{bmatrix} = \begin{bmatrix} l_1 + r_1(u_1 - l_1) \\ l_2 + r_2(u_2 - l_2) \\ \vdots \\ l_n + r_n(u_n - l_n) \end{bmatrix} \tag{5.2}$$

where $r_i \in [0, 1]$, $i = 1, 2, \ldots, n$ are random numbers. Equation (5.2) thus creates a number of random n-dimensional points within the hyperbox. The algorithm keeps track of the point with the lowest objective function value. The higher the number of generated random points, the better the approximation to the optimal solution is. This approach can, however, require a formidable execution time if a single objective function evaluation requires large simulation time or if the dimensionality of the problem n is relatively large.

A possible implementation of the random jump approach is shown in the MATLAB listing M5.1.

Example 5.1: Find the minimum of the function $f(x_1, x_2) = x_1 - x_2 + 2x_1^2 + 2x_1x_2 + x_2^2$ using the random jump method. Verify your answer using an analytical approach.

Solution: Using the listing M5.1 with 1000 random points, the following sampled output is obtained:

IterationCounter = 100 OldPoint = [1.1815 1.9512] OldValue = −1.0172
IterationCounter = 1000 OldPoint = [−0.9644 1.4277] OldValue = −1.2474

```
%M5.1
%This program carries out the random jump algorithm
NumberOfParameters=2; %This is n for this problem
UpperValues=[10   10]'; %upper values
LowerValues=[-10  -10]'; %lower values
OldPoint=0.5*(UpperValues+LowerValues); %select center of interval as old point
OldValue=getObjective(OldPoint); %Get the objective function at the old point
MaximumNumberOfIterations=1000; %maximum number of allowed iterations
IterationCounter=0; %iteration counter
while(IterationCounter<MaximumNumberOfIterations) %repeat until maximum number of iteration
                                                   %is achieved
  RandomVector=rand(NumberOfParameters,1); %get a vector of random variables
  NewPoint=LowerValues+RandomVector.*(UpperValues-LowerValues); %Get new random point
  NewValue=getObjective(NewPoint); %get new value
  if(NewValue<OldValue) %is there an improvement?. Then store the new point and value
    OldPoint=NewPoint;
    OldValue=NewValue;
  end
  IterationCounter=IterationCounter+1; %increment the iteration counter
end
OldPoint
OldValue
```

Using this code, we see that the obtained approximation to the optimal point is $\bar{x} = [-0.9644 \quad 1.4277]^T$. To verify this answer, we apply the necessary optimality conditions to get:

$$\nabla f = \begin{bmatrix} 1 + 4x_1 + 2x_2 \\ -1 + 2x_1 + 2x_2 \end{bmatrix} = \begin{bmatrix} 0 \\ 0 \end{bmatrix} \Rightarrow \boldsymbol{x}^* = \begin{bmatrix} -1.0 \\ 1.5 \end{bmatrix} \tag{5.3}$$

Because the function is quadratic, its Hessian is a constant and is given by:

$$\boldsymbol{H} = \begin{bmatrix} 4 & 2 \\ 2 & 2 \end{bmatrix} \tag{5.4}$$

which can be shown to be a positive definite matrix. It follows that $\boldsymbol{x}^*$ is indeed a local minimum. Notice that the solution obtained using the random jump method required 1000 iterations to get close enough to the optimal solution. An even higher number of function evaluations would be required if the dimensionality of the problem is higher.

5.4 The random walk method

This technique generates a random search direction $\boldsymbol{s}^{(k)}$ at the kth iteration. Starting from the current iteration $\boldsymbol{x}^{(k)}$, the algorithm then carries a line search along the line $\boldsymbol{x}^{(k)} + \lambda\boldsymbol{s}^{(k)}$ to get a better solution $\boldsymbol{x}^{(k+1)}$. The new solution $\boldsymbol{x}^{(k+1)}$ is accepted only if $f(\boldsymbol{x}^{(k+1)}) < f(\boldsymbol{x}^{(k)})$. Otherwise, this random direction is discarded and a new random direction $\boldsymbol{s}^{(k)}$ is generated. Another line search is then carried out until an improvement is achieved or the algorithm terminates. The same steps are repeated at the new point $\boldsymbol{x}^{(k+1)}$. The algorithm may terminate after N iterations have been carried out or if $\left|f\left(\boldsymbol{x}^{(k+1)}\right) - f\left(\boldsymbol{x}^{(k)}\right)\right| < \varepsilon$, i.e., no significant change is achieved in the objective function value. The MATLAB listing M5.2 illustrates a possible implementation of the random walk technique. Notice that this listing uses a Golden Section line search function developed based on the theory of Chapter 4.

```
%M5.2
%Implementation of the random walk method
NumberOfParameters=2; %This is n for this problem
OldPoint=[0 0]'; %This is the starting point
OldValue=getObjective(OldPoint); %Get the objective function at the old point
OnesVector=ones(NumberOfParameters,1); %get a vector of ones
negativeOnesVector=-1.0*OnesVector; %get a vector of -ve ones
MaximumNumberOfIterations=100; %maximum number of allowed iterations
IterationCounter=0; %iteration counter
LambdaMax=3; %maximum value of line search parameter
Tolerance=0.001;
while(IterationCounter<MaximumNumberOfIterations) %repeat until maximum
                                                   %number of iteration is reached
  RandomVector=rand(NumberOfParameters,1); %get a vector of random variables
  u=negativeOnesVector+RandomVector.*(OnesVector-negativeOnesVector); %make
                         % the random vector between -1 and 1 for each component
  LambdaOptimal = GoldenSection('getObjective',Tolerance,OldPoint,u,LambdaMax);
                                          %get the optimal line search parameter
  NewPoint=OldPoint+LambdaOptimal*u; %Get new random point
  NewValue=getObjective(NewPoint); %get new value
  if(NewValue<OldValue) % is there an improvement? store the new point and value
    OldPoint=NewPoint;
    OldValue=NewValue;
  end
  IterationCounter=IterationCounter+1; %increment the iteration counter
end
```

Example 5.2: Repeat Example 5.1 using the random walk method.

Solution: Utilizing the MATLAB listing M5.2, we obtain the following samples of the output:

OldPoint = [0 0]′ OldValue = 0 LambdaOptimal = 0.9969
u = [0.0442 0.3353]′ NewPoint = [0.0441 0.3343]′ NewValue = −0.1451

OldPoint = [0.0441 0.3343]′ OldValue = −0.1451 LambdaOptimal = 1.8230
u = [−0.1383 0.4191]′ NewPoint = [−0.2081 1.0983]′ NewValue = −0.4707

OldPoint = [−0.2081 1.0983]′ OldValue = −0.4707 LambdaOptimal = 0.3375
u = [−0.9456 −0.9202]′ NewPoint = [−0.5273 0.7877]′ NewValue = −0.9692

At the final solution, the code output is given by:

OldPoint = [−1.0000 1.4999]′ OldValue = −1.2500 LambdaOptimal = 4.2009e-004

u = [−0.2312 0.3371]′ NewPoint = [−1.0001 1.5000]′ NewValue = −1.2500

which gives an approximation to the optimal solution of $\bar{x} = [-1.0001 \quad 1.5000]^T$. The optimal solution, as explained in Example 5.1, is $x^* = [-1.0 \quad 1.5]^T$. Notice that fewer number of iterations was needed than in the random jump technique to obtain a better approximation to the optimal solution x^*. This is because the integrated line search helps achieve more reduction in the objective function per iteration.

5.5 Grid search method

The grid search method has some resemblance to the random jump technique. A possibly large number of points within an n-dimensional hyperbox are generated.

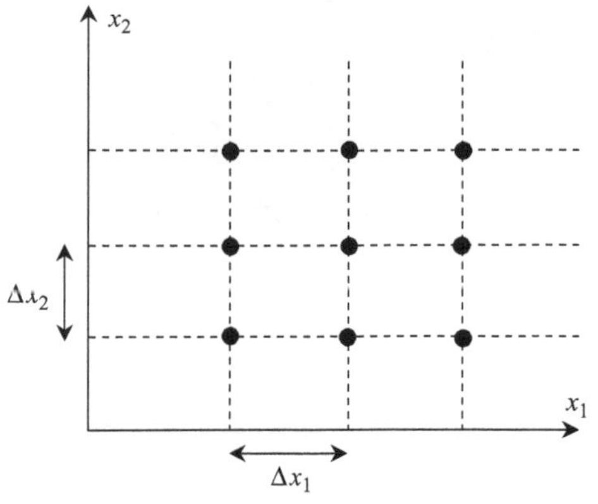

Figure 5.1 An illustration of the grid search method for the two-dimensional case; a large number of points are created on a grid and the corresponding function values are evaluated. The optimal solution is approximated by the grid point with the smallest objective function value

The main difference between the two approaches is that the points are created on a grid and not in a random way as shown in Figure 5.1 for a two-dimensional problem. The point with the smallest value of the objective function over the grid is picked as an approximation to the optimal solution. The main difficulty with this approach is the computational cost associated with it. If we have p points in every dimension, the total number of required objective function evaluations is p^n. For a large n, or for time-intensive simulations, the associated computational time may be formidable.

5.6 The univariate method

This method solves the n-dimensional optimization problem by solving a sequence of single-dimensional optimization problems. At the kth iteration, the algorithm solves the one-dimensional optimization problem:

$$\begin{aligned} \boldsymbol{x}^{(k+1)} &= \boldsymbol{x}^{(k)} + \lambda_k \boldsymbol{e}_k \\ \lambda_k &= \min_{\lambda} f(\boldsymbol{x}^{(k)} + \lambda \boldsymbol{e}_k) \end{aligned} \tag{5.5}$$

where $\boldsymbol{e}_k$ is the kth coordinate direction. The algorithm cycles through the coordinate systems one by one as follows:

$$\begin{aligned} &\boldsymbol{e}_k = \begin{bmatrix} 1 \\ 0 \\ \vdots \\ 0 \end{bmatrix}, \quad k = 1, n+1, 2n+1, \ldots \quad \boldsymbol{e}_k = \begin{bmatrix} 0 \\ 1 \\ \vdots \\ 0 \end{bmatrix}, \quad k = 2, n+2, 2n+2, \ldots \\ &\qquad\vdots \\ &\boldsymbol{e}_k = \begin{bmatrix} 0 \\ 0 \\ \vdots \\ 1 \end{bmatrix}, \quad k = n, 2n, 3n, \ldots \end{aligned} \tag{5.6}$$

An illustration of the univariate approach is shown in Figure 5.2. A possible implementation of the univariate approach in MATLAB is shown in listing M5.3.

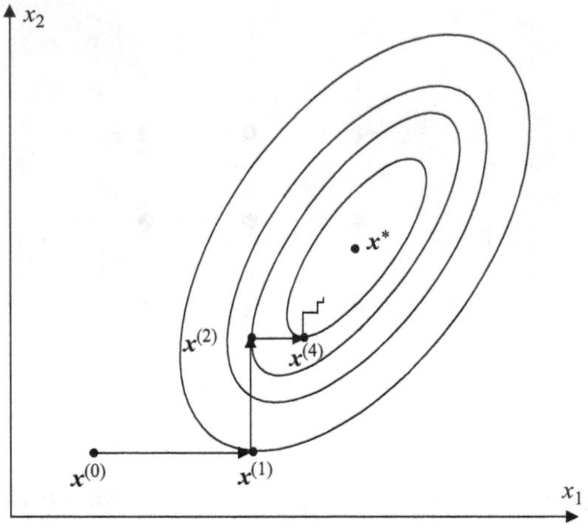

Figure 5.2 An illustration of the univariate optimization approach for a two-dimensional problem; notice how the iterations zigzag along the two coordinate directions before approaching the optimal point. The line connecting $x^{(0)}$ and $x^{(2)}$ offers a faster reduction in the objective function

```
%M5.3
%The univariate optimization approach
NumberOfParameters=3; %This is n for this problem
OldPoint=[3 -3 5]'; %This is the starting point
OldValue=getObjective(OldPoint); %Get the objective function at the old point
Identity =eye(NumberOfParameters); %Get identity matrix of size n
MaximumNumberOfIterations=100; %maximum number of allowed iterations
IterationCounter=0; %iteration counter
LambdaMax=3; %maximum value of Lambda
Tolerance=0.001; %terminating tolerance for line search
Epsilon=0.001; %exploration step
while(IterationCounter<MaximumNumberOfIterations) %repeat for a number of iterations
  for i=1:NumberOfParameters
   up=Identity(:,i); %get the vector ei
   un=-1.0*up; %get the vector -ei
   %we do first exploration in the -ve and +ve directions
   fp=feval('getObjective',OldPoint+Epsilon*up); %get +ve exploratory function value
   if(fp<OldValue) %positive direction is promising
     u=up; %choose the positive coordinate direction
   else
     u=un; %choose the negative coordinate direction
   end
   %get the optimal value
   LambdaOptimal =  GoldenSection('getObjective',Tolerance,OldPoint,u,LambdaMax);
   NewPoint=OldPoint+LambdaOptimal*u; %Get new rpoint
   NewValue=getObjective(NewPoint); %get new value
   if(NewValue<OldValue) %is there an improvement?. Then store the new point and value
     OldPoint=NewPoint;
     OldValue=NewValue;
   end
  end
  IterationCounter=IterationCounter+1; %increment the iteration counter
end
```

Notice that the MATLAB listing M5.3 uses an exploratory small step in the direction of each coordinate direction. If this exploratory step does not give a reduction in the objective function, we search instead in the direction $-e_k$. The following example illustrates the univariate optimization approach.

Example 5.3: Utilize the univariate approach to find the minimum of the function $f(\boldsymbol{x}) = x_1^2 + 3x_2^2 + 6x_3^2$ starting from the point $\boldsymbol{x}^{(0)} = [3.0 \quad -3.0 \quad 5.0]^T$.

Solution: Utilizing the MATLAB listing M5.3, the algorithm gives the following output after 100 iterations:

IterationCounter = 100 OldPoint = 1.0e-003* [−0.0000 −0.4201 0.0711]′
OldValue = 5.5979e-007

The obtained solution is thus $\bar{\boldsymbol{x}} = 1.0\text{e-}3 \ [0 \quad -0.4201 \quad 0.0711]^T$. The given function is a sum of squares and it reaches the minimum value when all terms are zeros. It follows that the optimal solution is $\boldsymbol{x}^* = [0 \quad 0 \quad 0]^T$. The solution obtained by the univariate approach is very close to the exact optimal solution. However, 100 simulations are considered a large number for achieving the minimum of a quadratic function of only three variables.

The univariate approach is known to be slow in convergence for objective functions with sharp contours. The algorithm may carry out a large number of iterations without making significant improvement in the value of the objective function as shown in Figure 5.2. We will discuss later techniques that are capable of solving similar problems with a much smaller number of function calculations.

5.7 The pattern search method

Figure 5.2 shows how the iterations of the univariate approach very slowly approach the optimal solution $\boldsymbol{x}^*$. One may observe that the line connecting two iterations with index difference of n, say $(\boldsymbol{x}^{(2)} - \boldsymbol{x}^{(0)})$, offers a faster reduction in the objective function. This is the main concept behind pattern search techniques. A number of pattern search techniques have been developed over the years for accelerated convergence.

One of the important pattern search techniques is the Hooke and Jeeves method. This technique is better illustrated by the steps shown in Figure 5.3. Starting with an initial point $\boldsymbol{x}^{(0)}$, sequential steps are taken along all coordinate directions to obtain the point $\boldsymbol{x}^{(1)}$. The algorithm then carries out a line search starting from the point $\boldsymbol{x}^{(1)}$ along the line $\boldsymbol{s}^{(0)} = \boldsymbol{x}^{(1)} - \boldsymbol{x}^{(0)}$ to obtain the new point:

$$\boldsymbol{x}^{(2)} = \boldsymbol{x}^{(1)} + \lambda_1^* \boldsymbol{s}^{(0)} \tag{5.7}$$

The new point is then taken as $\boldsymbol{x}^{(2)}$. The whole process is repeated from that point again until no significant reduction can be achieved in the objective function.

For the general n-dimensional case, the steps of the algorithm are as follows. Starting from the current point $\boldsymbol{x}^{(k)}$, a new point $\boldsymbol{x}^{(k+1)}$ is achieved by taking steps along all coordinate directions that would result in a reduction of the objective function. The new $\boldsymbol{x}^{(k+1)}$ is thus given by:

$$\boldsymbol{x}^{(k+1)} = \boldsymbol{x}^{(k)} + \beta^{(1)}\Delta^{(1)}\boldsymbol{e}^{(1)} + \beta^{(2)}\Delta^{(2)}\boldsymbol{e}^{(2)} + \cdots + \beta^{(n)}\Delta^{(n)}\boldsymbol{e}^{(n)}, \tag{5.8}$$

where $\Delta^{(i)}$ is the step taken in the direction of the ith coordinate parameter. $\beta^{(i)}$ is the sign of the step in ith coordinate direction with $\beta^{(i)} \in \{1, -1\}$. The step taken

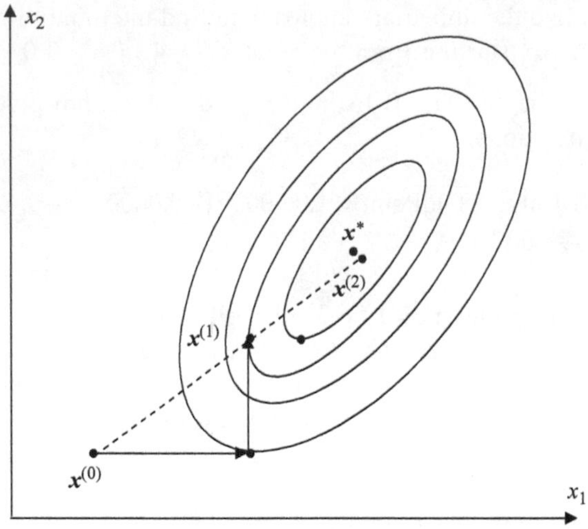

Figure 5.3 An illustration of one iteration of the Hooke and Jeeves method; the algorithm first takes sequential steps in all coordinate directions followed by a pattern search. Notice how the pattern search in the line connecting two consecutive iterations move the pattern very close to the optimal design

in the direction of a coordinate direction must result in a reduction in the objective function. The direction of the step should thus be reversed if a drop in the objective function is not possible. If $f(\boldsymbol{x}^{(k+1)}) > f(\boldsymbol{x}^{(k)})$, this implies that the steps taken in the direction of different coordinates are large. In this case, a new point $\boldsymbol{x}^{(k+1)}$ is evaluated using (5.8) with reduced step sizes. If the step sizes drop below a certain threshold, then this means that the current point $\boldsymbol{x}^{(k)}$ is the best possible solution.

Once a point $\boldsymbol{x}^{(k+1)}$ is obtained satisfying $f(\boldsymbol{x}^{(k+1)}) < f(\boldsymbol{x}^{(k)})$, the algorithm proceeds to the pattern search stage. The search direction $\boldsymbol{s}^{(k)} = \boldsymbol{x}^{(k+1)} - \boldsymbol{x}^{(k)}$ is then calculated. A point with a better value of the objective function is obtained by carrying out a line search step along that direction starting from $\boldsymbol{x}^{(k+1)}$ to get:

$$\boldsymbol{x}^{(k+2)} = \boldsymbol{x}^{(k+1)} + \lambda_k^* \boldsymbol{s}^{(k)} \tag{5.9}$$

The algorithm then repeats the sequential coordinate search (5.8) followed by the pattern move step (5.9) with $\boldsymbol{x}^{(k)}$ replaced by $\boldsymbol{x}^{(k+2)}$. The MATLAB listing M5.4 shows a possible implementation of this approach.

Note that in this MATLAB listing the variable "YOld" represents the best point obtained by carrying out the sequential steps in all coordinate directions as in (5.8). If after carrying out all sequential steps this point did not move from the starting point, this implies that the step taken is large and it should be reduced. The following example applies the Hooke and Jeeves method to a three-dimensional problem.

Example 5.4: Utilize the MATLAB listing M5.4 to solve the optimization problem:

minimize $\quad f(\boldsymbol{x}) = x_1^2 + 3x_2^2 + 6x_3^2 \quad$ (5.10)

starting from the point $\quad \boldsymbol{x}^{(0)} = [3.0 \quad -3.0 \quad 6.0]^T$

Solution: The output from the code M5.4 for the first few iterations is given by:

Iteration 0: OldPoint = $[3.0 \quad -3.0 \quad 5.0]'$ OldValue = 186
Iteration 1: OldPoint = $[-0.1999 \quad 0.1999 \quad 1.8001]'$ OldValue = 19.6020
Iteration 2: OldPoint = $[0.9601 \quad -0.9601 \quad 0.6401]'$ OldValue = 6.1455

```
%M5.4
%A possible implementation of the Hooke and Jeeves pattern search method
NumberOfParameters=3; %This is n for this problem
OldPoint=[3 -3 5]' %This is the starting point
Step=0.2; %the step taken in the direction of all coordinate parameters
OldValue=getObjective(OldPoint) %Get the objective function at the starting point
Identity =eye(NumberOfParameters); %Get identity matrix of size n
LambdaMax=15; %maximum value of Lambda
Tolerance=0.001; %terminating tolerance for line search
StepNorm=1000; %initialize stepNorm
MinimumDistance=1.0e-4; %termination condition
while(StepNorm>MinimumDistance) %repeat until maximum number of iteration is achieved
  %first we do search in the directions of the coordinate axes
  %YOld is the point from which steps are taken in coordinate directions.
  YOld=OldPoint; %start exploring from the current point
  YOldValue=OldValue; %store also the old value
  for i=1:NumberOfParameters %repeat for all coordinates
    up=Identity(:,i); %get the unity vector in the ith coordinate direction (ei)
    un=-1.0*up; %get the vector -ei
    %we do first exploration in the +ve ith coordinate direction
    fp=feval('getObjective',YOld+Step*up); %get +ve pertubed function value
    if(fp<YOldValue) %positive direction is promising
      YNew=YOld+Step*up; %Get new  exploration point
      YNewValue=fp; %value at step taken in the +ve ith coordinate direction
    else
      fn=feval('getObjective',YOld+Step*un); %get -ve pertubed function value
      if(fn<YOldValue) %is there an improvement in objective function value
        YNew=YOld+Step*un; %Get new  exploration point
        YNewValue=fn; %value at step taken in the -ve ith coordinate direction
      else %no improvement possible for the ith coordinate.  Do not move from start point
        YNew=YOld;
        YNewValue=YOldValue;
      end

    end
    YOld=YNew;  %move to the next coordinate direction
    YOldValue=YNewValue; %update the value
  end

  %YOLD contains now the point achieved after taking sequential steps in
  %all coordinates.
  if(YOld==OldPoint) % No change was made from the starting point
    Step=Step*0.8; %reduce the step taken in each coordinate direction
  else %successful reduction is possible
    PatternDirection=YOld-OldPoint; %determine the pattern direction
    %now we do a line search in this direction starting from YOld
     %get the optimal value in the direction of u starting from YOld
    LambdaOptimal =GoldenSection('getObjective',Tolerance,YOld,
                                              PatternDirection,LambdaMax);
    NewPoint=YOld+LambdaOptimal*PatternDirection; %Get new point
    NewValue=feval('getObjective',NewPoint); %Get the New Value
    StepNorm=norm(NewPoint-OldPoint); %get the norm of the step
    OldPoint=NewPoint %update the current point
    OldValue=NewValue %update the current value
    Step=0.2; %recover original step value after a successful step
  end
end
OldPoint
OldValue
```

Iteration 3: OldPoint = [0.1919 −0.1919 −0.1280]′ OldValue = 0.2457
Iteration 4: OldPoint = [−0.0082 0.0082 0.0721]′ OldValue = 0.0314

It can be seen from these numbers that the objective function value dropped from 186 at the starting point to only 0.0314 after four iterations only. The exact solution of this problem, as evident from the objective function nature, is $\boldsymbol{x}^* = [0 \quad 0 \quad 0]^T$ with an objective function value of 0.

5.8 The Simplex method

The Simplex method for unconstrained optimization is different from the linear programming Simplex method discussed in Chapter 2. The Simplex method, as the name implies, utilizes a simplex to search the space for a local minimum. The simplex in an n-dimensional space consists of $(n+1)$ points $\{\boldsymbol{x}_i,\ i = 1, 2, \ldots, n+1\}$. In a two-dimensional space, it is a triangle. In a three-dimensional space, it is a tetrahedron. The algorithm then proceeds to calculate the objective function at all these points. The point with the highest value of the objective function $\boldsymbol{x}_h$ is determined. The centroid of all other points is then determined and is given by:

$$\boldsymbol{x}_o = \frac{1}{n} \sum_{\substack{i=1, \\ i \neq h}}^{n+1} \boldsymbol{x}_i \tag{5.11}$$

Three basic operations are carried out within the Simplex method, namely, reflection, expansion, and contraction. These operations are illustrated in Figure 5.4. This figure shows an initial simplex given by the three points $\{\boldsymbol{x}_1, \boldsymbol{x}_2, \boldsymbol{x}_h\}$. The centroid

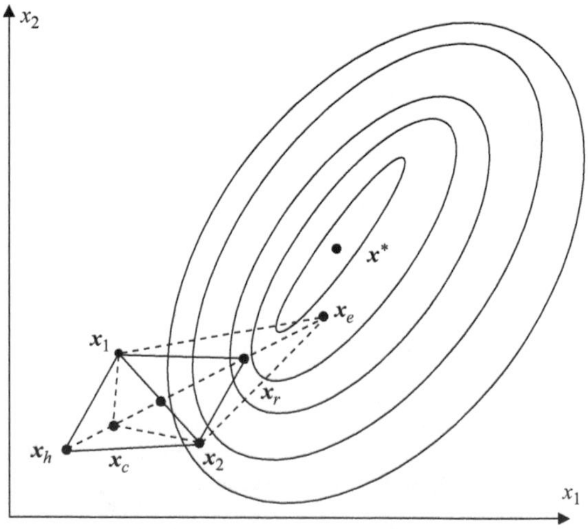

Figure 5.4 An illustration of the Simplex method; the point with the highest function value $\boldsymbol{x}_h$ is reflected to get the point $\boldsymbol{x}_r$. If the objective function at $\boldsymbol{x}_r$ is better than all other simplex points, an expansion results in the point $\boldsymbol{x}_e$. If the point $\boldsymbol{x}_r$ has a higher objective function than $\boldsymbol{x}_h$, the simplex is contracted and $\boldsymbol{x}_h$ is replaced by $\boldsymbol{x}_c$

point is at the middle of the line connecting $\boldsymbol{x}_1$ and $\boldsymbol{x}_2$. As the point $\boldsymbol{x}_h$ has the highest value of the objective function, reflecting this point across the centroid point would likely result in a better objective function value than $f(\boldsymbol{x}_h)$. This reflected point is given by:

$$\boldsymbol{x}_r = (1+\alpha)\boldsymbol{x}_o - \alpha\boldsymbol{x}_h \tag{5.12}$$

where α is the reflection coefficient with $0 \le \alpha \le 1$. Notice that if $\alpha = 0$, the reflected point $\boldsymbol{x}_r$ is the same as the centroid point. If $\alpha = 1$, the reflected point $(2\boldsymbol{x}_o - \boldsymbol{x}_h)$ has the same opposite distance from the centroid as the point $\boldsymbol{x}_h$. The point $\boldsymbol{x}_r$ then replaces the point $\boldsymbol{x}_h$ if $f(\boldsymbol{x}_r) < f(\boldsymbol{x}_h)$. In some cases, the point $\boldsymbol{x}_r$ shows very good reduction in the objective function such that $f(\boldsymbol{x}_r) < f(\boldsymbol{x}_i)$, $\forall i$. This implied that the direction $(\boldsymbol{x}_r - \boldsymbol{x}_h)$ is a promising direction for reducing the objective function. The simplex can thus be expanded further to the expansion point $\boldsymbol{x}_e$ given by:

$$\boldsymbol{x}_e = (1+\beta)\boldsymbol{x}_r - \beta\boldsymbol{x}_o, \tag{5.13}$$

where β is the expansion coefficient with $0 \le \beta \le 1$. As explained earlier for (5.12), the expansion point lies on the line between the two points $\boldsymbol{x}_r$ and $(2\boldsymbol{x}_r - \boldsymbol{x}_o)$. The expanded point replaces the point $\boldsymbol{x}_h$ if $f(\boldsymbol{x}_e) < f(\boldsymbol{x}_r)$. The reflection and expansion steps are illustrated in Figure 5.4 for a two-dimensional problem. The new simplex for both operations is either $\{\boldsymbol{x}_1, \boldsymbol{x}_2, \boldsymbol{x}_r\}$ or $\{\boldsymbol{x}_1, \boldsymbol{x}_2, \boldsymbol{x}_e\}$, respectively.

The last possible operation is contraction. This step is used when the reflected point is poor satisfying $f(\boldsymbol{x}_r) > f(\boldsymbol{x}_i)$, $\forall i$. This implies that the size of the simplex is too big for the objective function behavior. In this case, the simplex should be contracted. In the contraction step, the point $\boldsymbol{x}_h$ is replaced by the contraction point $\boldsymbol{x}_c$ given by:

$$\boldsymbol{x}_c = (1-\gamma)\boldsymbol{x}_o + \gamma\boldsymbol{x}_h \tag{5.14}$$

where γ is the contraction factor. The point $\boldsymbol{x}_c$ lies on the line connecting the centroid $\boldsymbol{x}_o$ and the point $\boldsymbol{x}_h$ as shown in Figure 5.4. The new simplex replaces $\boldsymbol{x}_h$ by $\boldsymbol{x}_c$ and the steps of the algorithm are repeated for the new simplex. The algorithm continues with the new steps until the size of the simplex has shrunk beyond a certain threshold. Figure 5.5 shows the first few iterations of the Simplex method for a two-dimensional problem.

A possible implementation of the Simplex algorithm for unconstrained optimization is given in the listing M5.5. The following example illustrates the application of the Simplex method.

Example 5.5: Solve the optimization problem:

$$\text{minimize}\, f(\boldsymbol{x}) = (x_1 - 1)^2 + (x_2 - 5)^2 + (x_3 - 4)^2 \tag{5.15}$$

using the Simplex method. Starting with the initial Simplex given by the points:

$$S = \left\{ \begin{bmatrix} 0 \\ 0 \\ 0.3 \end{bmatrix}, \begin{bmatrix} 0 \\ 0.4 \\ 0 \end{bmatrix}, \begin{bmatrix} 0.6 \\ 0 \\ 0 \end{bmatrix}, \begin{bmatrix} 0 \\ 0 \\ 0 \end{bmatrix} \right\} \tag{5.16}$$

Utilize the MATLAB listing M5.5.

```
%M5.5
%This program carries out the simplex method
NumberOfParameters=3; %This is n for this problem
Simplex=zeros(NumberOfParameters,NumberOfParameters+1); %create storage for simplex
SimplexValues=zeros(1,NumberOfParameters+1); % create storage for function values
%choose some initial points for the simplex
Simplex(:,1)=[0 0 0.3]';
Simplex(:,2)=[0 0.4 0]';
Simplex(:,3)=[0.6 0 0]';
Simplex(:,4)=[0 0 0]';
%Now we get all initial function values
for i=1:(NumberOfParameters+1)
  SimplexValues(1,i)=feval('getObjective',Simplex(:,i));
end
%set the values for the different types of coefficients
ReflectionCoefficient=1.0;
ExpansionCoefficient=0.5;
ContractionCoefficient=0.6;
IterationCounter=0;
while(IterationCounter<100) %repeat
   [MaxValue MaxIndex]=max(SimplexValues); %Get the maximum value and its index
   [MinValue MinIndex]=min(SimplexValues); %Get the minimum value and its index
   CenterLower=GetCenter(Simplex,MaxIndex); %Get the centroid
   %now we start by doing reflection
   ReflectedPoint=(1.0+ReflectionCoefficient)*CenterLower-
                                    ReflectionCoefficient*Simplex(:,MaxIndex);
   ReflectedValue=feval('getObjective',ReflectedPoint); %Get the reflected value
   if(ReflectedValue<MaxValue) %is there an improvement
     %now let see if there is a room for expansion
     if(ReflectedValue<MinValue) %if this point is better than all points do expansion
       ExpandedPoint=(1+ExpansionCoefficient)*ReflectedPoint-
                                   ExpansionCoefficient*CenterLower; %expanded point
       ExpandedValue=feval('getObjective',ExpandedPoint); %get the expanded value
       if(ExpandedValue<ReflectedValue)
         ReflectedPoint=ExpandedPoint; %store the better values in the reflected point
         ReflectedValue=ExpandedValue;
       end
     end
     Simplex(:,MaxIndex)=ReflectedPoint; %store the reflected point or the expanded point
     SimplexValues(1,MaxIndex)=ReflectedValue;
   else                           %Reflection failed. Simplex must contract
     ContractionPoint=ContractionCoefficient*Simplex(:,MaxIndex)+(1-
                                         ContractionCoefficient)*CenterLower;
     ContractionValue=feval('getObjective',ContractionPoint);
     %now we replace the highest value
     Simplex(:,MaxIndex)=ContractionPoint;
     SimplexValues(1,MaxIndex)=ContractionValue;
   end
   IterationCounter=IterationCounter+1; %increase the iteration counter
end
MinIndex
MinValue
Simplex
```

Solution: The output from the MATLAB listing M5.5 for the problem (5.15) shows the point with minimum objective function in the simplex and its associated objective function value. This output is given by:

Iteration 0: MinValue = 38.1600 Min = [0 0.4000 0]′
Iteration 1: MinValue = 36.0903 Min = [0.5000 0.3333 0.2500]′
Iteration 2: MinValue = 34.0060 Min = [−0.4833 0.0111 0.4503]′

after 20 iterations: MinValue = 0.0975 Min = [0.9848 3.2921 4.1044]′

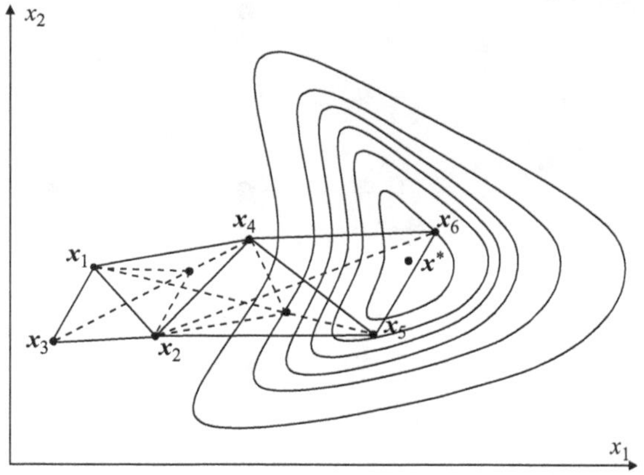

Figure 5.5 An illustration of several iterations of the Simplex method for a two-dimensional problem; in the first iteration, the point $\boldsymbol{x}_3$ is reflected and then expanded to the point $\boldsymbol{x}_4$ resulting in the new simplex $\{\boldsymbol{x}_1, \boldsymbol{x}_2, \boldsymbol{x}_4\}$. In the second iteration, the point $\boldsymbol{x}_1$ is reflected and then expanded to the point $\boldsymbol{x}_5$ resulting in the simplex $\{\boldsymbol{x}_2, \boldsymbol{x}_4, \boldsymbol{x}_5\}$. In the third iteration, the point $\boldsymbol{x}_2$ is reflected to the new point $\boldsymbol{x}_6$ to give the new simplex $\{\boldsymbol{x}_4, \boldsymbol{x}_5, \boldsymbol{x}_6\}$

The exact solution for this problem is $\boldsymbol{x}^* = [1.0 \quad 5.0 \quad 4.0]^T$ with a minimum value of the objective function of 0.

5.9 Response surface approximation

Another class of techniques for solving derivative-free unconstrained optimization problems is the response surface approximation. The basic idea in this class of techniques is to approximate the objective function, which is usually known only numerically, with another low-order function (usually a polynomial). The minimum of the surface approximation is then used to approximate the minimum of the objective function. If the response obtained through this approach is unsatisfactory, an improved surface approximation is constructed. These techniques are similar in their core concept to the quadratic and cubic interpolation techniques discussed in the previous chapter for the one-dimensional case. There are many variations of the response surface approximation approach that are beyond the scope of this book. We thus focus on only one approach and illustrate it with examples.

We approximate the objective function $f(\boldsymbol{x})$ within a certain subdomain of the parameter space by the multi-dimensional quadratic function:

$$A(\boldsymbol{x}) = a + \boldsymbol{b}^T\boldsymbol{x} + \boldsymbol{x}^T\boldsymbol{D}\boldsymbol{x} \tag{5.17}$$

where a is a scalar, $\boldsymbol{b} \in \Re^{n\times 1}$, and $\boldsymbol{D} \in \Re^{n\times n}$ with $\boldsymbol{D}$ being a diagonal matrix. By using this diagonal matrix, the approximation (5.17) neglects the mixed second-order derivatives of the quadratic approximation. This reduces the number of unknowns of the model (5.17).

The model (5.17) can be expanded in the form:

$$A(\boldsymbol{x}) = a + \sum_{i=1}^{n} b_i x_i + \sum_{i=1}^{n} d_i x_i^2 \tag{5.18}$$

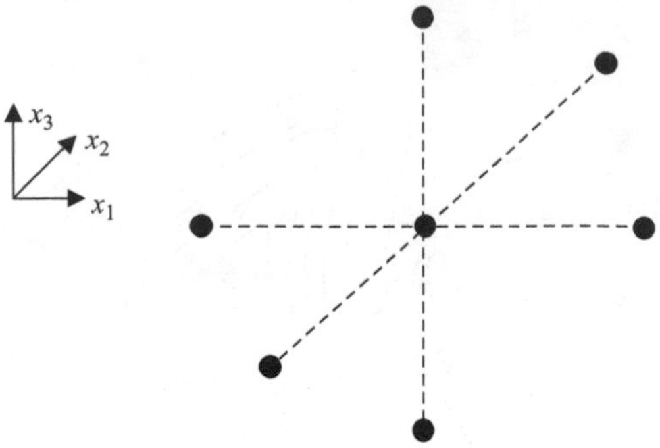

Figure 5.6 An illustration of the star distribution used for fitting models

where $\boldsymbol{b} = [b_1 \ \ b_2 \ \ \dots \ \ b_n]^T$ and $\boldsymbol{D} = \text{diag}(d_1, d_2, \dots, d_n)$. The approximation (5.18) has $(2n+1)$ unknown coefficients. It follows that $N = 2n+1$ independent pieces of information are needed to solve for these coefficients.

The approach illustrated in this section utilizes the objective function values at N points in the parameter space. As this approximation is intended to be only locally valid, we utilize the star distribution of points illustrated in Figure 5.6 for the three-dimensional case. At the kth iteration, this star distribution utilizes $\boldsymbol{x}^{(k)}$ as its center point. The points of this distribution at the kth iteration are thus given by:

$$S^{(k)} = \{\boldsymbol{x}^{(k)}, \boldsymbol{x}^{(k)} \pm \Delta_i \boldsymbol{e}_i, \quad i = 1, 2, \dots, n\} \tag{5.19}$$

In (5.19), points are created by using the perturbations $\pm\Delta_i$ in the direction of the ith coordinate. This set has $2n+1$ points. The values of the objective function are then evaluated at all N points. These values are given by:

$$S_f = \{f_i | f_i = f(\boldsymbol{x}_i), \quad i = 1, 2, \dots, N\} \tag{5.20}$$

where $\boldsymbol{x}_i$ is the ith point of the star distribution (5.19). The unknown coefficients of the model (5.18) are then solved for by forcing this model to have the same value as the objective function at all N points. This results in the following system of equations:

$$\begin{bmatrix} 1 & x_{1,1} & x_{1,2} & \dots & x_{1,n} & x_{1,1}^2 & x_{1,2}^2 & \dots & x_{1,n}^2 \\ 1 & x_{2,1} & x_{2,2} & \dots & x_{2,n} & x_{2,1}^2 & x_{2,2}^2 & \dots & x_{2,n}^2 \\ \vdots & \vdots & \vdots & \vdots & \vdots & \vdots & \vdots & \vdots & \vdots \\ 1 & x_{N,1} & x_{N,2} & \dots & x_{N,n} & x_{N,1}^2 & x_{N,2}^2 & \dots & x_{N,n}^2 \end{bmatrix} \begin{bmatrix} a \\ b_1 \\ b_2 \\ \vdots \\ b_n \\ d_1 \\ d_2 \\ \vdots \\ d_n \end{bmatrix} = \begin{bmatrix} f_1 \\ f_2 \\ \vdots \\ f_N \end{bmatrix} \tag{5.21}$$

where $x_{i,j}$ is the jth component of the ith point in the star distribution. The system (5.21) has N equations in the N unknown coefficients. It is then solved for the quadratic model coefficients. Once the coefficients are determined, the minimum of the quadratic model is obtained by enforcing the necessary optimality conditions on (5.17) to obtain:

$$\frac{\partial A}{\partial \boldsymbol{x}} = \boldsymbol{0} \Rightarrow \boldsymbol{b} + 2\boldsymbol{D}\bar{\boldsymbol{x}} = 0 \Rightarrow \bar{\boldsymbol{x}} = -\frac{1}{2}\boldsymbol{D}^{-1}\boldsymbol{b} = \begin{bmatrix} -\dfrac{b_1}{2d_1} \\ -\dfrac{b_2}{2d_2} \\ \vdots \\ -\dfrac{b_n}{2d_n} \end{bmatrix} \tag{5.22}$$

If this solution is inside the hyperbox including the star distribution, it is accepted as an approximation to the solution. Otherwise, another star distribution is constructed with $\bar{\boldsymbol{x}}$ being the center point and the steps are repeated again. A possible implementation of the approach explained in this section is shown in the MATLAB listing M5.6.

Example 5.6: Repeat Example 5.4 using the response surface implementation M5.6.

Solution: The function in Example 5.4 is quadratic and one should expect the algorithm to converge in a fast rate. In the first iteration, the recovered coefficients of the quadratic approximation are:

$\boldsymbol{b} = [\,0 \quad 0 \quad 0]'$
$\boldsymbol{d} = [1 \quad 3 \quad 6]'$

which gives the exact optimal solution $\boldsymbol{x} = [0 \quad 0 \quad 0]^T$. Notice how the vectors $\boldsymbol{b}$ and $\boldsymbol{d}$ accurately estimate the gradient and the Hessian matrices of the quadratic function.

Example 5.7: Apply the response surface method to the optimization problem:

$$\text{minimize}\, f(\boldsymbol{x}) = (x_1 - 1)^2 + (x_2 - 5)^3 + 2(x_3 - 4)^2 \tag{5.23}$$

Starting with the point $\boldsymbol{x}^{(0)} = [3.0 \quad -1.0 \quad 5.0]^T$. Utilize the MATLAB listing M5.6.

Solution: Appropriate changes are made in the initialization section of the MATLAB code to utilize the given starting point and evaluate the objective function (5.23). The function (5.23) is cubic and it tests how the algorithm behaves with non-quadratic functions. The algorithm took four iterations. The center of the star and the objective function value at each iteration are given by:

Iteration 0: OldPoint = [3.0 1.0 5.0]' oldPointValue = 210
Iteration 1: newPoint = [1.0000 2.0069 4.0000]' newPointValue = 26.8129
Iteration 2: newPoint = [1.0000 3.5174 4.0000]' newPointValue = 3.259
Iteration 3: newPoint = [1.0000 4.2868 4.0000]' newPointValue = 0.3628
Iteration 4: newPoint = [1.0000 4.7018 4.0000]' newPointValue = 0.0265

The algorithm took more iterations because the function is cubic. The optimal solution of this problem is $\boldsymbol{x}^* = [1.0 \quad 5.0 \quad 4.0]^T$. The code M5.6 can be improved by adjusting the size of the perturbation used to construct the star distribution to improve accuracy. The number of points required at every iteration can also be reduced by reusing some of the previous point or by using a linear model instead of a quadratic model.

```
%M5.6
%An Implementation of the Response Surface Method
numberOfParameters=3; %number of parameters
numberOfStarPoints=2*numberOfParameters+1; %number of points in star
systemMatrix=zeros(numberOfStarPoints,numberOfStarPoints); %system matrix
systemVector=zeros(numberOfStarPoints,1); %this is the right hand side vector
OldPoint=[3  -3  5]'; %the starting point of the algorithm
oldPointValue=getObjective(OldPoint);
inBox=false; % flag to check if solution is inside hyperbox
Delta=0.5; % size of the Star arm
starObjectiveValues=zeros(numberOfStarPoints,1); %storage for star values
while(inBox==false) %repeat until solution is within the modeling hyperbox
  currentStar=getStarPoints(OldPoint, Delta); %every row correspond to a point in the Star
  for i=1:numberOfStarPoints
    starObjectiveValues(i,1)=getObjective(currentStar(i,:)'); %get value at the ith point
  end
  %build the system of equations
  systemMatrix(:,1)=1; %column of 1s
  systemMatrix(:,2:numberOfParameters+1)=currentStar; %columns 2 to (n+1)
  %square of Star points
  systemMatrix(:,(numberOfParameters+2):(numberOfStarPoints))=currentStar.^2;
  systemVector=starObjectiveValues; %right hand side vector
  %solve the system matrix
  Coefficients=inv(systemMatrix)*systemVector; %these are the quadratic coefficients
  b=Coefficients(2:(numberOfParameters+1)); %the vector b
  d=Coefficients((numberOfParameters+2):(numberOfStarPoints)); %the diagonals of D
  newPoint=-0.5*b./d; %divide element by element
  newPointValue=getObjective(newPoint);
  inBox=IsInBox(newPoint,OldPoint, Delta); %check if solution is within hyperbox
  if(inBox==false)
    OldPoint=newPoint; %move to new point
  end
end
 %this function generates the star distribution points
function starPoints = getStarPoints(OldPoint, Delta)
numberOfParameters=size(OldPoint,1); % number of parameters
starPoints=zeros(2*numberOfParameters,numberOfParameters); % storage for points
starPoints(1,:)=OldPoint';%store center of star
Identity=eye(numberOfParameters); %idenitity matrix
for i=1:numberOfParameters
  starPoints(2*i,:)=(OldPoint+Delta*Identity(:,i))'; %perturb in +ve i
  starPoints(2*i+1,:)=(OldPoint-Delta*Identity(:,i))'; %perturb in -ve i
end
 %This function check if a given point is within a hyperbox centered at the given point
function inBox = IsInBox(newPoint, OldPoint, Delta)
numberOfParameters=size(newPoint,1); %recover number of parameters
onesVector=ones(numberOfParameters,1); %get vector of all ones
lowerBound=OldPoint-Delta*onesVector; %lower bounds on box
upperBound=OldPoint+Delta*onesVector; %upper bounds on box
if((lowerBound <=newPoint)& (newPoint<=upperBound)) %is new point within hyperbox
  inBox=true;
else
  inBox=false;
end
```

A5.1 Electrical application: impedance transformers

In many electrical engineering applications, it is required to connect different circuit components that have different impedances, different dimensions, or different material properties. Because of the different properties of these components, traveling electromagnetic waves may reflect back with little energy delivered to the rest of the circuit. Impedance transformers aim at addressing this problem. By designing smooth transitions from one component to another, the amount of reflection can be reduced significantly. The reflection over a desired frequency band is minimized by adjusting the dimensions of the transitions to their optimal values [5].

Figure 5.7 shows two different types of impedance transformers utilizing two different technologies. In Figure 5.7(a), a two-section microstrip impedance transformer is shown. In this case, the lengths and widths of two microstrip lines are designed to minimize the energy reflected when connecting the wide microstrip line to the much thinner microstip line. In this case, the optimization parameters are $\boldsymbol{x} = [L_1 \quad W_1 \quad L_2 \quad W_2]^T$. The energy reflection can be measured through the widely used S-parameters over the band of interest. The modulus of these complex quantities serves as a possible objective function.

Figure 5.7(b) shows a two-section waveguide impedance transformer. It is used to minimize reflection when connecting a wide microwave waveguide to a much narrower waveguide. Similar to the microstrip case, the dimensions of the transition sections are designed to minimize energy reflection over the frequency band of interest. The S-parameters may be used to formulate the objective function to be minimized.

To illustrate this approach, we carry out the design of the two-section transmission line impedance transformer shown in Figure 5.8. The target of this problem is to reduce reflection from a load resistance of value R_L when connected to a source with a different source resistance R_s. An appropriate objective function is given by

$$f(\boldsymbol{x}) = \sum_i |S_{11}(freq_i)|^2 \tag{5.24}$$

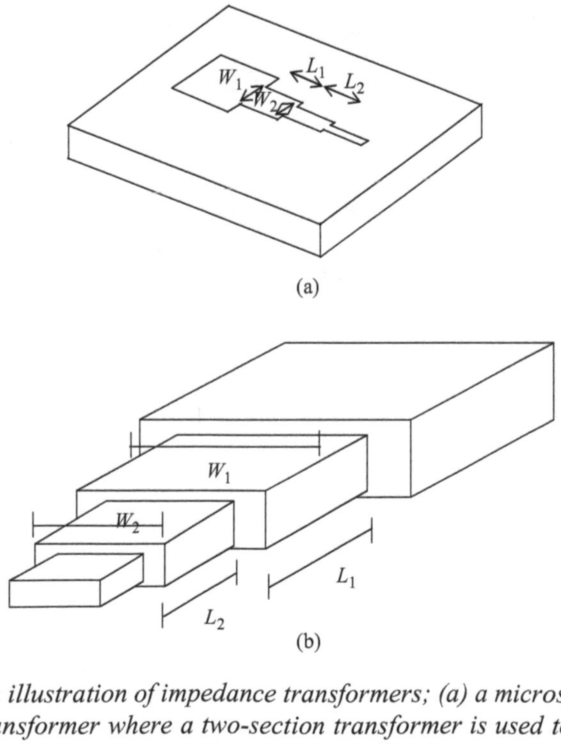

Figure 5.7 An illustration of impedance transformers; (a) a microstrip impedance transformer where a two-section transformer is used to connect two microstrip lines of different widths and (b) a two-section waveguide transformer is used to connect two different waveguides of different cross sections

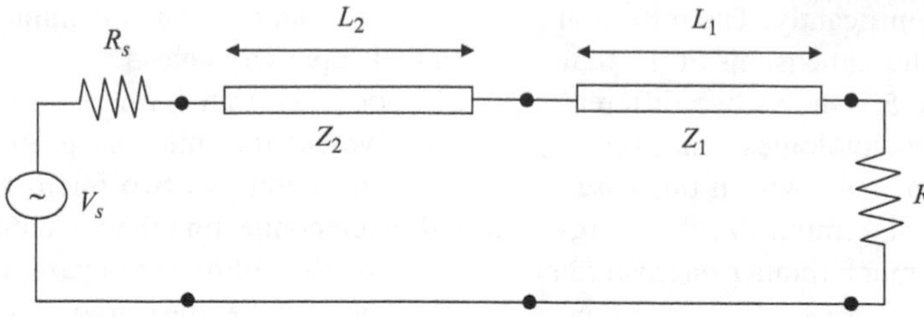

Figure 5.8 The two-section impedance transformer of application A5.1

where S_{11} is the S-parameter seen by the source port and $freq_i$ is the ith considered frequency. Energy reflection at a certain frequency vanishes if we can drive the modulus of S_{11} to zero. This may be possible at only one frequency. However, for a number of frequencies, we just aim at minimizing the reflection as much as possible.

The optimization variables for the problem in Figure 5.8 are the characteristic impedances and physical lengths of the two transmission lines with $\boldsymbol{x} = [Z_1 \quad L_1 \quad Z_2 \quad L_2]^T$. Because the parameters have different orders, scaling is used within the optimization code. The parameters are scaled back to their physical values before every call to the simulator.

To solve this problem, we utilize the MATLAB listing M5.4 with the following initialization section:

```
%M5.7
%The initialization section for the Hooke and Jeeves algorithm
sourceResistance=50;
loadResistance=75;
NumberOfParameters=4; %This is n for this problem
OldPoint=[1.0 0.75 1.0 0.75]' %This is the starting point (100 ohms and 7.5 cms)
Step=0.1; %the step taken in the direction of all coordinate parameters
LambdaMax=0.2; %maximum value of Lambda
```

For this application, we have $R_s = 50\ \Omega$ and $R_L = 75\ \Omega$. The starting point for optimization is $\boldsymbol{x}^{(0)} = [Z_1 \quad L_1 \quad Z_2 \quad L_2]^T = [100\,\Omega \quad 7.5\,\text{cm} \quad 100\,\Omega \quad 7.5\,\text{cm}]^T$. Notice how scaling is used in the MATLAB listing M5.7.

The objective function is evaluated by summing the square of the modulus of the reflection coefficient over three frequencies {0.75, 1.0, 1.25} GHz. The objective function is calculated as given by the MATLAB listing M5.8.

The reflection coefficient at the initial parameters over the band of interest is shown in Figure 5.9. The algorithm carried out five iterations until reflection was sufficiently reduced. The output of the algorithm is given by:

$$\boldsymbol{x}^{(0)} = [1.0000 \quad 0.7500 \quad 1.0000 \quad 0.7500]^T \quad f(\boldsymbol{x}^{(0)}) = 0.3024$$
$$\boldsymbol{x}^{(1)} = [1.1031 \quad 0.6469 \quad 0.8969 \quad 0.7500]^T \quad f(\boldsymbol{x}^{(1)}) = 0.2155$$
$$\boldsymbol{x}^{(2)} = [0.9832 \quad 0.6469 \quad 0.7769 \quad 0.7500]^T \quad f(\boldsymbol{x}^{(2)}) = 0.1238$$
$$\boldsymbol{x}^{(3)} = [0.8632 \quad 0.5269 \quad 0.6570 \quad 0.7500]^T \quad f(\boldsymbol{x}^{(3)}) = 0.0508$$
$$\boldsymbol{x}^{(4)} = [0.7632 \quad 0.4269 \quad 0.6570 \quad 0.7500]^T \quad f(\boldsymbol{x}^{(4)}) = 0.0235$$
$$\boldsymbol{x}^{(5)} = [0.7632 \quad 0.4269 \quad 0.6069 \quad 0.7500]^T \quad f(\boldsymbol{x}^{(5)}) = 0.0149$$

The final parameters are thus $[Z_1 \quad L_1 \quad Z_1 \quad L_1]^T = [76.32\,\Omega \quad 4.269\text{ cm} \quad 60.69\,\Omega \quad 7.50\text{ cm}]^T$. The modulus of the reflection coefficient at the final design over the band of interest is shown in Figure 5.10.

```
%M5.8
%calculation of objective function for application A5.1
function value=getObjective(xScaled)
scaleMatrix=diag([100 0.1 100 0.1] ); %scaling matrix
x=scaleMatrix*xScaled; %scale parameters back
Z1=x(1); %get first parameter
L1=x(2); %get second parameter
Z2=x(3); %get third parameter
L2=x(4); %get fourth parameter
Zs=50; %source resistance
ZL=75; %load resistance
numberOfFrequencies=3; %number of frequencies
Frequencies=[0.75e9 1.0e9 1.25e9]';
value=0; %initialize objective function value
for i=1:numberOfFrequencies %calculate objective function for the ith frequency
  Frequency=Frequencies(i); %get the ith frequency
  %calculate reflection coefficient at the ith frequency
  reflectionCoefficient=getReflectionCoefficient(Z1, L1, Z2, L2, Frequency, ZL, Zs);
  value=value+(abs(reflectionCoefficient))^2; %update objective function
end

 %This function returns S11 at a specific frequency for a multi-Section impedance transformer.
function reflectionCoefficient=getReflectionCoefficient(Z1, L1, Z2, L2, Frequency, ZL, Zs)
C=3.0e8; %speed of light is assumed for all transmission lines
Lambda=C/Frequency; %get wavelength at frequency
i=sqrt(-1);
%get the electric lengths of transmission lines at these frequencies
L1Electrical=L1/Lambda;
L2Electrical=L2/Lambda;
%calculate the input impedance seen by each section. The transmission line impedance formula is
%used
Zin1=Z1*(ZL+i*Z1*tan(2*pi*L1Electrical))/(Z1+i*ZL*tan(2*pi*L1Electrical));
Zin2=Z2*(Zin1+i*Z2*tan(2*pi*L2Electrical))/(Z2+i*Zin1*tan(2*pi*L2Electrical));
reflectionCoefficient=(Zin2-Zs)/(Zin2+Zs); %get reflection coefficient
```

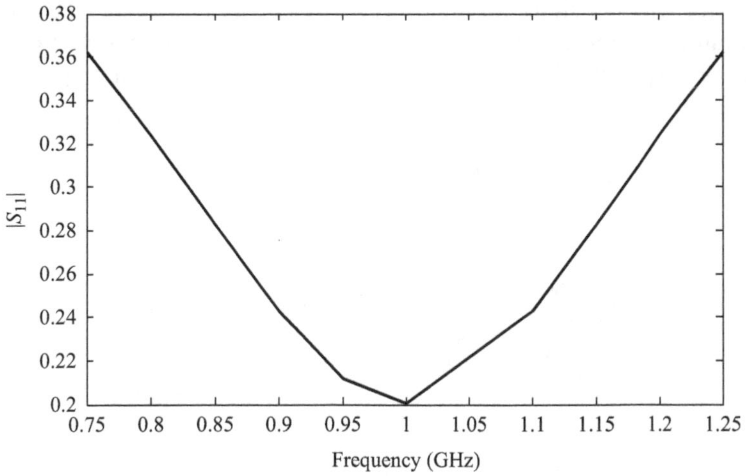

Figure 5.9 The response of the two-section impedance transformer at the initial design

A5.2 Electrical application: the design of photonic devices

The design of high-frequency structures is a very important application of optimization theory. Changing the different material properties or dimensions of different discontinuities changes the electromagnetic response of a circuit/structure.

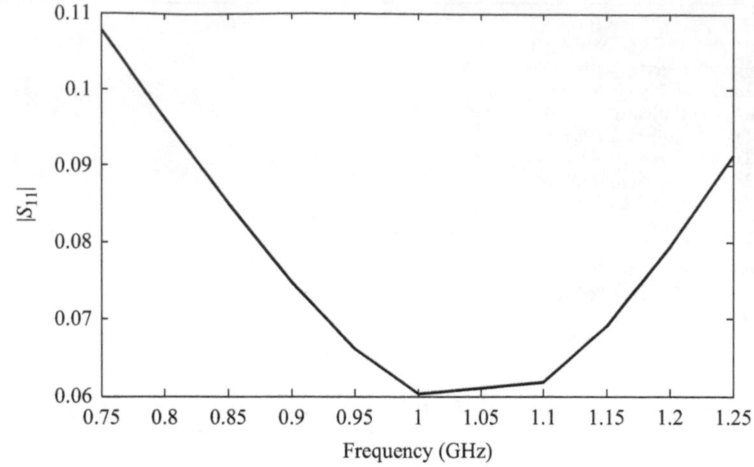

Figure 5.10 The response of the two-section impedance transformer at the final design

Interesting functionalities can be achieved by properly designing the different parameters.

Photonic devices are an important class of high-frequency structures. They are concerned with the generation and manipulation of light with applications in various areas. Optimization techniques have been utilized in the design of optical filters, grated structures, nanoantennas, and many other photonic structures.

We illustrate the application of optimization techniques to photonic devices by considering the plasmonic slit array shown in Figure 5.11. In this structure, the light incident from the left side is focused using the array of slits into a focal point by changing the width of all the metallic slits $w = [w_1 \quad w_2 \quad \ldots \quad w_N]^T$. Our target in this optimization problem is to move the focal point as far as possible from the slit array while keeping the beam width below 1.1 μm. The optimization problem may thus be formulated by:

$$\min_{w} -\alpha F_L - \log\left(1 - \frac{\text{BW}}{1.1}\right) \tag{5.25}$$

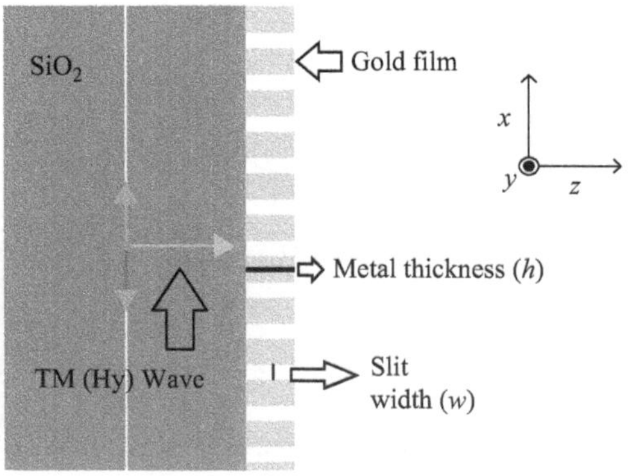

Figure 5.11 The structure of a plasmonic slit array

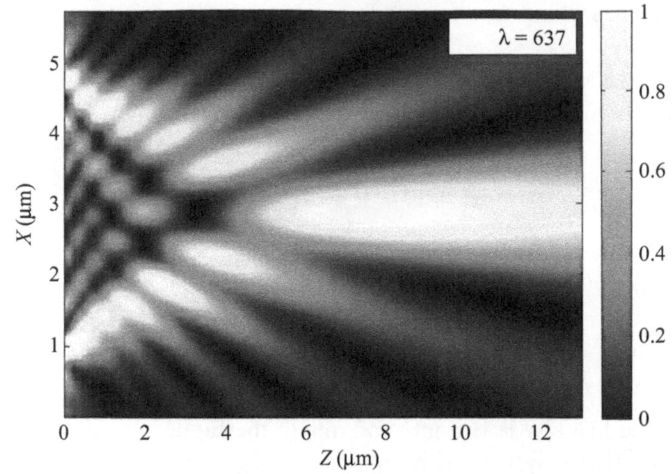

Figure 5.12 Intensity plot at the optimal design of the slit plasmonic array

where F_L is the focal length and BW is the beam width. The values of the focal length and beam width corresponding to every possible $\boldsymbol{w}$ are calculated through a computational electromagnetic approach. The number of optimization parameters can be reduced by assuming symmetry for the slits resulting in only $(N+1)/2$ parameters if N is an odd number. In this example, we utilize a structure with 19 metallic slits. Symmetry was employed and we have only 10 independent width values.

Pattern search optimization is utilized in solving this problem. The initial values give an unfocused light of beam width 3.74 μm. After 33 iterations, the widths at the optimal design are given by $\boldsymbol{w} = [0.19 \quad 0.08 \quad 0.06 \quad 0.05 \quad 0.04 \quad 0.04 \quad 0.03 \quad 0.03 \quad 0.03 \quad 0.03 \quad 0.03 \quad 0.03 \quad 0.03 \quad 0.04 \quad 0.04 \quad 0.05 \quad 0.06 \quad 0.08 \quad 0.19]^T$ μm. The optimal beam intensity is shown in Figure 5.12. This design gives the focal length and beam width at this design of 7.68 and 0.78 μm. The finite-difference time-domain (FDTD) solver lumerical [6] was used in simulating and optimizing this structure.

References

1. Singiresu S. Rao, *Engineering Optimization Theory and Practice*, Third Edition, John Wiley & Sons Inc, New York, 1996
2. Andreas Antoniou and Wu-Sheng Lu, *Practical Optimization Algorithms and Engineering Applications*, Springer, New York, 2007
3. Edwin K.P. Chong and Stanislaw H. Zak, *An Introduction to Optimization*, Second Edition, John Wiley & Sons Inc, New Jersey, 2001
4. Jorge Nocedal and Stephen J. Wright, *Numerical Optimization*, Second Edition, Springer, New York, 2006
5. John W. Bandler and R.M. Biernacki, *ECE3KB3 Courseware on Optimization Theory*, McMaster University, Canada, *1997*
6. F.D.T.D Lumerical, Lumerical Solutions Inc., http://www.lumerical.com

Problems

5.1 Modify the MATLAB listing M5.1 to implement the grid search technique. Using this code, solve the optimization problem:

$$\text{minimize } f(x_1, x_2) = \frac{-0.5}{x_1^2 + x_2^2 + 1.0}$$

in the two-dimensional interval $[-2.0 \quad -2.0]^T \leq \boldsymbol{x} \leq [2.0 \quad 2.0]^T$. Utilize a grid size of 0.02 in both dimensions. What is exact optimal solution of this problem?

5.2 A system of equations $\boldsymbol{y}(\boldsymbol{x}) = \boldsymbol{0}$, with $\boldsymbol{y} \in \Re^m$, can be solved as an optimization problem by minimizing the objective function $f(\boldsymbol{x}) = \sum_{j=1}^{m} y_j^2(\boldsymbol{x})$. By utilizing the MATLAB listing M5.2, apply the random walk approach to the solution of the system of equations $x_1 - 2x_2 = 5.0$, $x_1 + x_2 = -1.0$. Verify your answer by directly solving these two equations.

5.3 Repeat Problem 5.2 for the system of nonlinear equations $x_1^2 + x_2^2 = 9.0$, $x_2 = x_1^2 - 2.0$. Verify your solution by solving the problem graphically. How many solutions does this problem have?

5.4 Consider the electric circuit shown in Figure 5.13.
   (a) Write the power delivered to the load as a nonlinear function of the real and imaginary parts of the source impedance R_s and X_s.
   (b) Find the optimal complex source impedance that would maximize the power delivered to the load using the MATLAB listing M5.3 after making the necessary changes to the initialization section.
   (c) Verify your answer analytically to obtain R_s^* and X_s^* by applying the necessary optimality conditions.

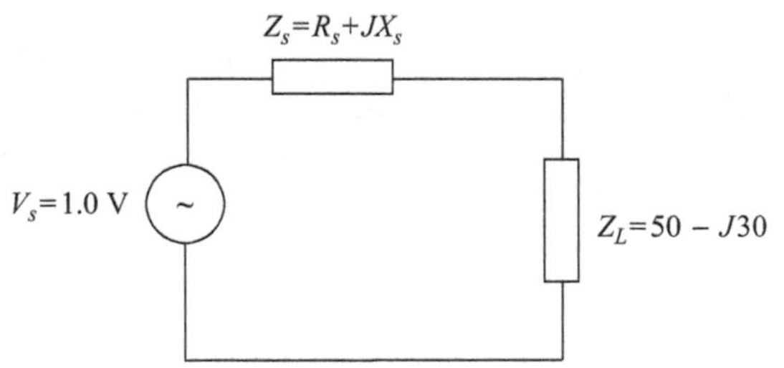

Figure 5.13 The circuit of Problem 5.4

5.5 Minimize the four-dimensional function:

$$f(\boldsymbol{x}) = (x_1 + 10x_2)^2 + 5(x_3 - x_4)^2 + (x_2 - 2x_3)^4 + 10(x_1 - x_4)^4.$$

Utilize the pattern search listing M5.4 starting from the point $\boldsymbol{x}^{(0)} = [3.0 \quad -1.0 \quad 0 \quad 1.0]^T$.

5.6 Repeat Problem 5.5 using the Simplex method. Utilize the MATLAB listing M5.5 with the initial simplex as the points $\boldsymbol{x}^{(0)}$, $\boldsymbol{x}^{(0)} + \Delta\boldsymbol{e}_1$, $\boldsymbol{x}^{(0)} + \Delta\boldsymbol{e}_2$, $\boldsymbol{x}^{(0)} + \Delta\boldsymbol{e}_3$, and $\boldsymbol{x}^{(0)} + \Delta\boldsymbol{e}_4$, where $\Delta = 0.3$.

5.7 Minimize the objective function:

$$f(\boldsymbol{x}) = (x_1^2 + x_2^2 - 1)^2 + (x_1 + x_2 - 2)^2$$

using the response surface method. Exploit the MATLAB listing M5.6 with a starting point $[8.0 \quad 8.0]^T$. Utilize a star distribution of size $\Delta = 0.5$.

Chapter 6

First-order unconstrained optimization techniques

6.1 Introduction

As explained in the previous chapter, there are techniques that do not require any sensitivity information. They only utilize the values of the objective function at a number of points in the parameter space to approach the optimal solution. These algorithms use the function values to infer the behavior of the function and predict descent directions.

Sensitivity-based optimization algorithms utilize the available first- or second-order sensitivities to guide the optimization iterations. These algorithms have robust convergence proofs. They can be guaranteed to generate a better solution at every iteration until the optimal point is sufficiently approached. These techniques, however, come at a cost. Estimating the sensitivities can be achieved through finite difference approximations. These finite approximations, as explained in Chapter 1, use repeated evaluations of the objective function with perturbed parameter values. For a system with n-parameter, at least n extra simulations are needed to estimate the first-order sensitivities. This overhead can be significant if n is large or if the objective function is estimated through a time-intensive numerical simulation. Estimating second-order adjoint sensitivities would require $O(n^2)$ extra simulations, which may be a formidable cost. Recent advances in first- and second-order adjoint sensitivities reduce this overhead significantly. Adjoint sensitivity approaches are addressed in Chapter 10.

We discuss in this chapter first-order unconstrained optimization approaches. These approaches assume the existence of first-order derivatives either through a finite difference approximation or through an adjoint sensitivity approach. We focus in this chapter on the steepest descent approach and conjugate gradient approaches.

6.2 The steepest descent method

As explained in Chapter 3, the gradient of the objective function ∇f is the direction of maximum function increase. The antiparallel of this direction $(-\nabla f)$ is the direction of maximum function decrease or steepest descent direction. At the kth iteration, starting from the current point $\boldsymbol{x}^{(k)}$, a reduction in the objective function can be achieved by searching along this direction to get the new point:

$$\boldsymbol{x}^{(k+1)} = \boldsymbol{x}^{(k)} + \lambda^*(-\nabla f(\boldsymbol{x}^{(k)})) \tag{6.1}$$

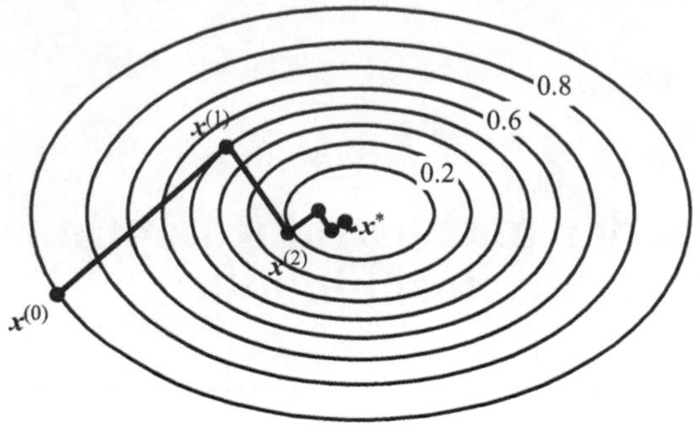

Figure 6.1 An illustration of the steepest descent method; at each iteration the steepest descent direction is the antiparallel of the gradient. A line search is carried out in this direction until the minimum value is reached along that direction

Through a line search approach, the optimal value of search parameter λ^* is determined. The line search, as explained in Chapter 4, is given by:

$$\lambda^* = arg \min_{\lambda} f(\boldsymbol{x}^{(k)} + \lambda(-\nabla f(\boldsymbol{x}^{(k)}))) \tag{6.2}$$

As mentioned earlier, the gradient $\nabla f(\boldsymbol{x}^{(k)})$ is normal to the objective function contour passing through the point $\boldsymbol{x}^{(k)}$. It is also normal to the gradient at the point $\boldsymbol{x}^{(k+1)}$ given by (6.1) and tangential to the contour at that point. This is shown in Figure 6.1. This figure also illustrates an important property of the steepest descent method. Far from the optimal solution, the algorithm shows fast reduction in the objective function. However, as the iterations approach the optimal point, the steps taken become smaller and smaller and the algorithm starts to zigzag. This zigzagging behavior can become severe for problems with sharp constant value surfaces. Because of the fast reduction in the objective function in the initial steps, the steepest descent method can be used to approach the solution. A switch to a second-order approach may then take place to accelerate convergence to the optimal point. I will explain in Chapter 7 approaches that combine the desired properties of first- and second-order methods.

The following example illustrates the steepest descent method.

Example 6.1: Use the method of steepest descent to find the minimizer of the function:

$$f(\boldsymbol{x}) = f(x_1, x_2, x_3) = (x_1 - 4)^4 + (x_2 - 3)^2 + 4(x_3 + 5)^4 \tag{6.3}$$

starting from the point $\boldsymbol{x}^{(0)} = [4.0 \quad 2.0 \quad -1.0]^T$. Carry out only two iterations.

Solution: We first evaluate the gradient of the objective function for any arbitrary point $\boldsymbol{x}$. This gradient is given by:

$$\nabla f = \left[4(x_1 - 4)^3 \quad 2(x_2 - 3) \quad 16(x_3 + 5)^3\right]^T \tag{6.4}$$

At the starting point, this gradient has the value:

$$\nabla f(\boldsymbol{x}^{(0)}) = [0 \quad -2 \quad 1024]^T \tag{6.5}$$

Line search is then carried out in the direction of steepest function descent $s^{(0)} = -\nabla f(x^{(0)}) = [0 \quad 2.0 \quad -1024]^T$. The new single dimension function is given by:

$$f(\lambda) = f(x^{(0)} + \lambda s^{(0)}) = (0)^4 + (2 + 2\lambda - 3)^2 + 4(-1 - 1024\lambda + 5)^4 \quad (6.6)$$

The minimum of this one-dimensional functional can be estimated using any of the techniques discussed in Chapter 4. Using the Secant method, we obtain:

$$\lambda_0^* = 3.967 \times 10^{-3} \Rightarrow x^{(1)} = [4.000 \quad 2.008 \quad -5.062]^T \quad (6.7)$$

Notice that the value of the line search λ is too small in this case. This could have been avoided if the gradient in (6.5) was first normalized to a unity norm. This normalization would make the same line search code useful for any problem regardless of the norm of the gradient.

For the second iteration, we first estimate the gradient at the point $x^{(1)}$. This gradient is given by:

$$\begin{aligned} \nabla f(x_1) &= \left[4(x_1 - 4) \quad 2(x_2 - 3) \quad 16(x_3 + 5)^3\right]^T\Big|_{x=x^{(1)}} \\ &= [0 \quad -1.984 \quad -0.003875]^T \end{aligned} \quad (6.8)$$

Line search is then carried out in the direction of steepest function descent $s^{(1)} = -\nabla f(x^{(1)})$. The single dimension function at the second iteration is given by:

$$\begin{aligned} f(\lambda) &= f(x^{(1)} + \lambda s^{(1)}) \\ &= (0)^4 + (2.088 + 1.984\lambda - 3)^2 + 4(-5.062 + 0.003875\lambda + 5)^4 \end{aligned} \quad (6.9)$$

Using a Secant method, we obtain the optimal value for the line search parameter $\lambda_1^* = 0.500$. This gives the new point:

$$x^{(2)} = x^{(1)} - \lambda_1^* \nabla f(x^{(1)}) = [4.000 \quad 3.000 \quad -5.060]^T \quad (6.10)$$

The gradient at the new point $x^{(2)}$ is given by:

$$\nabla f(x^{(2)}) = [0 \quad 0 \quad -0.0035]^T \quad (6.11)$$

Notice that the norm of the gradient is very close to zero, which indicates near first-order optimality. Actually, the point $x^{(2)}$ is very close to the actual minimum of the function (6.3), which is $x^* = [4.0 \quad 3.0 \quad -5.0]^T$. The function (6.3) is a sum of positive terms and it is minimized when each term is individually minimized to zero. This function has well-rounded contours and the steepest descent method works fine in this case. The MATLAB® listing M6.1 shows a possible implementation of the steepest descent approach.

```
% M6.1
%This program carries out the Steepest Descent direction
NumberOfParameters=3; %This is n for this problem
OldPoint=[3 -3 5]' %This is the starting point
OldValue=getObjective(OldPoint) %Get the objective function at the old point
Tolerance=0.001; %terminating tolerance for line search
Epsilon=0.001; %exploration step
LambdaMax=4.0; %maximum value of lambda for line search
StepNorm=1000; %initialize stepNorm
MinimumDistance=1.0e-4; %This is the terminating distance
while(StepNorm>MinimumDistance) %repeat until step size is small enough
   Gradient=getGradient('getObjecive',OldPoint, Epsilon); %get the gradient at the old point
   NormalizedNegativeGradient=-1.0*Gradient/norm(Gradient); %normalize the gradient
   LambdaOptimal =   GoldenSection('getObjective',Tolerance,OldPoint,
                      NormalizedNegativeGradient,LambdaMax); %get the optimal value
   NewPoint=OldPoint+LambdaOptimal*NormalizedNegativeGradient; %Get new point
   NewValue=feval('getObjective',NewPoint); %Get the New Value
   StepNorm=norm(NewPoint-OldPoint); %get the norm of the step
   OldPoint=NewPoint %update the current point
   OldValue=NewValue %update the current value
end
```

Example 6.2: Utilize the MATLAB listing M6.1 to find the minimum of the function:

$$f(\boldsymbol{x}) = f(x_1, x_2, x_3) = (x_1 - 1)^2 + (x_2 - 5)^2 + (x_3 - 4)^2 \tag{6.12}$$

starting from the point $\boldsymbol{x}^{(0)} = [3.0 \quad -3.0 \quad 5.0]^T$.

Solution: Executing the MATLAB code, we get the following output:

$\text{OldPoint} = [3.0 \quad -3.0 \quad 5.0]' \qquad \text{OldValue} = 69$
$\text{OldPoint} = [2.0368 \quad 0.8517 \quad 4.5183]' \qquad \text{OldValue} = 18.5518$
$\text{OldPoint} = [1.0735 \quad 4.7034 \quad 4.0365]' \qquad \text{OldValue} = 0.0947$
$\text{OldPoint} = [0.9994 \quad 5.0000 \quad 3.9994]' \qquad \text{OldValue} = 6.9495e-007$

The optimal solution is found in only three iterations. This function also have well-rounded constant value surfaces. This is why only three iterations are required to reach the optimal solution.

6.3 The conjugate directions method

The two earlier examples are well-behaved examples. The steepest descent technique was able to reach the optimal solution in n iterations. However, in other problems, this is not the case. Figure 6.2 shows the iterations taken in another quadratic problem with sharper contours. It is clear that the algorithm starts to show a zigzagging behavior as the optimal solution is approached. Sixteen iterations were needed to reach an accuracy of 0.01 ($\|\boldsymbol{x}^{(k+1)} - \boldsymbol{x}^{(k)}\| \leq 0.01$). Over 100 iterations would have been needed to reach an accuracy of 0.001. This slow convergence near the solution motivated the search for other gradient-based approaches that can eliminate this behavior.

A family of techniques called conjugate techniques is proposed that show excellent convergence for quadratic problems. They can solve an n-dimensional problem with a quadratic objective function in up to n iterations!

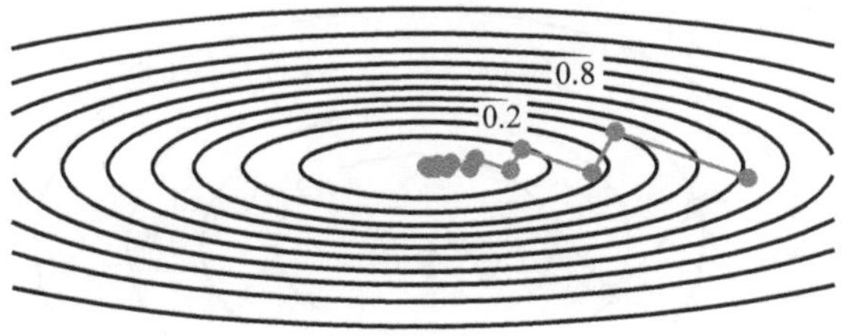

Figure 6.2 *An illustration of the slow convergence of the steepest descent method; Sixteen iterations are needed for this simple quadratic function with poorly scaled coordinates to reach an accuracy of ε = 0.01. Over a hundred iterations are needed to reach an accuracy of 0.001*

Conjugacy-based techniques can be generally classified into conjugate direction methods and conjugate gradient methods. Both techniques share the same basic concepts but they differ at how the steps are taken.

6.3.1 Definition of conjugacy

The conjugacy of two directions is defined in terms of a Hessian matrix. Consider a general quadratic function of the form:

$$f(\boldsymbol{x}) = a + \boldsymbol{g}^T\boldsymbol{x} + \frac{1}{2}\boldsymbol{x}^T\boldsymbol{H}\boldsymbol{x} \tag{6.13}$$

Equation (6.13) represents a second-order Taylor expansion around the origin $\boldsymbol{x}=0$. The expansion coefficients are $a \in \Re^1$, the gradient $\boldsymbol{g} \in \Re^{n\times 1}$, and the Hessian matrix $\boldsymbol{H} \in \Re^{n\times n}$. Any quadratic functions can be represented by (6.13) with a constant Hessian matrix for the whole domain. Non-quadratic functions can be approximated by (6.13) around a small neighborhood of the expansion point. In this case, both the gradient and the Hessian change with position.

Two vectors $\boldsymbol{s}_i$ and $\boldsymbol{s}_j$ are said to be conjugate with respect to a matrix $\boldsymbol{H}$ if they satisfy:

$$\boldsymbol{s}_i^T\boldsymbol{H}\boldsymbol{s}_j = 0, \quad i \neq j \tag{6.14}$$

Equation (6.14) implies that these two vectors are normal to each other using an $\boldsymbol{H}$-norm. Orthogonal vectors are a special case of (6.14) with $\boldsymbol{H}=\boldsymbol{I}$, the identity matrix.

The property (6.14) can be used to prove the following theorem:

Theorem 6.1: Consider minimizing an n-dimensional quadratic function of the form (6.13). The minimum of this function can be reached in at most n iterations if all the search directions are conjugate to each other.

Theorem 6.1 gives a "roadmap" for efficient optimization of a quadratic function. The algorithm starts from an initial point and an initial search direction is used. A line search is then carried out along that direction. At every subsequent iteration, a search direction is determined such that it is conjugate to all previous search directions. This guarantees convergence to the solution in at most n iterations if the function is quadratic.

A general non-quadratic function can be minimized using a similar approach. Such a function can be approximated over each subdomain as a quadratic function.

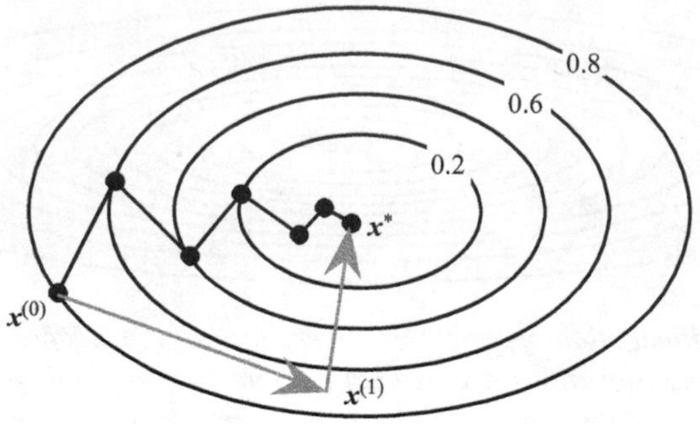

Figure 6.3 An illustration of conjugacy-based techniques; the algorithm converges in only two iterations for a two-dimensional quadratic problem while the steepest descent method shows a zigzagging behavior

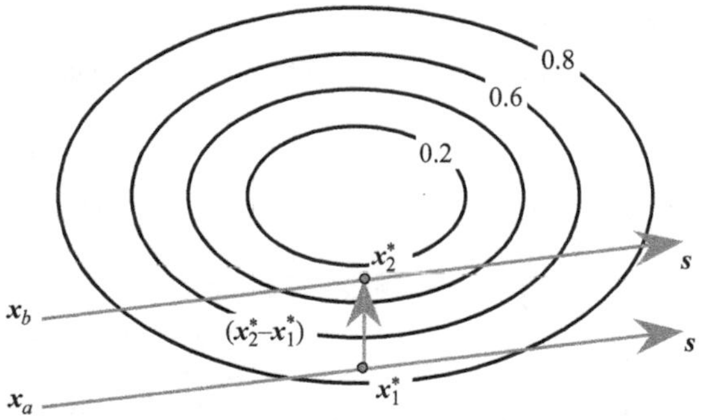

Figure 6.4 An illustration of the parallel subspace theorem for generating conjugate directions; starting from two arbitrary points $\boldsymbol{x}_a$ *and* $\boldsymbol{x}_b$ *and carrying out a line search along the direction* **s***, the direction connecting the two minimizers* $(x_2^* - x_1^*)$ *is conjugate to* **s**

It follows that after n iterations, the conjugate search directions are reset and a new set of other n search directions are generated until the optimal solution is sufficiently reached. Figure 6.3 illustrates how a technique using conjugate directions converges for the case $n = 2$ in only two iterations while the steepest descent method shows, as explained earlier, a much slower convergence.

The question that remains is how to generate these conjugate directions. There are a number of developed techniques to achieve this target. We give two of these approaches in the following two subsections:

6.3.2 Powell's method of conjugate directions

Powell's method of conjugate directions is based on the parallel subspace theorem stated below and illustrated in Figure 6.4.

Theorem 6.2: Consider a quadratic function $f(\boldsymbol{x})$. Starting from any point $\boldsymbol{x}_a$ and carrying out a line search along a direction $\boldsymbol{s}$, one obtains the minimum $\boldsymbol{x}_1^*$. Starting from a different point $\boldsymbol{x}_b$ and carrying out a line search along the same direction, one obtains another minimum $\boldsymbol{x}_2^*$. The direction $(x_2^* - x_1^*)$ is conjugate to $\boldsymbol{s}$.

Proof: The proof of this theorem is informative, so we will go through it. The quadratic objective function is assumed to have the form (6.13). The gradient of this function at an arbitrary point $\boldsymbol{x}$ is given by:

$$\nabla f(\boldsymbol{x}) = \boldsymbol{H}\boldsymbol{x} + \boldsymbol{g} \tag{6.15}$$

The gradients at the two points $\boldsymbol{x}_1^*$ and $\boldsymbol{x}_2^*$ are thus given by:

$$\nabla f(\boldsymbol{x}_1^*) = \boldsymbol{H}\boldsymbol{x}_1^* + \boldsymbol{g} \quad \text{and} \quad \nabla f(\boldsymbol{x}_2^*) = \boldsymbol{H}\boldsymbol{x}_2^* + \boldsymbol{g} \tag{6.16}$$

Because $\boldsymbol{x}_1^*$ and $\boldsymbol{x}_2^*$ are both minima along the search direction $\boldsymbol{s}$, they should satisfy:

$$\boldsymbol{s}^T\nabla f(\boldsymbol{x}_1^*) = \boldsymbol{s}^T\nabla f(\boldsymbol{x}_2^*) = 0 \Rightarrow \boldsymbol{s}^T(\nabla f(\boldsymbol{x}_2^*) - \nabla f(\boldsymbol{x}_1^*)) = 0 \tag{6.17}$$

Expanding the gradient in (6.17) using (6.16), we get the final result:

$$\boldsymbol{s}^T\boldsymbol{H}(\boldsymbol{x}_2^* - \boldsymbol{x}_1^*) = 0 \tag{6.18}$$

It follows that the direction $(\boldsymbol{x}_2^* - \boldsymbol{x}_1^*)$ is conjugate to $\boldsymbol{s}$.

Theorem 6.2 shows a possible way for a conjugate direction-based algorithm. We illustrate this algorithm first for the three-dimensional case ($n=3$). The target at each iterations is to create a direction conjugate to the previous conjugate direction. We assume in the first iteration ($k=0$) that we have a starting point $\boldsymbol{x}^{(0)}$. We carry out a line search from this point in the direction $\boldsymbol{e}_3$, the unit vector in the direction of the third parameter where:

$$\boldsymbol{e}_i = \begin{bmatrix} 0 & \cdots & \underset{i\text{th componnet}}{1} & \cdots & 0 \end{bmatrix}^T \tag{6.19}$$

This initial line search results in a better point $\boldsymbol{z}_1$. Starting from the point $\boldsymbol{z}_1$, we then carry out a sequence of line searches along the directions $\boldsymbol{e}_1$, $\boldsymbol{e}_2$, and $\boldsymbol{e}_3$ to obtain a better point $\boldsymbol{z}_2$. Using Theorem 6.2, the direction $\boldsymbol{p}_1 = \boldsymbol{z}_2 - \boldsymbol{z}_1$ is conjugate to the direction $\boldsymbol{e}_3$. We then carry out a third line search starting from the point $\boldsymbol{z}_2$ in the direction of $\boldsymbol{p}_1$ to reach the new point $\boldsymbol{x}^{(1)}$. This first iteration is illustrated for the two-dimensional case as shown in Figure 6.5.

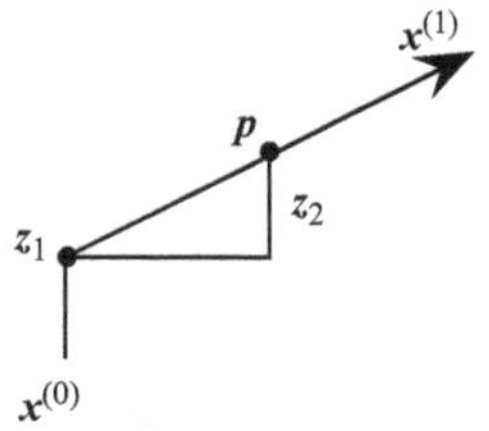

Figure 6.5 An illustration of one step of a conjugate direction algorithm for the two-dimensional case; starting from the point $\boldsymbol{x}^{(0)}$ *we carry out a line search along* $\boldsymbol{e}_2$ *to obtain the intermediate point* $\boldsymbol{z}_1$. *Starting from* $\boldsymbol{z}_1$, *we carry out sequential line searches along* $\boldsymbol{e}_1$ *and* $\boldsymbol{e}_2$ *to obtain* $\boldsymbol{z}_2$. *The direction* $\boldsymbol{p} = (\boldsymbol{z}_2 - \boldsymbol{z}_1)$ *is conjugate to* $\boldsymbol{e}_2$. *A final line search is then carried out starting from* $\boldsymbol{z}_2$ *in the direction* $\boldsymbol{p}$ *to obtain the better solution* $\boldsymbol{x}^{(1)}$. *In the following iteration* $\boldsymbol{e}_2$ *is replaced by* $\boldsymbol{p}$ *and the step is repeated staring from* $\boldsymbol{x}^{(1)}$

In the second iteration, the new direction $\boldsymbol{p}_1$ becomes the 3rd search direction while the other two directions are shifted to have the new set of directions $\{\boldsymbol{e}_2, \boldsymbol{e}_3, \boldsymbol{p}_1\}$. Because the first iteration ended with carrying out a line search along the direction $\boldsymbol{p}_1$, a new sequential line search is then carried out along the directions $\boldsymbol{e}_2$, $\boldsymbol{e}_3$, and $\boldsymbol{p}_1$ to get a new temporary point $\boldsymbol{z}$. According to the parallel subspace theorem, the direction $\boldsymbol{p}_2 = (\boldsymbol{z} - \boldsymbol{x}^{(1)})$ is conjugate to the direction $\boldsymbol{p}_1$. A line search is then carried out starting from the point $\boldsymbol{z}$ in the direction $\boldsymbol{p}_2$ to get a new point $\boldsymbol{x}^{(2)}$. The third iteration utilizes the set of directions $\{\boldsymbol{e}_3, \boldsymbol{p}_1, \boldsymbol{p}_2\}$. The algorithm then repeats until the optimal point is reached. For a non-quadratic function, the search directions are reset to the coordinate directions every three iterations.

For a general n-dimensional problem, a possible conjugate direction procedure is given by the following algorithm:

1. Initialization: set the starting point $\boldsymbol{x}^{(0)}$. Set the initial search directions to $\boldsymbol{p}_j = \boldsymbol{e}_j, j = 1, 2, \ldots, n$. Carry out an initial line search to minimize f along the direction $\boldsymbol{p}_n$ starting from $\boldsymbol{x}^{(0)}$ to get the point $\boldsymbol{x}^{(1)} = \boldsymbol{x}^{(0)} + \lambda^* \boldsymbol{p}_n$. Set iteration counter $k = 1$.
2. Set the temporary point $\boldsymbol{z}_1 = \boldsymbol{x}^k$.
3. Carry out sequential line searches from $\boldsymbol{z}_1$ in all directions to get the new temporary point $\boldsymbol{z}_{n+1} = \boldsymbol{z}_1 + \lambda_1^* \boldsymbol{p}_1 + \cdots + \lambda_n^* \boldsymbol{p}_n$.
4. Creation of a new search direction: a new conjugate direction is given by $\boldsymbol{s} = \boldsymbol{z}_{n+1} - \boldsymbol{z}_1$.
5. Get the new iterate: Starting from the point $\boldsymbol{z}_{n+1}$ carry out a line search along $\boldsymbol{s}$ to get the new better solution $\boldsymbol{x}^{(k+1)} = \boldsymbol{z}_{n+1} + \lambda^* \boldsymbol{s}$.
6. Update search directions: Set $\boldsymbol{p}_j = \boldsymbol{p}_{j+1}, j = 1, 2, \ldots, (n-1)$ and $\boldsymbol{p}_n = \boldsymbol{s}$.
7. If termination condition is satisfied, *stop*. Otherwise, set $k = k + 1$, go to step 2.

The MATLAB listing M6.2 illustrates a possible implementation of Powell's conjugate direction algorithm.

```
%M6.2
%This program implements Powell's method
NumberOfParameters=3; %This is n for this problem
OldPoint=[0 0 0]' %This is the starting point
OldValue=getObjective(OldPoint) %Get the objective function at the old point
Directions =eye(NumberOfParameters); %Get identity matrix of size n
LambdaMax=5; %maximum value of Lambda
Tolerance=0.00001; %terminating tolerance for line search
Epsilon=0.001; %exploration step
StepNorm=100; %initialize the norm of the step
%first we minimize along the nth direction
up=Directions(:,NumberOfParameters); %get the vector ei
un=-1.0*up; %get the vector -ei
%we do first exploration in the -ve and +ve directions
fp=feval('getObjective',OldPoint+Epsilon*up); %get +ve perturbed function value
if(fp<OldValue) %positive direction is promising
  u=up; %choose the positive coordinate direction
else
  u=un; %choose the negative coordinate direction
end
LambdaOptimal = GoldenSection('getObjective',Tolerance,OldPoint,u,LambdaMax); %get the
                                  %optimal value in the direction of u starting from OldPoint
NewPoint=OldPoint+LambdaOptimal*u; %get new point
NewValue=getObjective(NewPoint);
%make the new point the current point
OldPoint=NewPoint;
OldValue=NewValue;
```

```
while(StepNorm>1.0e-4) %repeat until termination condition satisfied
  %first we do search in the directions of the coordinate axes
  YOld=OldPoint; %start exploring from the current point
  YOldValue=OldValue; %store also the old value
  %sequential line search
  for i=1:NumberOfParameters %repeat for all coordinates
    up=Directions(:,i); %get the vector ei
    un=-1.0*up; %get the vector -ei
    %we do first exploration in the -ve and +ve directions
    fp=feval('getObjective',YOld+Epsilon*up); %get +ve pertubed function value
    if(fp<OldValue) %positive direction is promising
      u=up; %choose the positive coordinate direction
    else
      u=un; %choose the negative coordinate direction
    end
    LambdaOptimal = GoldenSection('getObjective',Tolerance,YOld,u,LambdaMax); %get the
                                 %optimal value in the direction of u starting from YOld
    YNew=YOld+LambdaOptimal*u; %Get new  exploration point
    YNewValue=feval('getObjective',YNew); %get new exploration value
    %store the new exploration point
    YOld=YNew;
    YOldValue=YNewValue;
  end
  ConjugateDirection=YOld-OldPoint; % the new Conjugate direction
  %get the optimal value in the direction of the new conjugate direction starting from YOld
  LambdaOptimal =
            GoldenSection('getObjective',Tolerance,YOld,ConjugateDirection,LambdaMax);
  NewPoint=YOld+LambdaOptimal*ConjugateDirection; %Get new point
  NewValue=feval('getObjective',NewPoint); %Get the New Value
  StepNorm=norm(NewPoint-OldPoint); %get the norm of the step
  %Now we update the search directions to add the new pattern search direction
  for j=1:(NumberOfParameters-1)
    Directions(:,j)=Directions(:,(j+1)); %create room for new direction
  end
  Directions(:,NumberOfParameters)=ConjugateDirection; %store the new pattern search direction
  OldPoint=NewPoint %update the current point
  OldValue=NewValue %update the current value
end
% print result
OldPoint
OldValue
```

It should be noted that in the MATLAB listing M6.2, line search is carried out either in the direction $\boldsymbol{e}_i$ or $-\boldsymbol{e}_i$ depending on an exploratory step. Also, the line search parameter is limited to a maximum value of 5. The following two examples illustrate the performance of the conjugate direction method.

Example 6.3: Utilize the conjugate directions method to minimize the function:

$$f(\boldsymbol{x}) = f(x_1, x_2, x_3) = 4(x_1 - 3)^2 + 3(x_2 - 5)^2 + (x_3 - 4)^2 \tag{6.20}$$

starting from the initial point $\boldsymbol{x}^{(0)} = [0 \quad 0 \quad 0]^T$.

Solution: Utilizing the MATLAB listing M6.2, we obtain the following iterations:

OldPoint = [0 0 0]′ OldValue = 95
OldPoint = [3.0001 5.0003 4.0001]′ OldValue = 2.1657e − 007

We could see that the solution to this three-dimensional problem was sufficiently approached in only one iteration! As stated earlier, a maximum of n iterations could be needed if exact line search is used. In practice, convergence can be achieved in fewer iterations!

Example 6.4: Utilize the conjugate directions method to minimize the function:

$$f(\boldsymbol{x}) = f(x_1, x_2, x_3) = 1.5x_1^2 + 2x_2^2 + 1.5x_3^2 + x_1x_3 + 2x_2x_3 - 3x_1 - x_3 \quad (6.21)$$

starting from the initial point $\boldsymbol{x}^{(0)} = [0 \quad 0 \quad 0]^T$.

Solution: Utilizing the same MATLAB code M6.2 and changing the call to the objective function evaluation to evaluate (6.21), we get the iterations:

OldPoint = [0 0 0]' OldValue = 0
OldPoint = [0.8889 0.0000 0.3333]' OldValue = −1.3518
OldPoint = [0.8889 −0.1667 0.3333]' OldValue = −1.4074
OldPoint = [0.9583 −0.1667 0.1250]' OldValue = −1.4653
OldPoint = [1.0000 0.0000 −0.0000]' OldValue = −1.5000

One more iteration was needed because of using line search with upper limit on the search parameter. The solution in the third iteration was, however, very close to the optimal solution.

The solution obtained from the MATLAB code for the problem (6.21) can also be evaluated analytically by enforcing first-order optimality conditions. The gradient at an arbitrary point $\boldsymbol{x}$ is given by:

$$\nabla f(\boldsymbol{x}) = \begin{bmatrix} 3x_1 + x_3 - 3 \\ 4x_2 + 2x_3 \\ x_1 + 2x_2 + 3x_3 - 1 \end{bmatrix} \quad (6.22)$$

Setting all components of the gradient (6.22) to zero, we obtain the system of linear equations:

$$\begin{aligned} 3x_1 + 0x_2 + x_3 &= 3 \\ 0x_1 + 4x_2 + 2x_3 &= 0 \\ x_1 + 2x_2 + 3x_3 &= 1 \end{aligned} \quad (6.23)$$

Solving for the unknown coordinates, we obtain $\boldsymbol{x}^* = [1.0 \quad 0 \quad 0]^T$, which corresponds to an objective function value $f(\boldsymbol{x}^*) = -1.5$. This result agrees with the result obtained using the conjugate directions method.

6.4 Conjugate gradient methods

The conjugate directions method, explained in the previous section, does not use sensitivity information. There is a class of techniques that utilize gradient information to create the conjugate directions for accelerated optimization. Instead of using the steepest descent direction, these techniques modify this direction to satisfy conjugacy to previous search directions. Once a new conjugate direction is determined, line search is then carried out to obtain the minimum along that direction.

One of the most used conjugate gradient techniques is the Fletcher–Reeves method. In the first iteration of this technique, the search direction $\boldsymbol{s}^{(0)}$ is the steepest descent direction $\boldsymbol{s}^{(0)} = -\nabla f(\boldsymbol{x}^{(0)})$. A line search is carried out in the steepest descent direction to obtain the new iterate $\boldsymbol{x}^{(1)}$. The new conjugate direction at the point $\boldsymbol{x}^{(1)}$ is obtained from the gradient at this point through the formula:

$$\boldsymbol{s}^{(1)} = -\nabla f(\boldsymbol{x}^{(1)}) + \frac{\|\nabla f(\boldsymbol{x}^{(1)})\|^2}{\|\nabla f(\boldsymbol{x}^{(0)})\|^2}\boldsymbol{s}^{(0)} \tag{6.24}$$

It can be shown that the direction $\boldsymbol{s}^{(1)}$ is conjugate to the direction $\boldsymbol{s}^{(0)}$. A line search is carried out along this new direction to obtain a better design. In general, for the $(k+1)$st iteration, the new conjugate direction is given in terms of the previous conjugate direction and current and previous gradients through the formula:

$$\boldsymbol{s}^{(k+1)} = -\nabla f(\boldsymbol{x}^{(k)}) + \frac{\|\nabla f(\boldsymbol{x}^{(k+1)})\|^2}{\|\nabla f(\boldsymbol{x}^{(k)})\|^2}\boldsymbol{s}^{(k)} \tag{6.25}$$

The MATLAB listing M6.3 gives a possible implementation of the Fletcher–Reeves method.

```
%M6.3
%implementation of the Fletcher Reeves method
NumberOfParameters=3; %This is n for this problem
OldPoint=[3 -7 0]' %This is the starting point
OldValue=getObjective(OldPoint) %Get the objective function at the old point
Epsilon=1.0e-3; %this is the perturbation
LambdaMax=5; %maximum lambda for line search
Tolerance=1.0e-4; %termination condition for line search
Gradient=getGradient('getObjective', OldPoint, Epsilon); %get the gradient at the first point
NormGradientOld=norm(Gradient); %get the norm of the gradient at the current point
ConjugateDirection=-1.0*Gradient; %initialize the first direction
IterationCounter=0;
while(IterationCounter<NumberOfParameters) %do only n iterations
  LambdaOptimal = GoldenSection('getObjective',Tolerance,OldPoint,
                                            ConjugateDirection,LambdaMax);%do line search
  NewPoint=OldPoint+LambdaOptimal*ConjugateDirection; %get new point
  NewValue=feval('getObjective',NewPoint); %get new objective function value
  NewGradient=getGradient('getObjective', NewPoint, Epsilon); %get new gradient
  NormGradientNew=norm(NewGradient); %get norm of the new gradient
  %now we determine the new conjugate direction
  NewConjugateDirection=-1.0*NewGradient+((NormGradientNew*NormGradientNew)/
                                (NormGradientOld*NormGradientOld))*ConjugateDirection;
  %now we make the new point the current point
  OldPoint=NewPoint
  OldValue=NewValue
  Gradient=NewGradient;
  NormGradientOld=NormGradientNew;
  ConjugateDirection=NewConjugateDirection;
  IterationCounter=IterationCounter+1;
end
```

The Fletcher–Reeves approach is illustrated through the following example.

Example 6.5: Utilize the Fletcher–Reeves method to minimize the objective function:

$$f(\boldsymbol{x}) = f(x_1, x_2, x_3) = 4(x_1-1)^2 + 3(x_2-5)^2 + (x_3-4)^2 \tag{6.26}$$

starting from the point $\boldsymbol{x}^{(0)} = [3.0 \quad -7.0 \quad 0]^T$. Utilize the MATLAB listing M6.3.

Solution: The output from the MATLAB code is given by:

OldPoint = [3 −7 0]′ OldValue = 464
OldPoint = [0.3529 4.9085 1.3231]′ OldValue = 8.8661
OldPoint = [1.3885 5.1720 2.4461]′ OldValue = 3.1071
OldPoint = [0.9996 4.9990 3.9992]′ OldValue = 4.6080e − 006

The solution is sufficiently reached in three iterations.

A6.1 Solution of large systems of linear equations

A problem faced often in the solution of practical electrical engineering problems is to solve large systems of linear equations. These systems may arise, e.g., from the solution of linear circuits with large number of components. The system of interest has the form:

$$\boldsymbol{A}\boldsymbol{x} = \boldsymbol{b} \tag{6.27}$$

If the matrix $\boldsymbol{A}$ has a large size, then its inversion may be time-intensive. An alternative way of solving the problem (6.27) is to instead minimize the quadratic function:

$$f(\boldsymbol{x}) = \frac{1}{2}\boldsymbol{x}^T\boldsymbol{A}\boldsymbol{x} - \boldsymbol{b}^T\boldsymbol{x} \tag{6.28}$$

The gradient of this objective function $\nabla f = \boldsymbol{A}\boldsymbol{x} - \boldsymbol{b} = -\boldsymbol{r}(\boldsymbol{x})$, where $\boldsymbol{r}$ is the residue of the linear system of equations (6.27). It follows that the point at which $\boldsymbol{r} = 0$ (the system (6.27) is satisfied) is an extreme point of the objective function (6.28). The Hessian of the objective function is the matrix $\boldsymbol{A}$. This implies that this conversion from the linear system of equations (6.27) to the optimization problem (6.28) is valid only if the matrix $\boldsymbol{A}$ is symmetric and positive semi-definite to guarantee convergence to a local minimum. We will show later how we can handle an arbitrary matrix.

Because a set of conjugate directions spans the whole n-dimensional space, any n-dimensional vector can be expanded in terms of these conjugate directions. It follows that the optimal solution of (6.27) can be expanded as:

$$\boldsymbol{x}^* = \sum_{j=1}^{n} \alpha_j \boldsymbol{p}^{(j)} \tag{6.29}$$

where $\boldsymbol{p}^{(k)}$, $k = 1, 2, \ldots, n$ is a possible sequence of conjugate directions with respect to the matrix $\boldsymbol{A}$. This characteristic simplifies the application of conjugate gradients to the solution of (6.28).

The problem (6.28) can be solved iteratively using the Fletcher–Reeves method discussed in Section 6.4. Staring with an initial guess $\boldsymbol{x}^{(0)}$, the initial conjugate gradient direction is given by $\boldsymbol{p}^{(0)} = -\nabla f(\boldsymbol{x}^{(0)}) = \boldsymbol{r}(\boldsymbol{x}^{(0)}) = \boldsymbol{r}^{(0)}$, the system residue at the starting point. In the Fletcher–Reeves method, a line search is carried out starting from $\boldsymbol{x}^{(0)}$ in the direction $\boldsymbol{p}^{(0)}$. However, because of the special properties of the linear system, no line search is needed. This can be seen by multiplying both sides of (6.29) with the matrix $\boldsymbol{A}$ and then subtracting $\boldsymbol{b}$, to get:

$$Ax^* - b = \sum_{j=1}^{n}(\alpha_j A p^{(j)}) - b = 0 \tag{6.30}$$

Multiplying both sides of (6.30) by an arbitrary $p^{(k)}$ and exploiting the concept of conjugacy, we get:

$$\alpha_k p^{(k)T} A p^{(k)} - p^{(k)T} b \Rightarrow \alpha_k = \frac{p^{(k)T} b}{p^{(k)T} A p^{(k)}} - \frac{p^{(k)T}(r^{(k)} + Ax^{(k)})}{p^{(k)T} A p^{(k)}} = \frac{p^{(k)T} r^{(k)}}{p^{(k)T} A p^{(k)}} \tag{6.31}$$

It follows that the line search step is known once we know the conjugate direction and the residue at the current point. Notice that $x^{(k)}$ is a linear combination of the conjugate directions $p^{(0)}, \ldots, p^{(k-1)}$ and is thus conjugate to the direction $p^{(k)}$.

The Fletcher–Reeves method can be thus used for solving linear systems of equations with symmetric positive definite matrices. This approach can be implemented using only vector scalar multiplications and matrix vector multiplications only. There is no need to invert the system matrix. The steps of this algorithm are given by the following steps:

Initialization: Given $x^{(0)} = 0$, A and b. Set $r^{(0)} = b - Ax^{(0)}$. The initial conjugate gradient direction is given by $p^{(0)} = r^{(0)}$. Set $k = 0$.

Step 1: Calculate the line search step α_k using (6.31).

Step 2: Determine new point $x^{(k+1)} = x^{(k)} + \alpha^{(k)} p^{(k)}$.

Step 3: Calculate the new residue $r^{(k+1)} = r^{(k)} - \alpha^{(k)} A p^{(k)}$.

Comment: This step can be derived by substituting the step taken in Step 2 back into the system of equations.

Step 4: If termination condition is satisfied, stop.

Comment: This condition may impose a certain value on the norm of the residue, i.e., $\|r^{(k+1)}\|^2 \le \varepsilon \Rightarrow r^{(k+1)T} r^{(k+1)} \le \varepsilon$.

Step 5: Determine the new conjugate direction using the Fletcher–Reeves formula (6.25) to get $p^{(k+1)} = -\nabla f(x^{(k+1)}) + \dfrac{\|\nabla f(x^{(k+1)})\|^2}{\|\nabla f(x^{(k)})\|^2} p^{(k)} = r^{(k+1)} + \dfrac{r^{(k+1)T} r^{(k+1)}}{r^{(k)T} r^{(k)}} p^{(k)}$.

Step 6: Set $k = k + 1$. Go to Step 1.

We illustrate this approach for solving linear systems of equations through the following circuit example:

Example 6.6: Consider the circuit shown in Figure 6.6. Apply loop analysis to formulate the loop equations. Solve for the unknown currents using the conjugate gradient method.

Solution: This circuit has 4 loops. The loop equations for all loops are given by:

$$\begin{aligned} 14I_1 - 3I_2 - 3I_3 + 0 &= 0 \\ -3I_1 + 10I_2 + 0 - 3I_4 &= 0 \\ -3I_1 + 0I_2 + 10I_3 - 3I_4 &= 0 \\ 0 - 3I_2 - 3I_3 + 14I_4 &= 0 \end{aligned} \tag{6.32}$$

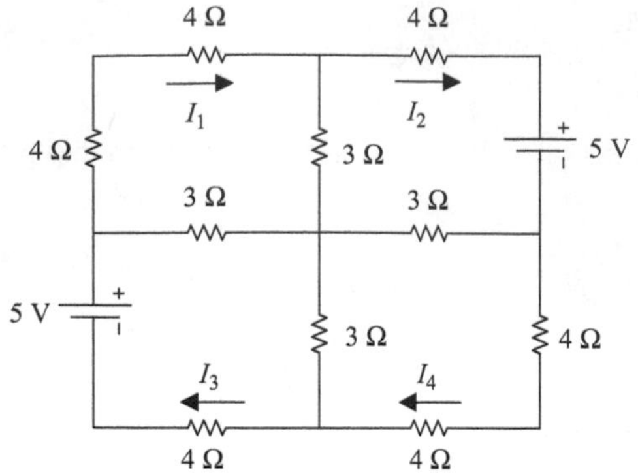

Figure 6.6 The circuit of Example 6.6. The resulting impedance matrix is symmetric and positive definite

or in matrix form:

$$\begin{bmatrix} 14 & -3 & -3 & 0 \\ -3 & 10 & 0 & -3 \\ -3 & 0 & 10 & -3 \\ 0 & -3 & -3 & 14 \end{bmatrix} \begin{bmatrix} I_1 \\ I_2 \\ I_3 \\ I_4 \end{bmatrix} = \begin{bmatrix} 0 \\ -5 \\ 5 \\ 0 \end{bmatrix} \tag{6.33}$$

The unknowns in this system are $\boldsymbol{x} = [I_1 \quad I_2 \quad I_3 \quad I_4]^T$. The matrix in (6.33) is symmetric and positive semi-definite as it is diagonally dominant with positive diagonals. The MATLAB listing M6.4 adapts the Fletcher–Reeves method to apply for solving linear system of equations with symmetric and positive definite matrices. Using this listing, the solution was found in only one iteration! The output from the algorithm at the starting and end points are:

Staring Point

$\text{OldPoint} = [0 \quad 0 \quad 0 \quad 0]^T$
$\text{Gradient} = [0 \quad 5.0 \quad -5.0 \quad 0]^T$

Final Point

$\text{OldPoint} = [0 \quad -0.5000 \quad 0.5000 \quad 0]^T$
$\text{Gradient} = [0 \quad 0 \quad 0 \quad 0]^T$

It should be noted that a zero gradient in the final result indicates that all four equations are exactly satisfied. By checking Figure 6.6, it could be verified that the reached solution satisfies all the circuit equations.

The technique discussed in (6.27)–(6.31) works only for symmetric positive semi-definite matrices. However, in general problems the system matrix may not have these properties. A more general approach is to solve the normal system of equations:

$$(\boldsymbol{A}^T\boldsymbol{A})\boldsymbol{x} = \boldsymbol{A}^T\boldsymbol{b} \tag{6.34}$$

This system is obtained by multiplying both sides of (6.27) by $\boldsymbol{A}^T$. The matrix $(\boldsymbol{A}^T\boldsymbol{A})$ is symmetric and positive semi-definite. The normal system of equations can

```
%M6.4
%Fletcher Reeves method for solving a system of linear equations
NumberOfParameters=4; %This is n for this problem
A =[-14 -3 -3  0;
  -3  10 0  -3;
  -3  0  10 -3;
  0  -3 -3  14];
b=[0 -5 5 0]';
OldPoint=[0  0  0  0]' %This is the starting point
OldValue=getObjective(OldPoint); %Get the objective function at the old point
OldResidue=getLinearSystemResidue(A,OldPoint,b); %get the residue
Gradient=-1.0*OldResidue; %get the gradient at the first point
Epsilon=1.0e-3; %this is the perturbation
NormGradientOld=norm(Gradient); %get the norm of the gradient at the current point
ConjugateDirection=-1.0*Gradient; %initialize the first direction
%now this is the main loop
while(NormGradientOld>Epsilon) %check norm of gradient
  %get line search value through formula
  LambdaOptimal =(ConjugateDirection'*OldResidue)/ConjugateDirection'*A*ConjugateDirection);
  NewPoint=OldPoint+LambdaOptimal*ConjugateDirection; %get new point
  NewValue=0.5*NewPoint'*A*NewPoint-b'*NewPoint; %evaluate new objective function value
  NewResidue=getLinearSystemResidue(A,NewPoint,b); %get new residue
  NewGradient=-1.0*NewResidue; %gradient is the negative of residue
  NormGradientNew=norm(NewGradient); %get norm of the new gradient
  %now we determine the new conjugate direction
  NewConjugateDirection=-1.0*NewGradient+((NormGradientNew*NormGradientNew)/
                                   NormGradientOld*NormGradientOld))*ConjugateDirection;
  %now we make the new point the current point
  OldPoint=NewPoint
  OldValue=NewValue
  Gradient=NewGradient;
  NormGradientOld=NormGradientNew;
  ConjugateDirection=NewConjugateDirection;
end
```

also be implemented using only matrix-vector products and vector inner products. This approach is illustrated by the following example.

Example 6.7: Consider the circuit shown in Figure 6.7. Utilize nodal analysis to determine the voltage at each node. Apply the Fletcher–Reeves method staring with zero voltages.

Solution: This circuit contains dependent voltage and current sources. Thus, its nodal matrix is not symmetric. There are four nodes to consider in this example as shown in Figure 6.7. Writing the nodal equations at the first node, we get:

$$\frac{V_1}{2} - 5V_b + \frac{(V_1 - V_2)}{5} + \frac{(V_4 - V_3)}{5} = 0 \tag{6.35}$$

Substituting with $V_b = V_1 - V_2$ and reorganizing, we get:

$$-4.3V_1 + 4.8V_2 - 0.2V_3 + 0.2V_4 = 0 \tag{6.36}$$

At node 2, the nodal equation gives:

$$-0.2V_1 + 0.65V_2 - 0.2V_3 + 0 = 5 \tag{6.37}$$

At node 3, we have the nodal equation:

$$0 - 0.2V_2 + 0.4V_3 - 0.2V_4 = 5 \tag{6.38}$$

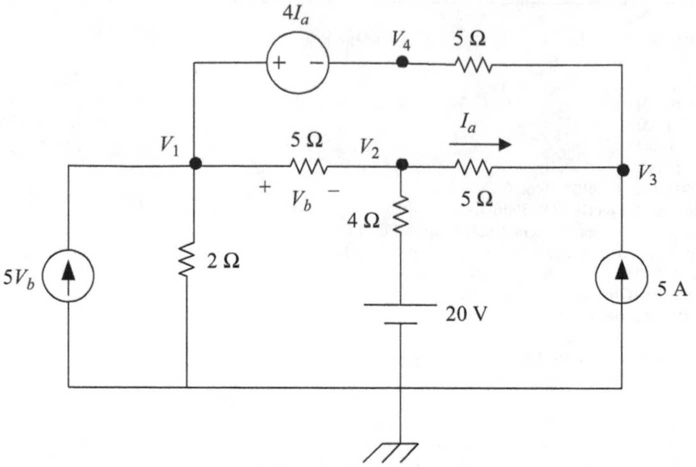

Figure 6.7 The circuit of Example 6.7; the resulting admittance matrix is not symmetric or positive definite because of the controlled sources. The normal equation should be utilized in the Fletcher–Reeves method

Finally, the nodal equation at node 4 is given by:

$$V_4 = V_1 - 4I_a = V_1 - 4\left(\frac{V_2 - V_3}{5}\right) = V_1 - 0.8V_2 + 0.8V_3 \tag{6.39}$$

The system of nodal equations, in matrix form, is thus given by:

$$\begin{bmatrix} -4.30 & 4.80 & -0.20 & 0.2 \\ -0.20 & 0.65 & -0.20 & 0 \\ 0 & -0.20 & 0.40 & -0.20 \\ 1.0 & -0.80 & 0.80 & -1.0 \end{bmatrix} \begin{bmatrix} V_1 \\ V_2 \\ V_3 \\ V_4 \end{bmatrix} = \begin{bmatrix} 0 \\ 5 \\ 5 \\ 0 \end{bmatrix} \tag{6.40}$$

The matrix in (6.40) is neither symmetric nor positive semi-definite. Applying the Fletcher–Reeves approach to the corresponding normal system, we get the following output from the MATLAB code:

Starting Point

OldPoint = [0   0   0   0]′
Gradient = [1.0000   −2.2500   −1.0000   1.0000]′

Second Point

OldPoint = [−0.0368   0.0827   0.0368   −0.0368]′
Gradient = [−1.3719   0.4080   −1.1461   1.1438]′

Third Point

OldPoint = [0.2032   0.4668   0.7768   −0.7758]′
Gradient = [−0.3833   −0.3234   0.0679   −0.2763]′

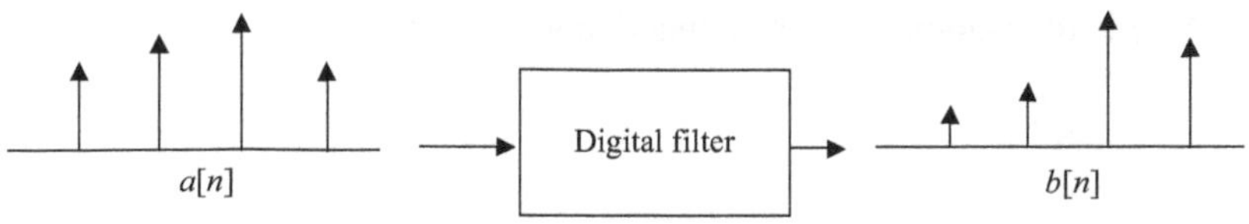

Figure 6.8 An illustration of digital filters; both the input and output are available at multiples of the sampling time step

Fourth Point

OldPoint = [4.9697 4.5749 1.3592 0.8552]′
Gradient = [0.0978 0.0822 −0.4228 −0.3357]′

Final Point

OldPoint = [49.3627 44.2157 69.3382 69.4608]′
Gradient = 1.0e − 008 *[0.0985 −0.1094 0.0086 −0.0091]′

The algorithm converges in exactly four iterations to the optimal point. It can be easily checked that the optimal solution $[V_1 \quad V_2 \quad V_3 \quad V_4]^T = [49.3627 \quad 44.2157 \quad 69.3382 \quad 69.4608]^T$ volts indeed satisfies both loop and nodal equations in Figure 6.7.

A6.2 The design of digital FIR filters

In previous chapters, optimization algorithms were applied to a number of analog circuits. In these circuits, the input and the output are both continuous functions of time. Another type of filters, called digital filters, receive the input at discrete instants of time and create the output at discrete instants of time. Figure 6.8 illustrates such a filter. The input sequence $a[n\Delta t]$, $n = 0, 1, \ldots, N_T$ is available only at discrete multiples of the time step Δt. Similarly, the output from the filter is $b[n\Delta t]$, $\forall n$. The time step is usually dropped from the notations and the sequence of input and output samples are referred to $a[n]$ and $b[n]$.

In finite impulse response (FIR) filters, the output is a function of the current input sample and a finite number of past input samples. Mathematically, this implies that:

$$b[k] = \sum_{n=0}^{M} h_n a[k-n] \tag{6.41}$$

The output from the filter (6.41) at the kth time step $b[k]$ depends on the input at the same time sample $a[k]$ and the input at previous M input values $a[k-1]$, $a[k-2], \ldots,$ and $a[k-M]$. The response of the filter is characterized by its coefficients $\boldsymbol{h} = [h_0 \quad h_1 \quad \ldots \quad h_M]^T$.

The z transform is used to convert the filter expression to the z domain [5]. The filter z domain expression is given by:

$$b[z] = H(z)a[z] \tag{6.42}$$

where *H*(*z*) is the *z*-domain transfer function and is given by:

$$H(z) = \sum_{n=0}^{M} h_n \; z^{-n} \tag{6.43}$$

The corresponding frequency domain response can be obtained by substituting $z = e^{-J\omega}$, $J = \sqrt{-1}$, with ω being the normalized frequency $0 \leq \omega \leq \pi$. The FIR filter response in the frequency domain is denoted by $H(\omega)$.

Occasionally, the desired filter response is given in the frequency domain at discrete set of frequencies $\{\omega_1, \omega_2, \ldots, \omega_N\}$. The desired response at these frequencies is given by $H_d(\omega_i)$, $i = 1, 2, \ldots, N$. If the filter order M is even, we have an odd number of coefficients. If we assume that the coefficients are symmetrical around the middle coefficient $h_{(M/2)}$, the filter transfer function can be written as:

$$H(\omega) = e^{-J\omega M/2} R(\omega) \tag{6.44}$$

where $R(\omega)$ is a real amplitude response function that is given by [6]:

$$R(\omega) = \sum_{n=0}^{M/2} c_n \cos(n\omega) \tag{6.45}$$

The coefficients are related to the original coefficients by $c_n = t h_n$, with $t = 1$ for $n = M/2$ and $t = 2$ for $n = 0, 1, \ldots, (M/2) - 1$.

The design problem is to determine the optimal filter coefficients $\boldsymbol{h} = [h_0 \;\; h_1 \;\; h_2 \ldots h_M]^T$ that would make the filter response $H(\boldsymbol{\omega})$ match the given response H_d at the given frequencies.

The optimal filter parameters that would generate the desired response may be obtained by minimizing the objective function:

$$f(\boldsymbol{h}) = \sum_{i=1}^{N} |H(\boldsymbol{h}, \omega_i) - H_d(\omega_i)|^p \tag{6.46}$$

where p is a positive number. Matching two complex numbers implies matching the real and the imaginary parts. Alternatively, the objective function may be formulated in terms of the amplitude response function (6.45). It follows that unconstrained optimization techniques with all real quantities can be used to solve this problem. The following example illustrates the design of FIR digital filters using optimization techniques.

Example 6.8: Design an FIR digital filter of order 20 that satisfies the following low-pass amplitude response:

$$R(\omega) = \begin{cases} 1, & \text{for } 0 \leq \omega \leq 0.40\pi \\ 0, & \text{for } 0.50\pi \leq \omega \leq \pi \end{cases} \tag{6.47}$$

Assume symmetrical filter coefficients.

Solution: This problem has 21 coefficients. Assuming symmetry, the unknowns are the coefficients $c_0, c_1, \ldots, c_{11}$ as given by (6.45). We utilize the steepest descent method in solving this problem. The MATLAB listing M6.1 was used after changing the initialization section to the following:

```
numberOfParameters=11; %This is (n/2)+1 for this problem
OldPoint=0.05*ones(numberOfParameters,1); %This is the starting point
```

The MATLAB listing M6.5 illustrates the objective function calculations used in solving:

```
%M6.5
%This function evaluates the objective function
function Objective=getObjective(c)
objectiveExponent=2; %exponent of objective function
load Data.mat;%load the frequencies and the target responses at these frequencies (Omegas, Targets)
Objective=0; %initialize the objective function value
Responses=getFIRMultipleFrequencyResponse(c, Omegas); %get the vector of responses
Errors=(Responses-Targets); %get vector of errors;
normErrors=norm(Errors); % this is the L2 error norm
Objective=normErrors^objectiveExponent; %get objective function value


%This function evaluates the response of an FIR filter assuming even order and
%linear phase at a number of frequencies by calling the single frequency response function
%inputs are the coefficients of the real amplitude function
function Responses=getFIRMultipleFrequencyResponse(c, Omegas)
numberOfFrequencies=size(Omegas,1); %this is number of frequencies
Responses=zeros(numberOfFrequencies,1); %vector of frequency responses
%repeat for all frequencies
for i=1:numberOfFrequencies
   Responses(i)=getFIRResponse(c,Omegas(i)); %get response at the ith frequency
end


%This function returns the response of a FIR filter for given coefficients
%and frequency assuming even order and linear phase assumptions
 function Response=getFIRResponse(c,Omega) %input is coefficient and normalized %frequency
numberOfCoefficients=size(c,1); %get number of FIR filter coefficients (M/2)+1
CosineArguments=Omega*linspace(0,(numberOfCoefficients-1), numberOfCoefficients)'; %n*omega
CosineTerms=cos(CosineArguments); % get cos(n*Omega), n=0,1,2,....,(M/2)+1
Response=c'*CosineTerms; %multiply each cosine term with its coefficient  and do summation
```

The coefficients are given the initial values $x^{(0)} = [0.05 \;\; 0.05 \;\; 0.05 \;\; 0.05 \;\; 0.05 \;\; 0.05 \;\; 0.05 \;\; 0.05 \;\; 0.05 \;\; 0.05 \;\; 0.05]^T$. The response of the filter at the starting design is shown in Figure 6.9. The objective function is the square of the norm of an error vector. This vector evaluates the difference between

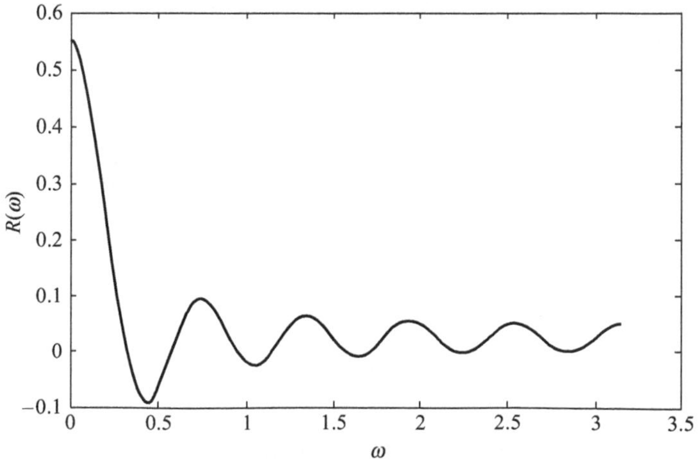

Figure 6.9 The initial response of the FIR digital filter of Example 6.8

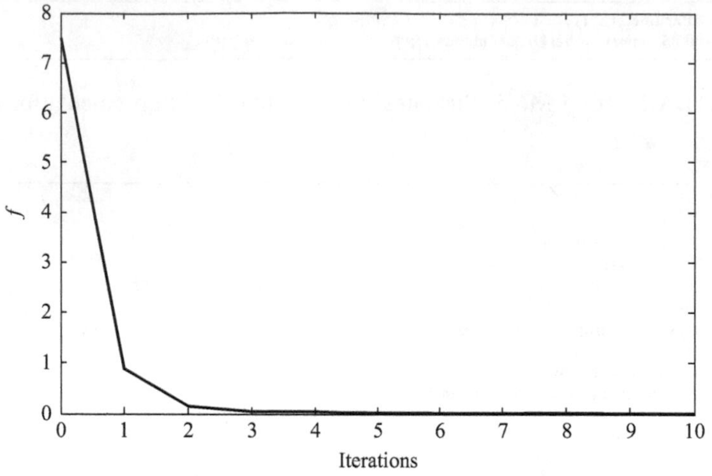

Figure 6.10 The objective function at every iteration for Example 6.8

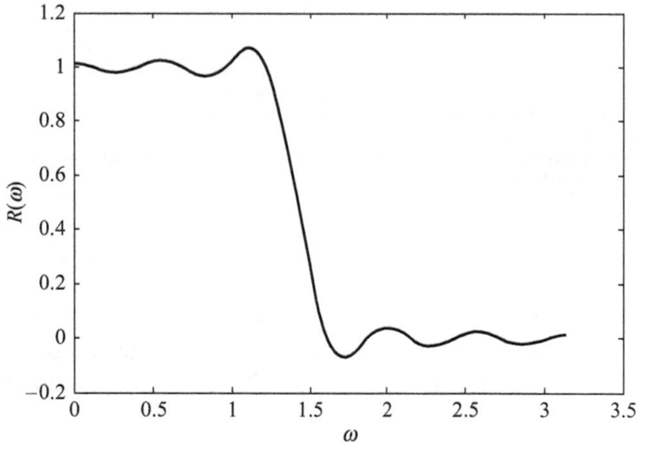

Figure 6.11 The response at the final design of the FIR digital filter of Example 6.8

the filter response and the target response at 20 frequencies. Nine of these frequencies are in the passband $\{0, 0.05\pi, 0.10\pi, 0.15\pi, 0.20\pi, 0.25\pi, 0.30\pi, 0.35\pi, 0.40\pi\}$ and the rest are in the stopband $\{0.5\pi, 0.55\pi, 0.6\pi, 0.65\pi, 0.70\pi, 0.75\pi, 0.80\pi, 0.85\pi, 0.90\pi, 0.95\pi, \pi\}$. We allowed for only ten iterations of the steepest descent algorithm. The change in the objective function is shown in Figure 6.10. The final design obtained by the algorithm is given by $\boldsymbol{x}^{(10)} = [0.4493 \quad 0.6275 \quad 0.0961 \quad -0.1856 \quad -0.0918 \quad 0.0854 \quad 0.0780 \quad -0.0371 \quad -0.0668 \quad 0.0092 \quad 0.0486]^T$. The response of the filter at the final design is shown in Figure 6.11. The filter approximates well the design specifications.

References

1. Singiresu S. Rao, *Engineering Optimization Theory and Practice*, Third Edition, John Wiley & Sons Inc, New York, 1996
2. John W. Bandler and R.M. Biernacki, *ECE3KB3 Courseware on Optimization Theory*, McMaster University, Canada, 1997
3. Jasbir S. Arora, *Introduction to Optimum Design*, Third Edition, Elsevier Inc., MA, USA, 2012
4. Gene H. Golub, Van Loan, and F. Charles, *Matrix Computations*, Johns Hopkins University Press, Baltimore, 1996
5. B.P. Lathi, *Signal Processing and Linear Systems*, Oxford University Press, New York, 1998
6. Andreas Antoniou and Wu-Sheng Lu, *Practical Optimization Algorithms and Engineering Applications*, Springer, New York, 2007
7. John O. Attia, *Electronics and Circuit Analysis Using Matlab*, CRC Press, Florida, 1999
8. Richard C. Jaeger, *Microelectronic Circuit Design*, WCB/McGraw-Hill, New York, 1997

Problems

6.1 Consider the objective functions $f_1(\boldsymbol{x}) = x_1^2 + x_2^2$, $f_2(\boldsymbol{x}) = x_1^2 + 10x_2^2$, and $f_3(\boldsymbol{x}) = x_1^2 + 100x_2^2$.
   (a) Apply necessary optimality conditions to obtain the minima of these functions.
   (b) Utilize the MATLAB listing M6.1 to apply the steepest descent method to these functions.

   Comment on the convergence of the algorithm in each case.

6.2 Solve the unconstrained optimization problem:

$$\text{minimize} \quad f(\boldsymbol{x}) = 2x_1^2 + 2x_2^2 + 2x_1x_2 + x_2 + 1$$

   starting from the point $\boldsymbol{x}^{(0)} = [0.1 \quad 0.1]^T$. Use the steepest descent code M6.1. Verify your answer by writing a MATLAB code that plots the contours of this two-dimensional function.

6.3 Repeat Problem 6.2 for the objective function $f(\boldsymbol{x}) = x_1{}^2 + 2x_2{}^2 + x_1x_2 + 2x_2 + 2$.

6.4 Write the objective functions in (6.2) and (6.3) in the standard quadratic form $f(\boldsymbol{x}) = \frac{1}{2}\boldsymbol{x}^T\boldsymbol{H}\boldsymbol{x} + \boldsymbol{b}^T\boldsymbol{x} + c$. Evaluate the analytical minima of these functions given by $\boldsymbol{x}^* = \boldsymbol{H}^{-1}\boldsymbol{b}$. Compare these analytical minima to the numerical solutions.

6.5 Minimize the function $f(\boldsymbol{x}) = x_1{}^2 e^{1-x_2{}^2-10(x_1-x_2)^2}$ using the conjugate directions listing M6.2. Use the starting point $\boldsymbol{x}^{(0)} = [0.1 \quad 0.1]^T$. How does the solution change by changing the starting point and why? Verify your answer by plotting the contours of this function.

6.6 Obtain a local minimum of the function:

$$f(\boldsymbol{x}) = (x_1 + 5x_2)^4 + 2(x_3 - x_4)^2 + (x_2 - 2x_3)^4 + 20(x_1 - x_4)^2$$

starting from the point $x^{(0)} = [3.0 \quad 2.0 \quad -1.0 \quad 2.0]^T$. Utilize the Fletcher–Reeves MATLAB listing M6.3. Verify your answer by considering the fact that this objective function is a sum of square terms.

6.7 Consider the Op-Amp circuit shown in Figure 6.12(a).

(a) Assuming ideal Op-Amps, obtain the transfer function $v_o = f(v_a, v_b)$. Evaluate the output for $v_a = 4.3$ V and $v_b = 5.5$ V.

(b) Utilize the more realistic Op-Amp model shown in Figure 6.12(b) to write the nodal system of equations $Av = b$, where $v = [v_1 \quad v_2 \quad v_3 \quad v_4 \quad v_5 \quad v_6 \quad v_0]^T$. Use the same values of the inputs as in (a).

(c) Utilize the MATLAB listing M6.4 to solve the corresponding normal equation $A^T A v = A^T b$ starting from the initial point $v^{(0)} = [1.0 \quad 1.0 \quad 1.0 \quad 1.0 \quad 1.0 \quad 1.0 \quad 1.0]^T$. Make any necessary changes to the code. Compare the answers in (a) and (c).

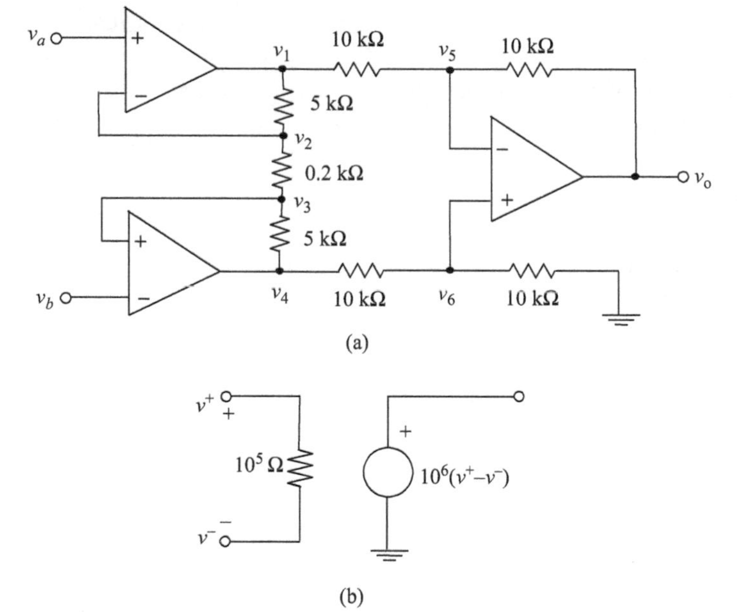

Figure 6.12 The circuit of Problem 6.7; (a) the Op-Amp circuit and (b) the equivalent circuit of the Op-Amp to be used in nodal analysis

6.8 Design an FIR digital filter of order 20 that satisfies the following bandpass amplitude response:

$$R(\omega) = \begin{cases} 0, & 0 \le \omega \le 0.25\pi \\ 1, & 0.35\pi \le \omega \le 0.65\pi \\ 0, & 0.75\pi \le \omega \le \pi \end{cases}$$

Utilize the MATLAB listings M6.3 and M6.4 after making the necessary changes. Utilize 10 frequency values in each band.

Chapter 7

Second-order unconstrained optimization techniques

7.1 Introduction

Taylor expansion shows us that the more derivatives we know about a function at a point, the more we can accurately predict its value over a larger neighborhood of the expansion point. Actually, if we know all the higher-order derivatives of a function at the expansion point, then we know how the function behaves over the whole space! This is a remarkable result that motivates using higher-order sensitivities. The main obstacle in estimating these sensitivities is their evaluation cost. First-order derivatives estimated using finite differences require $O(n)$ extra simulations as discussed in Chapter 1. Second-order sensitivities are estimated using $O(n^2)$ extra simulations. The cost of estimating third- or higher-order sensitivities is even higher. This is the main reason why most techniques utilize only first-order sensitivities.

Second-order optimization techniques utilize second-order sensitivities or an approximation of these derivatives to predict the steps taken within the optimization algorithm. They all enjoy a faster rate of convergence than that achieved using first-order sensitivities especially near the optimal solution. Their fast convergence rate near the optimal point comes at the high cost of estimating the second-order sensitivities.

Because second-order derivatives are the derivatives of first-order derivatives, they can be approximated if first-order derivatives are known at different space points. A whole class of techniques was developed that uses approximate formulas to predict the second-order derivatives using different values of the first-order ones. This approximation reduces the convergence rate but makes the cost of these techniques more acceptable.

We start this chapter by reviewing Newton's method, which is the main concept behind all these techniques. We then move to discuss techniques for improving this technique either by using a combination of first- and second-order derivatives or by using approximate second-order derivatives.

7.2 Newton's method

Newton's method exploits an expansion of the gradient of the objective function. At any point $\boldsymbol{x} \in \Re^n$ in the neighborhood of the point $\boldsymbol{x}^{(k)}$, the gradient can be expanded as:

$$\nabla f(\boldsymbol{x}) = \nabla f(\boldsymbol{x}^{(k)}) + \boldsymbol{H}(\boldsymbol{x}^{(k)})(\boldsymbol{x} - \boldsymbol{x}^{(k)}) \tag{7.1}$$

where $\boldsymbol{H}(\boldsymbol{x}^{(k)})$ is the Hessian at the expansion point $\boldsymbol{x}^{(k)}$. Our target is to find a minimizer of the function $f(\boldsymbol{x})$. Using necessary optimality conditions, we know that at the optimal point we should have $\nabla f(\boldsymbol{x}^*) = \boldsymbol{0}$. It follows that to determine the optimal point, we set the right-hand side of (7.1) to zero to get:

$$\nabla f(\boldsymbol{x}^{(k)}) + \boldsymbol{H}(\boldsymbol{x}^{(k)})(\boldsymbol{x}^* - \boldsymbol{x}^{(k)}) = \boldsymbol{0} \Rightarrow \boldsymbol{H}(\boldsymbol{x}^{(k)})(\boldsymbol{x}^* - \boldsymbol{x}^{(k)}) = -\nabla f(\boldsymbol{x}^{(k)}) \tag{7.2}$$

Equation (7.2) is a system of linear equations whose solution is the extreme point $\boldsymbol{x}^*$ at which the gradient vanishes. This extreme point may be a local minimum, a local maximum, or a saddle point. If the function is quadratic, the expansion (7.1) is exact. It follows that Newton's method converges in only one iteration for quadratic functions. However, if the function is not quadratic, which is the general case, the expansion (7.1) is not exact and the solution of (7.2) is a better approximation of the solution. This point is denoted by $\boldsymbol{x}^{(k+1)}$. The step (7.2) is then repeated at the new point $\boldsymbol{x}^{(k+1)}$ until a suitable termination condition is satisfied.

In addition to the cost of estimating the Hessian in (7.2), which can be formidable for problems with extensive simulation time, the convergence of Newton's method is also of concern. For a general non-quadratic function, if the starting point is not close enough to the solution, the algorithm may start to diverge. In other words, the norm $\|\boldsymbol{x}^{(k)} - \boldsymbol{x}^*\|$ grows with every iteration rather than decreases. Starting close to the solution guarantees that the Hessian matrix is positive semi-definite and thus the vector $(\boldsymbol{x}^* - \boldsymbol{x}^{(k)})$ has an acute angle with the steepest descent direction $(-\nabla f(\boldsymbol{x}^{(k)}))$ as explained in Chapter 1.

Newton's method may converge to a local maximum or a saddle point rather than a local minimum. If the Hessian matrix is negative semi-definite in (7.2), as is the case near a local maximum, thus the vector $(\boldsymbol{x}^* - \boldsymbol{x}^{(k)})$ will have an obtuse angle with the steepest descent direction $(-\nabla f(\boldsymbol{x}^{(k)}))$. This results in a new solution with even higher value of the objective function. The algorithm may converge as well to a saddle point which is neither a maximum nor a minimum.

Newton's method can be summarized in the following steps:

Initialization: Given the start point $\boldsymbol{x}^{(0)}$. Set $k = 0$.

Step 1: Evaluate $\nabla f(\boldsymbol{x}^{(k)})$ and $\boldsymbol{H}(\boldsymbol{x}^{(k)})$.
Step 3: Solve (7.2) for the new point $\boldsymbol{x}^{(k+1)}$.
Step 4: If the termination condition is satisfied, stop.
Step 5: Set $k = k + 1$. Go to Step 1.

The termination condition in Step 4 of the algorithm may include a condition on the norm of the step taken, e.g., $\|\boldsymbol{x}^{(k+1)} - \boldsymbol{x}^{(k)}\| \leq \varepsilon$, or alternatively, a condition may be imposed on the norm of the gradient at the new point.

The following example illustrates Newton's method.

Example 7.1: Use Newton's method to minimize the function:

$$f(\boldsymbol{x}) = (x_1 + 10x_2)^2 + 5(x_3 - x_4)^2 + (x_2 - 2x_3)^4 + 10(x_1 - x_4)^4 \tag{7.3}$$

starting with the initial point $\boldsymbol{x}^{(0)} = [3.0 \quad -1.0 \quad 0 \quad 1.0]^T$. Carry out only two iterations.

Solution: To apply Newton's method, we need first to evaluate both the gradient and the Hessian at the current point. At a general point $\boldsymbol{x} = [x_1 \quad x_2 \quad x_3 \quad x_4]^T$, the gradient and the Hessian are given by:

$$\nabla f(\boldsymbol{x}) = \begin{bmatrix} \dfrac{\partial f}{\partial x_1} \\ \dfrac{\partial f}{\partial x_2} \\ \dfrac{\partial f}{\partial x_3} \\ \dfrac{\partial f}{\partial x_4} \end{bmatrix} = \begin{bmatrix} 2(x_1 + 10x_2) + 40(x_1 - x_4)^3 \\ 20(x_1 + 10x_2) + 4(x_2 - 2x_3)^3 \\ 10(x_3 - x_4) - 8(x_2 - 2x_3)^3 \\ -10(x_3 - x_4) - 40(x_1 - x_4)^3 \end{bmatrix} \tag{7.4}$$

and

$$\boldsymbol{H}(\boldsymbol{x}) = \begin{bmatrix} 2 + 120(x_1 - x_4)^2 & 20 & 0 & -120(x_1 - x_4)^2 \\ 20 & 200 + 12(x_2 - 2x_3)^2 & -24(x_2 - 2x_3)^2 & 0 \\ 0 & -24(x_2 - 2x_3)^2 & 10 + 48(x_2 - 2x_3)^2 & -10 \\ -120(x_1 - x_4)^2 & 0 & -10 & 10 + 120(x_1 - x_4)^2 \end{bmatrix} \tag{7.5}$$

where every row of the Hessian in (7.5) is obtained by differentiating each component of the gradient in (7.4) relative to all four parameters.

At the starting point $\boldsymbol{x}^{(0)}$, we thus have:

$$f(\boldsymbol{x}^{(0)}) = 215, \quad \nabla f(\boldsymbol{x}^{(0)}) = [306 \quad -144 \quad -2 \quad -310]^T$$

$$\boldsymbol{H}(\boldsymbol{x}^{(0)}) = \begin{bmatrix} 482 & 20 & 0 & -480 \\ 20 & 212 & -24 & 0 \\ 0 & -24 & 58 & -10 \\ -480 & 0 & -10 & 490 \end{bmatrix} \tag{7.6}$$

By inverting the Hessian matrix in (7.6) and solving for the new step $(\boldsymbol{x}^{(1)} - \boldsymbol{x}^{(0)})$, we get:

$$\boldsymbol{x}^{(1)} - \boldsymbol{x}^{(0)} = -\left(\boldsymbol{H}(\boldsymbol{x}^{(0)})\right)^{-1} \nabla f(\boldsymbol{x}^{(0)}) = \begin{bmatrix} -1.4127 \\ 0.8413 \\ 0.2540 \\ 0.7460 \end{bmatrix} \tag{7.7}$$

This results in the new point $\boldsymbol{x}^{(1)} = [1.5873 \quad -0.1587 \quad 0.2540 \quad 0.2540]^T$. The function value, the gradient, and the Hessian are evaluated at this new point to get:

$$f(\boldsymbol{x}^{(1)}) = 31.8, \quad \nabla f(\boldsymbol{x}^{(1)}) = [94.81 \quad -1.179 \quad 2.371 \quad -94.81]^T$$

$$\boldsymbol{H}(\boldsymbol{x}^{(1)}) = \begin{bmatrix} 215.3 & 20 & 0 & -213.3 \\ 20 & 205.3 & -10.67 & 0 \\ 0 & -10.67 & 31.34 & -10 \\ -213.3 & 0 & -10 & 223.3 \end{bmatrix} \tag{7.8}$$

By inverting the Hessian matrix in (7.8) and solving for $(\boldsymbol{x}^{(2)} - \boldsymbol{x}^{(1)})$ using (7.2), we get:

$$\boldsymbol{x}^{(2)} - \boldsymbol{x}^{(1)} = -\left(\boldsymbol{H}(\boldsymbol{x}^{(1)})\right)^{-1} \nabla f(\boldsymbol{x}^{(1)}) = \begin{bmatrix} 0.5291 \\ -0.0529 \\ 0.0846 \\ 0.0846 \end{bmatrix} \tag{7.9}$$

which results in the new point $\boldsymbol{x}^{(2)} = [\,1.0582 \quad -0.1058 \quad 0.1694 \quad 0.1694\,]^T$. The objective function value at this new point is $f(\boldsymbol{x}^{(2)}) = 6.28$. Newton's method is obviously converging for this non-quadratic function.

7.3 The Levenberg–Marquardt method

To address the shortcomings of the original Newton method, several variations of the technique were suggested to guarantee convergence to a local minimum. One of the most important variations is the Levenberg–Marquardt method. This method effectively uses a step that is a combination between the Newton method given by (7.2) and the steepest descent step, which can be obtained from (7.2) with a Hessian matrix equal to unity. The step taken by this method is given by:

$$\boldsymbol{x}^{(k+1)} - \boldsymbol{x}^{(k)} = -\left(\boldsymbol{H}(\boldsymbol{x}^{(k)}) + \mu \boldsymbol{I}\right)^{-1} \nabla f(\boldsymbol{x}^{(k)}) \tag{7.10}$$

where μ is a positive scalar and $\boldsymbol{I} \in \Re^{n \times n}$ is the identity matrix. Notice that in (7.10) if μ is small enough, the Hessian matrix $\boldsymbol{H}(\boldsymbol{x}^{(k)})$ dominates and the step (7.10) becomes effectively a Newton's step. If the parameter μ is large enough, the matrix $\mu\boldsymbol{I}$ dominates and the step in (7.10) is approximately in the steepest descent direction. By increasing μ, the inverse matrix becomes small in norm and subsequently the norm of the step taken $\|\boldsymbol{x}^{(k+1)} - \boldsymbol{x}^{(k)}\|$ becomes smaller. It follows that the parameter μ controls also the step size.

One interesting mathematical property of this approach is that adding the matrix $\mu\boldsymbol{I}$ to the Hessian matrix increases each eigenvalue of this matrix by μ. If the matrix $\boldsymbol{H}(\boldsymbol{x}^{(k)})$ is not positive semi-definite then adding μ to each eigenvalue makes them more positive. The value of μ can be increased until all the eigenvalues are positive thus guaranteeing that the step (7.10) is a descent step.

The Levenberg–Marquardt approach starts each iteration with a very small value of μ, thus giving effectively the Newton's step. If an improvement in the objective function is achieved, the new point is accepted. Otherwise, the value of μ is increased until a reduction in the objective function is obtained.

The MATLAB® listing M7.1 shows a possible implementation of the Levenberg–Marquardt approach.

Example 7.2: Find the minimum of the function:

$$f(\boldsymbol{x}) = 1.5x_1^2 + 2x_2^2 + 1.5x_3^2 + x_1x_3 + 2x_2x_3 - 3x_1 - x_3 \tag{7.11}$$

starting from the point $\boldsymbol{x}^{(0)} = [\,3.0 \quad -7.0 \quad 0\,]^T$. Utilize the MATLAB listing M7.1.

```
%M7.1
%The Levenberg Marquardt Method
NumberOfParameters=3; %This is n for this problem
Epsilon=1.0e-3; %perturbation used in finite differences
OldPoint=[3 -7 0]' %This is the starting point
OldValue=getObjective(OldPoint); %initial function value
OldGradient=getGradient('getObjective', OldPoint, Epsilon); %initial gradient
OldGradientNorm=norm(OldGradient); %get the old gradient norm
OldHessian=getHessian('getObjective', OldPoint, Epsilon); %initial Hessian
Identity=eye(NumberOfParameters); %This is the identity matrix
while (OldGradientNorm>1.0e-5) %repeat until gradient is small enoug
  TrustRegionParameter=0.001; %initialize trust region parameter
  DescentFlag=0; %flag signaling if the new step is a descent step
  while(DescentFlag==0) %repeat until descent direction found
    MarquardtMatrix=OldHessian+TrustRegionParameter*Identity;
    NewStep=-1.0*inv(MarquardtMatrix)*OldGradient;
    NewPoint=OldPoint+NewStep %get the new trial point
    NewValue=feval('getObjective',NewPoint); %calculate new value
    if(NewValue<OldValue) %a descent step?
      DescentFlag=1.0; % set success flag
    else
      TrustRegionParameter=TrustRegionParameter*4; %Increase Mu
    end
  end
  Step=NewPoint-OldPoint; %get the new step
  StepNorm=norm(Step); %get the step norm
  NewGradient=getGradient('getObjective', NewPoint, Epsilon); %get new gradient
  NewHessian=getHessian('getObjective', NewPoint, Epsilon); %get new Hessian
  %now we swap parameters
  OldPoint=NewPoint;
  OldGradient=NewGradient;
  OldHessian=NewHessian;
  OldGradientNorm=norm(OldGradient);
end
```

Solution: Using this MATLAB listing, the following outputs are generated:

$$\begin{aligned}
&\text{OldPoint} = [3.0 \quad -7.0 \quad 0]' \quad \text{OldValue} = 102.5000\\
&\text{OldGradient} = [6.0015 \quad -27.9980 \quad -11.9985]'\\
&\text{OldHessian} = \begin{bmatrix} 3.0000 & 0 & 1.0000\\ 0 & 4.0000 & 2.0000\\ 1.0000 & 2.0000 & 3.0000 \end{bmatrix}\\
&\text{NewStep} = [-2.0004 \quad 6.9969 \quad 0.0017]'\\
&\text{NewPoint} = [0.9996 \quad -0.0031 \quad 0.0017]'\\
&\text{NewValue} = -1.5000
\end{aligned}$$

The algorithm terminated in only one iteration. The exact solution for this problem is $\boldsymbol{x}^* = [\,1.0 \quad 0 \quad 0\,]^T$ with a minimum value of $f(\boldsymbol{x}^*) = -1.50$.

7.4 Quasi-Newton methods

The original Newton's method uses the analytical Hessian in estimating the new step in each iteration as given by (7.2). An implementation for a general function estimates the Hessian through a finite difference approximation. This requires, as shown earlier, $O(n^2)$ extra simulations. This cost can be prohibitive for simulations with extensive simulation time or when n is large.

A number of techniques were developed that utilize an approximation $\boldsymbol{B}$ of the Hessian matrix. Formulas were developed to update the Hessian, its inverse, or both. These formulas differ in the rank of the update matrix and whether they maintain the symmetry and positive definiteness for the updated Hessian.

The basic concept of quasi-Newton methods is illustrated in Figure 7.1. At the kth iteration, we have the current point $\boldsymbol{x}^{(k)}$. We also assume that the gradient $\boldsymbol{g}^{(k)}$ and an approximation to the Hessian $\boldsymbol{B}^{(k)}$ are available at this point. Using Newton's method with the Hessian $\boldsymbol{H}^{(k)}$ replaced by $\boldsymbol{B}^{(k)}$, a better solution $\boldsymbol{x}^{(k+1)}$ is predicted. The gradient at this new point $\boldsymbol{g}^{(k+1)}$ is assumed to be available. Using the change in the gradient $\Delta\boldsymbol{g}^{(k)} = \boldsymbol{g}^{(k+1)} - \boldsymbol{g}^{(k)}$ and the change in position vector $\Delta\boldsymbol{x}^{(k)} = \boldsymbol{x}^{(k+1)} - \boldsymbol{x}^{(k)}$, we estimate a better approximation of the Hessian matrix $\boldsymbol{B}^{(k+1)}$ at the new point $\boldsymbol{x}^{(k+1)}$ through an update formula. This update formula differs from one quasi-Newton technique to another. These techniques usually start with the identity matrix as the initial Hessian approximation. The identity matrix is both symmetric and positive definite as desired in Hessian matrices. Quasi-Newton algorithms were shown to have a superlinear rate of convergence. This can be contrasted with the quadratic rate of convergence of Newton's method.

7.4.1 Broyden's rank-1 update

The simplest quasi-Newton approach utilizes Broyden's rank-1 formula given by:

$$\boldsymbol{A}^{(k+1)} = \boldsymbol{A}^{(k)} + \frac{\left(\Delta\boldsymbol{x}^{(k)} - \boldsymbol{A}^{(k)}\Delta\boldsymbol{g}^{(k)}\right)\left(\Delta\boldsymbol{x}^{(k)} - \boldsymbol{A}^{(k)}\Delta\boldsymbol{g}^{(k)}\right)^T}{\left(\Delta\boldsymbol{x}^{(k)} - \boldsymbol{A}^{(k)}\Delta\boldsymbol{g}^{(k)}\right)^T\Delta\boldsymbol{g}^{(k)}} = \boldsymbol{A}^{(k)} + \Delta\boldsymbol{A}^{(k)} \tag{7.12}$$

with the matrix $\boldsymbol{A}^{(k)}$ approximating the inverse of the Hessian matrix $(\boldsymbol{H}^{(k)})^{-1}$. Two things to notice about the formula (7.12); first, the update matrix $\Delta\boldsymbol{A}^{(k)}$ is a rank-1 update because it consists of the product of the vector $(\Delta\boldsymbol{x}^{(k)} - \boldsymbol{A}^{(k)}\,\Delta\boldsymbol{g}^{(k)})$ with its transpose. The columns of this update matrix are all multiples of this vector; thus the rank of the resulting matrix is 1. Second, this formula satisfies the quasi-Newton condition $\Delta\boldsymbol{x}^{(k)} = \boldsymbol{A}^{(k+1)}\,\Delta\boldsymbol{g}^{(k)}$, which relates the change in the gradient to the change in the position. All quasi-Newton methods satisfy this formula. The

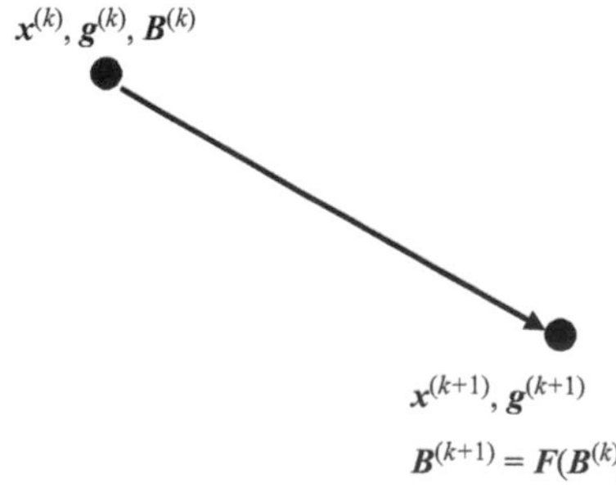

Figure 7.1 An illustration of quasi-Newton techniques; the new approximation to the Hessian matrix $\boldsymbol{B}^{(k+1)}$ *is approximated using the taken step, the difference in gradient, and the previous approximation of the Hessian*

Broyden's rank-1 update formula does not maintain symmetry or positive definiteness of the Hessian matrix. So, starting with a symmetric positive definite matrix, the updated matrix may not have these desired features.

A general algorithm for quasi-Netwon methods may thus be given by the following steps:

Initialization: Set $k=0$. Given $\boldsymbol{x}^{(0)}$, $\boldsymbol{g}^{(0)}$, and the initial approximation to the inverse of the Hessian matrix $\boldsymbol{A}^{(0)}=\boldsymbol{I}$.

Step 1: Get the Newton-like search direction using $\boldsymbol{s}^{(k)} = -\boldsymbol{A}^{(k)}\boldsymbol{g}^{(k)}$.

Comment: Because we are using an approximation of the Hessian rather than the Hessian itself, we cannot blindly trust this step. It is simply taken as a search direction.

Step 2: Carry out a line search in the direction $\boldsymbol{s}^{(k)}$ to get the new point $\boldsymbol{x}^{(k+1)}=\boldsymbol{x}^{(k)}+\lambda^{*}\boldsymbol{s}^{(k)}$.

Step 3: If termination condition is satisfied, stop.

Step 4: Calculate the new gradient $\boldsymbol{g}^{(k+1)}=\nabla f(\boldsymbol{x}^{(k+1)})$. Evaluate $\Delta\boldsymbol{x}^{(k)}=\boldsymbol{x}^{(k+1)}-\boldsymbol{x}^{(k)}$ and $\Delta\boldsymbol{g}^{(k)}=\boldsymbol{g}^{(k+1)}-\boldsymbol{g}^{(k)}$.

Step 5: Utilize the update formula (formula (7.12) for rank-1 Broyden update) to estimate $\boldsymbol{A}^{(k+1)}$.

Step 6: Set $k=k+1$. Go to Step 1.

The MATLAB listing M7.2 shows a possible implementation of the Broyden's rank-1 update.

Example 7.3: Repeat Example 7.2 using the MATLAB listing M7.2.

Solution: Using the MATLAB listing M7.2, we obtain the following output from the program:

```
First Iterate

OldPoint = [3.0  -7.0  0]'     OldValue = 102.5000     OldGradient = [6.0015
-27.9980   -11.9985]'
AOld = [1   0   0
        0   1   0
        0   0   1]

Second Iterate

OldPoint = [1.8133  -1.4638  2.3725]'   OldValue = 7.2047   OldGradient =
[4.8139  -1.1080  5.0049]'
AOld = [1.0000   0.0000   0.0000
        0.0000   0.4459  -0.3796
        0.0000  -0.3796   0.7399]

Third Iterate

OldPoint = [0.2566  -0.6897  1.0391]'   OldValue = -0.3060  OldGradient =
[-1.1895  -0.6785  0.9961]'
AOld = [0.4235   0.1218  -0.2328
        0.1218   0.4202  -0.3305
       -0.2328  -0.3305   0.6459]
```

```
%M7.2
%This program implements Broyeden's rank 1 formula
NumberOfParameters=3; %This is n for this problem
Epsilon=1.0e-3; %perturbation used in finite differences
Tolerance=1.0e-4; %termination condition for line search
OldPoint=[3 -7 0]' %This is the starting point
OldValue=getObjective(OldPoint) %Get the objective function at the old point
OldGradient=getGradient('getObjective', OldPoint, Epsilon) %get the gradient at the first point
OldGradientNorm=norm(OldGradient); %get the old gradient norm
LambdaMax=1.0; %limit the upper bound by the ideal Newton step
AOld=eye(NumberOfParameters); %initialize the approximation to the Hesian inverse by the identity
while(OldGradientNorm>0.5) %repeat under the gradient is small enoug
  QuasiNewtonDirection=-1.0*AOld*OldGradient; %get the current search direction
  LambdaOptimal =
    GoldenSection('getObjective',Tolerance,OldPoint,QuasiNewtonDirection,LambdaMax);
  NewPoint=OldPoint+LambdaOptimal*QuasiNewtonDirection; %get new point
  NewValue=getObjective(NewPoint); %get new objective function value
  NewGradient=getGradient('getObjective', NewPoint, Epsilon); %get the new gradient
  DeltaGradient=NewGradient-OldGradient; %this is the gradient difference
  DeltaPosition=NewPoint-OldPoint; %this is the position difference
  TempVector=DeltaPosition-AOld*DeltaGradient; %get the temporary vector in
                                               %Broyden's formula
  ANew=AOld+(TempVector*TempVector')/(TempVector'*DeltaGradient); %This is the
                                                                  %Broyden formula
  % make new values current values
  AOld=ANew;
  OldPoint=NewPoint;
  OldValue=NewValue;
  OldGradient=NewGradient;
  OldGradientNorm=norm(OldGradient); %get new norm
end
```

Fourth Iterate

OldPoint = [0.9998 −0.0002 −0.0004]′ OldValue = −1.5000 OldGradient = 1.0e−003 *[0.5181 0.4985 −0.1899]′

AOld = [0.4000 0.1000 −0.2000
0.1000 0.4000 −0.3000
−0.2000 −0.3000 0.6000]

The algorithm terminates in three iterations. The final point is $\boldsymbol{x}^{(3)} = [\,0.9998 \quad -0.0002 \quad -0.0004\,]^T$, which is very close to the optimal point $\boldsymbol{x}^* = [\,1.0 \quad 0 \quad 0\,]^T$. It can also be shown that the matrix $\boldsymbol{A}$ also converged to the inverse of the exact Hessian matrix at the solution.

7.4.2 The Davidon–Fletcher–Powell (DFP) formula

The DFP is another quasi-Newton method. It utilizes a rank-2 update formula. This update formula is the sum of two terms with each term including a product between a vector and its transpose. The resulting matrix update has thus only two independent columns. This formula is given by:

$$\boldsymbol{A}^{(k+1)} = \boldsymbol{A}^{(k)} + \frac{\Delta\boldsymbol{x}^{(k)}\Delta\boldsymbol{x}^{(k)T}}{\Delta\boldsymbol{x}^{(k)T}\Delta\boldsymbol{g}^{(k)}} - \frac{\boldsymbol{A}^{(k)}\Delta\boldsymbol{g}^{(k)}\Delta\boldsymbol{g}^{(k)T}\boldsymbol{A}^{(k)}}{\Delta\boldsymbol{g}^{(k)T}\boldsymbol{A}^{(k)}\Delta\boldsymbol{g}^{(k)}} = \boldsymbol{A}^{(k)} + \Delta\boldsymbol{A}^{(k)} \tag{7.13}$$

This formula is more complex than the much simpler Broyden's formula. However, it can be shown to preserve symmetry and positive definiteness of the new matrix $\boldsymbol{A}^{(k+1)}$ if the previous approximation $\boldsymbol{A}^{(k)}$ is symmetric and positive definite. A possible MATLAB implementation of the DFP formula is given in the listing M7.3.

The following example illustrates the DFP method.

Example 7.4: Apply the MATLAB listing M7.3 to minimize the objective function:

$$f(\boldsymbol{x}) = (x_1 + 10x_2)^2 + 5(x_3 - x_4)^2 + (x_2 - 2x_3)^4 + 100(x_1 - x_4)^4 \tag{7.14}$$

starting from the point $\boldsymbol{x}^{(0)} = [-2.0 \quad -1.0 \quad 1.0 \quad 2.0]^T$.

Solution: The output from the MATLAB listing M7.3 is as follows:

First Iteration

```
OldPoint=[-2.0    -1.0    1.0    2.0]'    OldValue=25830  OldGradient=
1.0e+004 *[-2.5614  -0.0348  0.0206  2.5620]'
AOld=[1    0    0    0
      0    1    0    0
      0    0    1    0
      0    0    0    1]
```

Second Iteration

```
OldPoint=[-0.6300     -0.9814     0.9890     0.6297]'  OldValue=438.1776
OldGradient=[-819.4021  -312.3930   211.142  796.8315]'
AOld=[0.5006  -0.0007  -0.0001   0.5000
     -0.0007   1.0000  -0.0000   0.0007
     -0.0001  -0.0000   1.0000   0.0001
      0.5000   0.0007   0.0001   0.4995]
```

Third Iteration

```
OldPoint=[-0.5913     0.0558     0.2848     0.6693]'    OldValue=253.3708
OldGradient=[-800.4861  -1.1169  -2.7474  806.1762]'
AOld=[ 0.4992  -0.0310   0.0207   0.4986
      -0.0310   0.3249   0.4639  -0.0304
       0.0207   0.4639   0.6812   0.0215
       0.4986  -0.0304   0.0215   0.4980]
```

Fourth Iteration

```
OldPoint=[-1.0170  0.3041  0.5910  0.2281]'  OldValue=245.6528
```

Fifth Iteration

```
OldPoint=[-3.1177  0.1884  0.1617  -1.9597]'  OldValue=203.8119
```

Sixth Iteration

```
OldPoint=[-2.8520  0.1498  0.1055  -1.7947]'  OldValue=144.8359
```

Seventh Iteration

```
OldPoint=[-1.0582  0.0487  0.0391  -0.3274]'  OldValue=29.5268
```

```
%M7.3
% Implementation of the DFP method
NumberOfParameters=4; %This is n for this problem
Epsilon=1.0e-3; %perturbation used in finite differences
Tolerance=1.0e-4; %termination condition for line search
OldPoint=[-2 -1 1 2]' %This is the starting point
OldValue=getObjective(OldPoint) %Get the objective function at the old point
OldGradient=getGradient('getObjective', OldPoint, Epsilon) %get the gradient at the first
                                                                    %point
OldGradientNorm=norm(OldGradient); %get the old gradient norm
LambdaMax=1.0; %limit the upper bound by the ideal Newton step
AOld=eye(NumberOfParameters); %initialize the approximation to the Hessian inverse
while(OldGradientNorm>0.5) %repeat under the gradient is small enoug
  QuasiNewtonDirection=-1.0*AOld*OldGradient; %get the current search direction
  LambdaOptimal =  GoldenSection('getObjective',Tolerance,OldPoint,
                                     QuasiNewtonDirection,LambdaMax);%line search
  NewPoint=OldPoint+LambdaOptimal*QuasiNewtonDirection %get new point
  NewValue=getObjective(NewPoint) %get new objective function value
  NewGradient=getGradient('getObjective', NewPoint, Epsilon) %get the new gradient
  DeltaGradient=NewGradient-OldGradient; %this is the gradient difference
  DeltaPosition=NewPoint-OldPoint; %this is the position difference
  TempVector=AOld*DeltaGradient; %one of the formula's terms
  ANew=AOld+((DeltaPosition*DeltaPosition')/(DeltaPosition'*DeltaGradient))-
                              ((TempVector*TempVector')/(TempVector'*DeltaGradient))
  %make new values current values
  AOld=ANew;
  OldPoint=NewPoint;
  OldValue=NewValue;
  OldGradient=NewGradient;
  OldGradientNorm=norm(OldGradient); %get new norm
end
```

Eighth Iteration

OldPoint = [−0.4868 0.0535 0.0425 0.0876]′ OldValue = 10.9043

Ninth Iteration

OldPoint = [−0.3013 0.0513 0.0061 0.1377]′ OldValue = 3.8436

Tenth Iteration

OldPoint = [−0.3057 0.0427 −0.0460 0.0274]′ OldValue = 1.2738

Eleventh Iteration

OldPoint = [−0.3340 0.0347 −0.0910 −0.0834]′ OldValue = 0.3968

Final Iteration

OldPoint = [0.0067 −0.0006 −0.0187 −0.0197] OldValue = 5.5776e-005

The convergence in the solution of this problem is slower than in quadratic functions because of the higher degree of nonlinearity of the function. The objective function value dropped from 25,830 at the starting point to only 0.00005 at the final point.

7.4.3 The Broyden–Fletcher–Goldfarb–Shanno method

One of the most widely used quasi-Newton methods is the Broyden–Fletcher–Goldfarb–Shanno (BFGS) method. It is named after the four scientists who, independently, arrived at this formula around the same time. This method has the advantage of offering two formulas for updating the Hessian matrix and its inverse

This saves the effort of having to invert the Hessian matrix at each iteration. It also preserves the symmetry and positive definiteness at each iteration. The BFGS formulas for updating the Hessian and its inverse are given by, respectively:

$$\boldsymbol{B}^{(k+1)} = \boldsymbol{B}^{(k)} + \frac{\Delta \boldsymbol{g}^{(k)} \Delta \boldsymbol{g}^{(k)T}}{\Delta \boldsymbol{g}^{(k)T} \Delta \boldsymbol{x}^{(k)}} + \frac{\left(\boldsymbol{B}^{(k)} \Delta \boldsymbol{x}^{(k)}\right)\left(\boldsymbol{B}^{(k)} \Delta \boldsymbol{x}^{(k)}\right)^T}{\left(\boldsymbol{B}^{(k)} \Delta \boldsymbol{x}^{(k)}\right)^T \Delta \boldsymbol{x}^{(k)}} = \boldsymbol{B}^{(k)} + \Delta \boldsymbol{B}^{(k)} \tag{7.15}$$

and

$$\boldsymbol{A}^{(k+1)} = \boldsymbol{A}^{(k)} + \frac{\Delta \boldsymbol{x}^{(k)} \Delta \boldsymbol{x}^{(k)T}}{\Delta \boldsymbol{x}^{(k)T} \Delta \boldsymbol{g}^{(k)}} \left(1 + \frac{\Delta \boldsymbol{g}^{(k)T} \boldsymbol{A}^{(k)} \Delta \boldsymbol{g}^{(k)}}{\Delta \boldsymbol{x}^{(k)T} \Delta \boldsymbol{g}^{(k)}}\right) - \frac{\boldsymbol{A}^{(k)} \Delta \boldsymbol{g}^{(k)} \Delta \boldsymbol{x}^{(k)T}}{\Delta \boldsymbol{x}^{(k)T} \Delta \boldsymbol{g}^{(k)}}$$
$$- \frac{\Delta \boldsymbol{x}^{(k)} \Delta \boldsymbol{g}^{(k)T} \boldsymbol{A}^{(k)}}{\Delta \boldsymbol{x}^{(k)T} \Delta \boldsymbol{g}^{(k)}} = \boldsymbol{A}^{(k)} + \Delta \boldsymbol{A}^{(k)} \tag{7.16}$$

where $\boldsymbol{B}^{(k)}$ approximates the Hessian at the kth iteration while $\boldsymbol{A}^{(k)}$ approximates its inverse.

7.4.4 The Gauss–Newton method

The Gauss–Newton method is a variation of Newton's method that is suited for solving nonlinear least squares problems. These problems consider a vector of nonlinear functions of the form:

$$\boldsymbol{f}(\boldsymbol{x}) = [f_1(\boldsymbol{x}) \quad f_2(\boldsymbol{x}) \quad \cdots \quad f_m(\boldsymbol{x})]^T \tag{7.17}$$

These functions usually represent error functions that should ideally be driven to zero at the optimal solution. The problem to be solved can then be cast in the form:

$$\boldsymbol{f}(\boldsymbol{x}) = [f_1(\boldsymbol{x}) \quad f_2(\boldsymbol{x}) \quad \cdots \quad f_m(\boldsymbol{x})]^T = \boldsymbol{0}^T \tag{7.18}$$

Most often the number of functions exceeds the number of parameters ($m > n$) and the system (7.18) does not have an exact solution. Instead, we aim at minimizing the square of the L_2 norm of the error vector which is given by:

$$U(\boldsymbol{x}) = f_1^2 + f_2^2 + \cdots + f_m^2 = \sum_{p=1}^{m} f_p^2 \tag{7.19}$$

The nonlinear least squares problem aims at finding the optimal point $\boldsymbol{x}^*$ that minimizes the objective function (7.19). The gradient of the objective function (7.19) can be found by differentiating each term with respect to the parameters $\boldsymbol{x}$ to obtain:

$$\nabla U(\boldsymbol{x}) = \sum_{p=1}^{m} 2 f_p \frac{\partial f_p}{\partial \boldsymbol{x}} = \sum_{p=1}^{m} 2 f_p \nabla f_p \tag{7.20}$$

It follows from (7.20) that the gradient of the objective function is a weighted sum of the gradients of all nonlinear error functions. As explained in Chapter 1, (7.20) can be expressed in terms of the Jacobian of the error vector to give the more compact vector form:

$$\nabla U(\boldsymbol{x}) = 2\boldsymbol{J}^T \boldsymbol{f} \tag{7.21}$$

To evaluate the Hessian of the least squares objective function $U(\boldsymbol{x})$, we differentiate each component of the gradient vector (7.20) with respect to all parameters. The ith component of the gradient vector (7.20) is given by:

$$\frac{\partial U}{\partial x_i} = \sum_{p=1}^{m} 2f_p \frac{\partial f_p}{\partial x_i} \tag{7.22}$$

Differentiating (7.22) with respect to the jth parameter x_j, we get:

$$\frac{\partial^2 U}{\partial x_i \partial x_j} = 2\sum_{p=1}^{m} \frac{\partial f_p}{\partial x_j}\frac{\partial f_p}{\partial x_i} + 2\sum_{p=1}^{m} f_p \frac{\partial^2 f_p}{\partial x_i \partial x_j} \tag{7.23}$$

Notice that the second term in the right-hand side of (7.23) requires the Hessian of each individual error function. These Hessian components are, however, multiplied by the value of individual error function f_p. If Newton's method starts with a point $\boldsymbol{x}^{(0)}$ close enough to the optimal point, the components of the vector $\boldsymbol{f}$ should be close enough to zero. In this case, the Hessian of the objective function can be approximated by:

$$\frac{\partial^2 U}{\partial x_i \partial x_j} \approx 2\sum_{p=1}^{m} \frac{\partial f_p}{\partial x_j}\frac{\partial f_p}{\partial x_i}, \quad i = 1, \ldots, n \quad \text{and} \quad j = 1, \ldots, n \tag{7.24}$$

In a more compact matrix form, the approximated Hessian of the objective function U is thus given by:

$$\boldsymbol{H}_u(\boldsymbol{x}) \approx 2\boldsymbol{J}^T\boldsymbol{J} \tag{7.25}$$

The Gauss–Newton method utilizes the Hessian approximation (7.25) in minimizing the objective function (7.19). The step taken in this method is given by:

$$\boldsymbol{s}^{(k)} = -\boldsymbol{H}_u^{(k)-1}\nabla U^{(k)} = -\left(\boldsymbol{J}^{(k)T}\boldsymbol{J}^{(k)}\right)^{-1}\boldsymbol{J}^{(k)T}\boldsymbol{f}(\boldsymbol{x}^{(k)}) \tag{7.26}$$

where the Jacobian matrix is estimated at the point $\boldsymbol{x}^{(k)}$. Because the step in (7.26) uses an approximation of the Hessian matrix of the objective function and because the objective function $U(\boldsymbol{x})$ may not be a quadratic one, the full Newton step should not be trusted. Instead, a line search should be carried out in the direction $\boldsymbol{s}^{(k)}$.

The same formula (7.26) can be derived in an alternative way using first-order Taylor expansion. The error vector function in (7.18) can be expanded to give:

$$\boldsymbol{f}(\boldsymbol{x}) \approx \boldsymbol{f}(\boldsymbol{x}^{(k)}) + \boldsymbol{J}^{(k)}(\boldsymbol{x} - \boldsymbol{x}^{(k)}) \tag{7.27}$$

To find the optimal point, the one with zero errors, we force the left-hand side of (7.27) to zero to get the over-determined system:

$$\boldsymbol{J}^{(k)}(\boldsymbol{x}^* - \boldsymbol{x}^{(k)}) = -\boldsymbol{f}(\boldsymbol{x}^{(k)}) \tag{7.28}$$

Because the matrix $\boldsymbol{J}^{(k)}$ is a rectangular matrix ($m > n$), it cannot be directly inverted to solve for the optimal point $\boldsymbol{x}^*$. Assuming that the matrix $\boldsymbol{J}^{(k)}$ has a full rank, the matrix $\boldsymbol{J}^{(k)T}\boldsymbol{J}^{(k)}$ is invertible. Both sides of (7.28) are thus multiplied by $\boldsymbol{J}^{(k)T}$ and then the resulting matrix is inverted:

$$\boldsymbol{J}^{(k)T}\boldsymbol{J}^{(k)}(\boldsymbol{x}^* - \boldsymbol{x}^{(k)}) = -\boldsymbol{J}^{(k)T}\boldsymbol{f}(\boldsymbol{x}^{(k)}) \Rightarrow \Delta\boldsymbol{x}^{(k)} = -\left(\boldsymbol{J}^{(k)T}\boldsymbol{J}^{(k)}\right)^{-1}\boldsymbol{J}^{(k)T}\boldsymbol{f}(\boldsymbol{x}^{(k)}) \tag{7.29}$$

which is identical to (7.26). Formula (7.29) is an extension of the Newton–Raphson method for solving nonlinear equations discussed in Chapter 1.

The linear least squares problem is a variation of the general least squares problem where the error functions are linear functions of the parameters. Each error function is given by a linear term of the form:

$$f_i(\boldsymbol{x}) = a_{i1}x_1 + a_{i2}x_2 + \cdots + a_{in}x_n - b_i = \boldsymbol{a}_i^T\boldsymbol{x} - b_i \tag{7.30}$$

The linear least squares problem thus aims to solve the over-determined system of equations:

$$\begin{bmatrix} \boldsymbol{a}_1^T\boldsymbol{x} - b_1 \\ \boldsymbol{a}_2^T\boldsymbol{x} - b_2 \\ \vdots \\ \boldsymbol{a}_m^T\boldsymbol{x} - b_m \end{bmatrix} = \begin{bmatrix} 0 \\ 0 \\ \vdots \\ 0 \end{bmatrix} \Rightarrow \boldsymbol{Ax} = \boldsymbol{b} \tag{7.31}$$

where $\boldsymbol{A} \in \Re^{m\times n}$ is a matrix whose ith row is $\boldsymbol{a}_i^T$ and $\boldsymbol{b} = [b_1 \quad b_2 \quad \ldots \quad b_m]^T$. The L_2 norm squared of the error vector (7.31) is thus given by:

$$U(\boldsymbol{x}) = \|\boldsymbol{f}\|_2^2 = (\boldsymbol{Ax} - \boldsymbol{b})^T(\boldsymbol{Ax} - \boldsymbol{b}) = \boldsymbol{x}^T\boldsymbol{A}^T\boldsymbol{Ax} - \boldsymbol{x}^T\boldsymbol{A}^T\boldsymbol{b} - \boldsymbol{b}^T\boldsymbol{Ax} + \boldsymbol{b}^T\boldsymbol{b} \tag{7.32}$$

Differentiating both sides of (7.32) with respect to $\boldsymbol{x}$ and setting it to zero, we get:

$$\frac{\partial U(\boldsymbol{x})}{\partial \boldsymbol{x}} = \boldsymbol{A}^T\boldsymbol{Ax} + \boldsymbol{A}^T\boldsymbol{Ax} - \boldsymbol{A}^T\boldsymbol{b} - \boldsymbol{A}^T\boldsymbol{b} = \boldsymbol{0} \Rightarrow \boldsymbol{A}^T\boldsymbol{Ax} = \boldsymbol{A}^T\boldsymbol{b} \tag{7.33}$$

If the matrix $\boldsymbol{A}$ is full rank, then the square matrix $\boldsymbol{A}^T\boldsymbol{A}$ is invertible and (7.33) gives the result:

$$\boldsymbol{x}^* = \left(\boldsymbol{A}^T\boldsymbol{A}\right)^{-1}\boldsymbol{A}^T\boldsymbol{b} \tag{7.34}$$

```
%M7.4
%Implementation of the Gauss-Newton Method
NumberOfParameters=3; %This is n for this problem
NumberOfResidues=5;
Epsilon=1.0e-3; %perturbation used in finite differences
Tolerance=1.0e-4; %termination condition for the line search
OldPoint=[1 1 1]' %This is the starting point
OldResidues=getResidues(OldPoint) %Get the vector of residues at the old point
OldValue=norm(OldResidues)
OldJacobian=getJacobian('getResidues', OldPoint, Epsilon) %get the Jacobian at the current point
LambdaMax=1.0; %maximum value of line search
IterationCounter=0; %this is the iteration counter
MaxIterations=20; %this is the maximum number of iterations
while(IterationCounter<MaxIterations) %repeat under the gradient is small enoug
  GaussNewtonDirection=-1.0*inv(OldJacobian'*OldJacobian)*OldJacobian'*OldResidues;
  LambdaOptimal = GoldenSection('getResiduesNorm',Tolerance,OldPoint,GaussNewtonDirection
                        ,LambdaMax);%do line search
  NewPoint=OldPoint+LambdaOptimal*GaussNewtonDirection %get new point
  NewResidues=getResidues(NewPoint) %get new vector of residues
  NewValue=norm(NewResidues); %get the norm of the residues
  NewJacobian=getJacobian('getResidues', NewPoint, Epsilon) %get the new gradient
  OldPoint=NewPoint;
  OldResidues=NewResidues;
  OldValue=NewValue;
  OldJacobian=NewJacobian;
end
```

It should be noted that the solution obtained in (7.34) is a minimum. This can be checked by differentiating (7.33) one more time with respect to $\boldsymbol{x}$ to get the positive semi-definite matrix Hessian matrix $\boldsymbol{H}_u = \boldsymbol{A}^T\boldsymbol{A}$.

MATLAB listing M7.4 shows a possible implementation of the Gauss–Newton method for solving least squares problems.

Example 7.5: Find the minimum of the sum of squares function:

$$U(\boldsymbol{x}) = (x_1 + 5)^2 + (x_2 + 8)^2 + (x_3 + 7)^2 + 2x_1^2x_2^2 + 4x_1^2x_3^2 \tag{7.35}$$

using the Gauss–Newton method and starting with the point $\boldsymbol{x}^{(0)} = [1.0 \quad 1.0 \quad 1.0]^T$. Exploit the MATLAB listing M7.4.

Solution: The output from the MATLAB listing M7.4 for the problem (7.35) is given by:

OldPoint = [1.0 1.0 1.0]′ OldValue = 13.6748
OldPoint = [1.7106 −0.7766 −0.4213]′ OldValue = 12.0871
OldPoint = [−0.6874 −1.8077 −0.9990]′ OldValue = 9.8959
OldPoint = [0.1968 −3.8691 −2.4145]′ OldValue = 8.1952
OldPoint = [−0.1411 −6.4462 −5.1237]′ OldValue = 5.7695
NewPoint = [−0.0158 −7.9924 −6.9887]′ OldValue = 4.9923

The final solution obtained from this code after few iterations is $\boldsymbol{x}^* = [-0.0158 \quad -7.9924 \quad -6.9887]^T$. The exact analytical solution of this problem is $\boldsymbol{x}^* = [0 \quad -8.0 \quad -7.0]^T$. As the starting point is relatively far from the actual solution, more iterations are needed than in Newton's method, which would have required only one iteration for such a quadratic problem.

We discuss a number of real-life engineering applications that are relevant to the algorithms discussed in this chapter. A special focus is put on the linear and nonlinear least squares problems.

A7.1 Wireless channel characterization

Figure 7.2 illustrates a wireless channel with a transmitter and a receiver. Any sequence of signals sent by the transmitter is received on the other side after it undergoes certain changes due to the channel. By using a certain sequence of transmitted signals and recording how they are received on the other side, the channel can be characterized.

As an example of such a system, we assume that the transmitter sends a sequence of 3 discrete signals (s_0, s_1, s_2) each of duration 3 time units. As shown in Figure 7.2, the transmitted signal takes two paths with attenuation factors a_1 and a_2 and delays $\tau_1 = 10$ time units and $\tau_2 = 12$ time units, respectively. Our target is to find the channel coefficients a_1 and a_2 given that the following sequences were sent and received $s_0 = 1$, $s_1 = 2$, $s_2 = 1$, $r_{10} = 4$, $r_{11} = 7$, $r_{12} = 8$, $r_{13} = 6$, $r_{14} = 3$, where the subscript indicates the time unit at which the signal was sent or received.

For this problem, the delay in the first path is smaller than the delay on the second path. The signal received after 10 time units must have been received only through the first path due to the signal sent 10 time units earlier. It follows that we have $r_{10} = a_1 \times s_0$ or $4 = a_1 \times 1$. Similarly, the signal r_{11} must be due to the signal

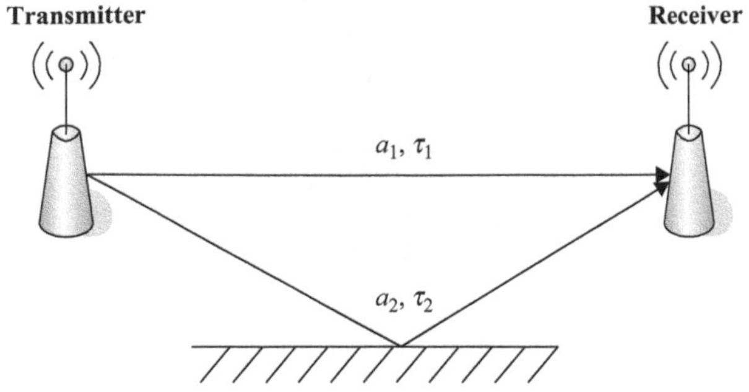

Figure 7.2 An illustration of a simplified wireless communication system; the signal sent by the transmitter reaches the receiver through two different paths each has its own attenuation and delay

s_1 transmitted along the first path 10 time units earlier. It follows that we have $r_{11} = a_1 \times s_1$ or $7 = a_1 \times 2$. The signal r_{12} is different because signals start arriving through the second path. This signal is thus a linear combination of the signal s_2 transmitted along the first path 10 time units earlier and the signal s_0 transmitted along the second path 12 time units earlier. In equation form this means that $r_{12} = a_1 \times s_2 + a_2 \times s_0$ or $8 = a_1 \times 1 + a_2 \times 1$. Writing the equations for all the available received signals, we obtain the system of linear equations:

$$\begin{bmatrix} 1 & 0 \\ 2 & 0 \\ 1 & 1 \\ 0 & 2 \\ 0 & 1 \end{bmatrix} \begin{bmatrix} a_1 \\ a_2 \end{bmatrix} = \begin{bmatrix} 4 \\ 7 \\ 8 \\ 6 \\ 3 \end{bmatrix} \tag{7.36}$$

The system (7.36) is an over-determined system with a full rank matrix. Utilizing the formulation (7.34), the optimal estimates of the channel parameters that would minimize the least squares errors are given by:

$$\begin{bmatrix} a_1^* \\ a_2^* \end{bmatrix} = \left(\begin{bmatrix} 1 & 2 & 1 & 0 & 0 \\ 0 & 0 & 1 & 2 & 1 \end{bmatrix} \begin{bmatrix} 1 & 0 \\ 2 & 0 \\ 1 & 1 \\ 0 & 2 \\ 0 & 1 \end{bmatrix} \right)^{-1} \begin{bmatrix} 1 & 2 & 1 & 0 & 0 \\ 0 & 0 & 1 & 2 & 1 \end{bmatrix} \begin{bmatrix} 4 \\ 7 \\ 8 \\ 6 \\ 3 \end{bmatrix} = \begin{bmatrix} \frac{133}{35} \\ \frac{112}{35} \end{bmatrix} \tag{7.37}$$

The same concept presented in this application can be extended to problems with multi-path sequences with more sent and received sequences.

A7.2 The parameter extraction problem

An interesting problem that appears very often in electrical engineering is the problem of parameter extraction. This problem is illustrated in Figure 7.3. The response $\boldsymbol{R}$ of a certain system is controlled by its parameters $\boldsymbol{x}$. This response may be the vector of S-parameters at a number of frequencies or field values at a number

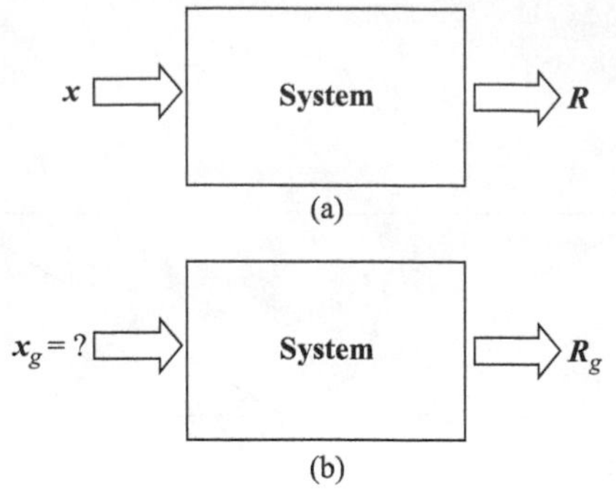

Figure 7.3 An illustration of the parameter extraction procedure; (a) the forward simulation is shown where given the parameter $\mathbf{x}$*, the response* $\mathbf{R}$ *is simulated or measured; and (b) the inverse problem where the response* $\mathbf{R}_g$ *is given and the target is to determine the parameters* $\mathbf{x}_g$ *that would have given rise to these responses*

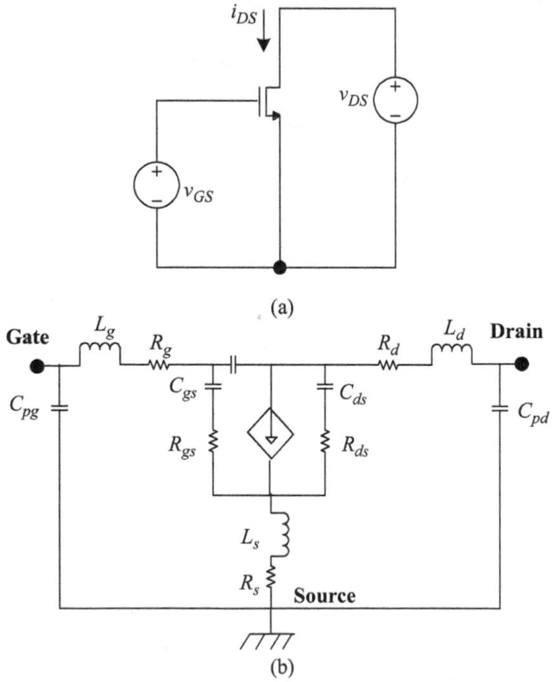

Figure 7.4 An illustration of the parameter extraction problem as applied to transistors; (a) the transistor is measured under different bias conditions and different frequencies to determine its response $\mathbf{R}_g$*; and (b) the measured response is then used to extract values for the equivalent circuit parameters through an optimization problem. The extracted parameters include inductors, capacitors, resistors, and the coefficients of dependent current sources*

of points. The system is usually simulated in the forward direction. The values of the parameters $\boldsymbol{x}$ are first chosen and then the vector $\boldsymbol{R}$ is determined through numerical modeling or through actual fabrication/assembly and measurement of the system.

In some cases, it is required to solve the inverse problem. A set of measurements $\boldsymbol{R}_g$ are given. The target is to determine the parameter $\boldsymbol{x}$ that would have given rise to this response. This is the case, for example, when determining the values of the components of a small signal model of a transistor. The transistor is first measured and then, using an optimization approach, the values of the small signal model that could have created this response are determined. The microwave imaging approach is illustrated in Figure 7.4.

Another application of this approach is microwave imaging. Here, a number of antennas are used to illuminate an object with unknown material properties distribution. By measuring the reflected field, which constitutes $\boldsymbol{R}$ in this case, one tries to recover the vector $\boldsymbol{x}$ of unknown material properties (permittivity and conductivity) of the object. This effectively creates an image of this object. This approach is illustrated in Figure 7.5.

In general, the target of a parameter extraction problem is to solve for the parameter values that would give rise to the given response $\boldsymbol{R}_g$. This problem can be cast as the following optimization problem:

$$\boldsymbol{x}^* = \arg\min_{x} \|\boldsymbol{R}(\boldsymbol{x}) - \boldsymbol{R}_g\| \tag{7.38}$$

By choosing to minimize the square of the L_2 norm, the parameter extraction problem is cast in the form:

$$\boldsymbol{x}^* = \arg\min_{x} U(\boldsymbol{x}) = \left(R_1 - R_{g,1}\right)^2 + \left(R_2 - R_{g,2}\right)^2 + \cdots + \left(R_m - R_{g,m}\right)^2 \tag{7.39}$$

This problem has the same form as (7.19) and can be solved using the Gauss–Newton method.

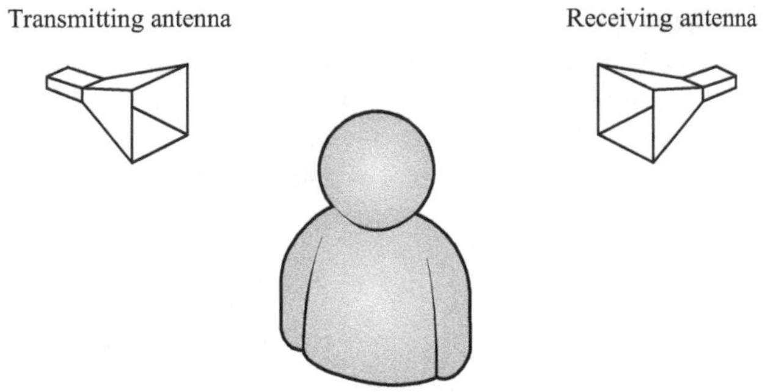

Figure 7.5 An illustration of the application of parameter extraction in microwave imaging; the change in human tissue properties (permittivity and conductivity) results in a change in the received signal by the receiving antenna

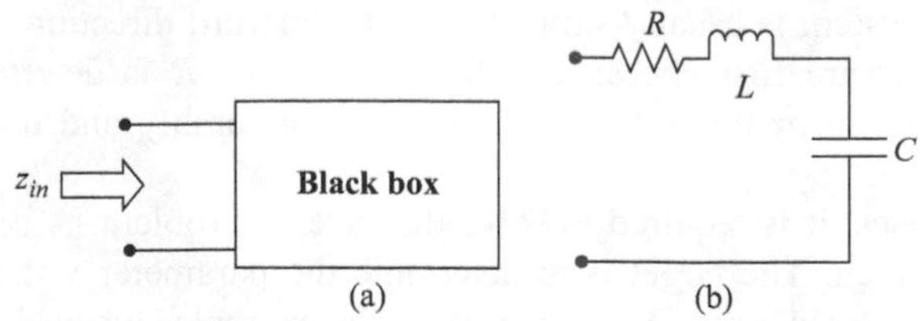

Figure 7.6 The input impedance of a circuit represented by a black box is approximated by a series RLC circuit with unknown parameters

As an example, consider the black box shown in Figure 7.6. The input impedance of this black box is given at a number of frequencies. These given values are:

$$\boldsymbol{Z}_g = 1.0e+003 * [0.0500 - 1.0233\text{i} \\ 0.0500 - 0.4551\text{i} \\ 0.0500 - 0.2406\text{i} \\ 0.0500 - 0.1145\text{i} \\ 0.0500 - 0.0237\text{i} \\ 0.0500 + 0.0494\text{i} \\ 0.0500 + 0.1123\text{i} \\ 0.0500 + 0.1690\text{i} \\ 0.0500 + 0.2214\text{i} \\ 0.0500 + 0.2709\text{i}] \tag{7.40}$$

at frequencies (100 Hz, 200 Hz, ..., 1.0 kHz).

It is believed that the response of this black box can be approximated by a series RLC circuit with unknown components $\boldsymbol{x} = [R \quad C \quad L]^T$. The target of the parameter extraction problem is thus to find the values of $\boldsymbol{x}$ that would generate the same measured values of the input impedance. In this case, the vector of measured response is complex. This vector can be divided into real and imaginary parts to give a vector of real target response with double the number of components.

We utilize the Gauss–Newton method with a starting point $\boldsymbol{x}^{(0)} = [20\,\Omega \quad 4.0\,\mu\text{F} \quad 1.0\,\text{mH}]^T$. Because the parameters have different orders of magnitude, scaling is applied to make all parameters of the same scale. When calling the circuit simulator to evaluate the impedance at any set of values and for any frequency, the parameters are scaled back to their physical values. The MATLAB code M7.4 is adapted for this problem as shown in the MATLAB listing M7.5.

The algorithm takes only 4 iterations. The output from the algorithm is as follows:

```
OldPoint = [2.0  4.0  1.0]'  OldValue = 1.1381e+003
NewPoint = [4.9998  59.9981  1.3337]'  NewValue = 164.7848
NewPoint = [5.0000  60.0011  1.4817]'  NewValue = 16.3573
NewPoint = [5.0000  60.0011  1.4998]'  NewValue = 0.2031
```

The final extracted set of values are thus $\boldsymbol{x}^{(3)} = [50.0\,\Omega \quad 60.0011\,\mu\text{F} \quad 1.4998\,\text{mH}]^T$. The least squares error function dropped from 1138 to 0.2031 indicating good convergence. The modulus of the input impedance of the circuit as compared to the modulus of the given impedance at every frequency for all iterations are shown in Figures 7.7–7.10. The input impedance gradually approaches the target response. Excellent match is achieved in the final iteration.

```
%M7.5
%Gauss-Newton method applied for the parameter extraction
NumberOfParameters=3; %This is n for this problem
NumberOfResidues=20;
Epsilon=1.0e-3; %perturbation used in finite differences
Tolerance=1.0e-4; %termination condition for the line search
OldPoint=[2 4 1]' %This is the starting point
OldResidues=getResidues(OldPoint) %Get the vector of residues at the old point
OldValue=norm(OldResidues)
OldJacobian=getJacobian('getResidues', OldPoint, Epsilon) %get the Jacobian at the current point
LambdaMax=1.0; %maximum value of line search
IterationCounter=0; %this is the iteration counter
MaxIterations=20; %this is the maximum number of iterations
while(IterationCounter<MaxIterations) %repeat under the gradient is small enoug
  GaussNewtonDirection=-1.0*inv(OldJacobian'*OldJacobian)*OldJacobian'*OldResidues;
  LambdaOptimal =
  GoldenSection('getResiduesNorm',Tolerance,OldPoint,GaussNewtonDirection,LambdaMax);
  NewPoint=OldPoint+LambdaOptimal*GaussNewtonDirection %get new point
  NewResidues=getResidues(NewPoint) %get new vector of residues
  NewValue=norm(NewResidues); %get the norm of the residues
  NewJacobian=getJacobian('getResidues', NewPoint, Epsilon) %get the new gradient
 % make new values current values
  OldPoint=NewPoint;
  OldResidues=NewResidues;
  OldValue=NewValue;
  OldJacobian=NewJacobian;
end
```

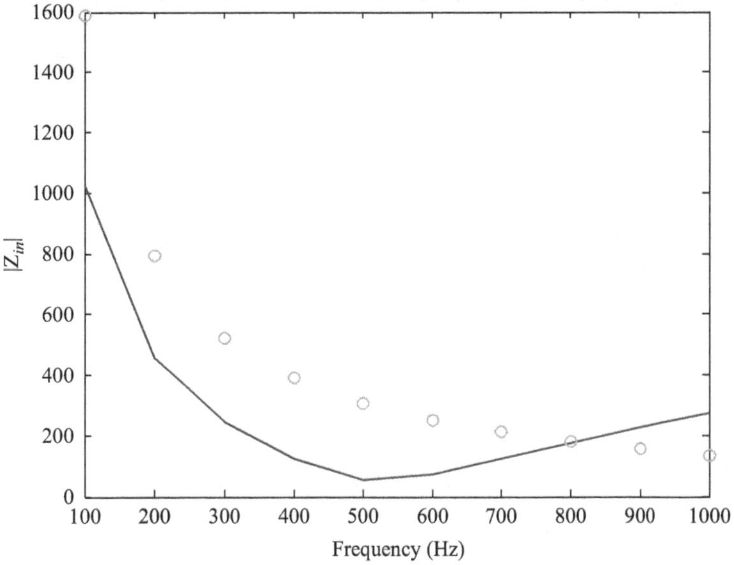

Figure 7.7 The modulus of the target input impedance (–) versus the modulus of the input impedance at the starting point $\mathbf{x}^{(0)}$ at all considered frequency points

A7.3 Artificial neural networks training

An artificial neural network (ANN) is a mathematical model that approximates a general mapping from the space of parameters $\Re^n$ to the space of responses $\Re^m$. The basic building block of an ANN is a mathematical model of the human neuron. The biological neurons fire when the input from all its nervous connections exceeds a

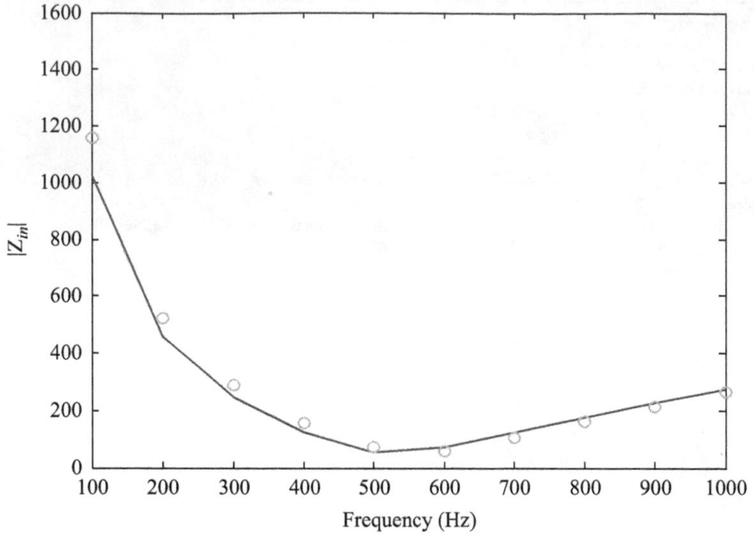

Figure 7.8 The modulus of the target input impedance (–) versus the modulus of the input impedance at the second point $\mathbf{x}^{(1)}$ at all considered frequency points

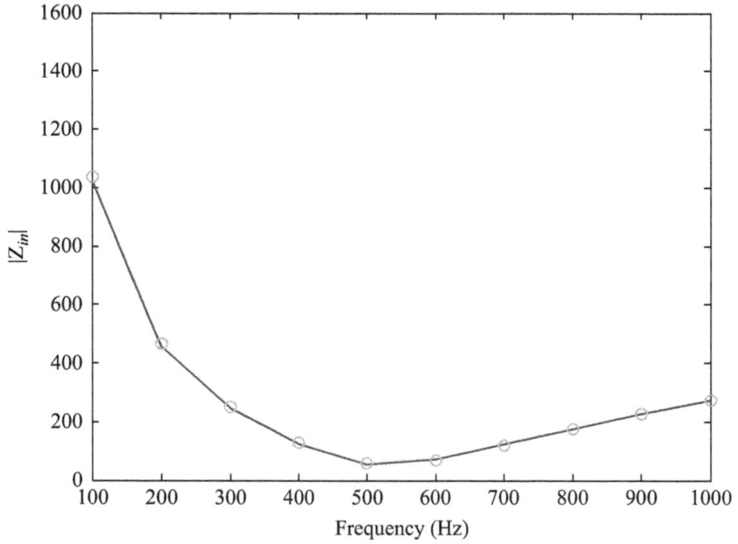

Figure 7.9 The modulus of the target input impedance (–) versus the modulus of the input impedance at the third point $\mathbf{x}^{(2)}$ at all considered frequency points

certain threshold. The mathematical neuron imitates this biological behavior. This neuron is shown in Figure 7.11. The signals from all neurons connected to the inputs of this neuron v_i, $i = 1, 2, \ldots, p$ are given different weights. The total stimulus of this neuron is the sum of all the weighted inputs and is given by:

$$z = \sum_{i=1}^{p} w_i v_i \tag{7.41}$$

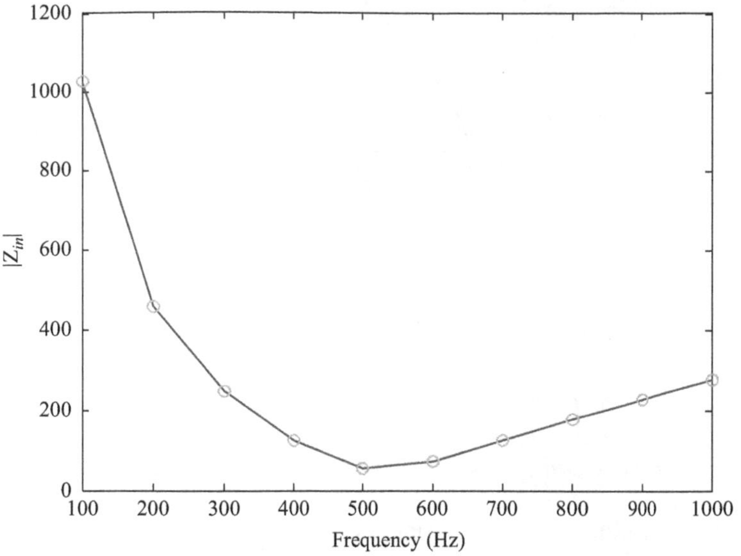

Figure 7.10 The modulus of the target input impedance (–) versus the modulus of the input impedance at the fourth point $\boldsymbol{x}^{(3)}$ at all considered frequency points

This sum triggers the activation function of the neuron to create the neuron output y, which is given by:

$$y(\boldsymbol{w}, \boldsymbol{v}) = \sigma(z) = \sigma\left(\sum_{i=1}^{p} w_i v_i\right) \tag{7.42}$$

where $\boldsymbol{w} = \begin{bmatrix} w_1 & w_2 & \dots & w_p \end{bmatrix}^T$ and $\boldsymbol{v} = \begin{bmatrix} v_1 & v_2 & \dots & v_p \end{bmatrix}^T$. The function $\sigma(z)$ is the neuron activation function. Several activation functions are used in the ANN literature. These include the following functions:

$$\begin{aligned} \text{Sigmoid function } \sigma(z) &= \frac{1}{1+e^{-z}} \\ \text{Arc-tangent function } \sigma(z) &= \left(\frac{2}{\pi}\right)\arctan(z) \\ \text{Hyperbolic-tangent function } \sigma(z) &= \frac{e^{z}-e^{-z}}{e^{z}+e^{-z}} \end{aligned} \tag{7.43}$$

All these functions saturate at a minimum value for very low stimulus. They also saturate at a maximum output for high stimulus values. In between, these functions increase monotonically with the increase in the stimulus of the neuron. As an example, the Sigmoid function is shown in Figure 7.12.

An ANN is a connection of neurons to model a mathematical mapping. We consider here only the feed forward ANN. In this configuration, the neurons are organized in layers and the outputs from the neurons of one layer are fed forward as inputs to the neurons on the next layer. The training process of the ANN aims at finding the optimal weights of the ANN so that it implements a certain mapping.

In particular, we focus on the single hidden layer ANN shown in Figure 7.13. The universal approximation theorem [5] states that this configuration is capable of

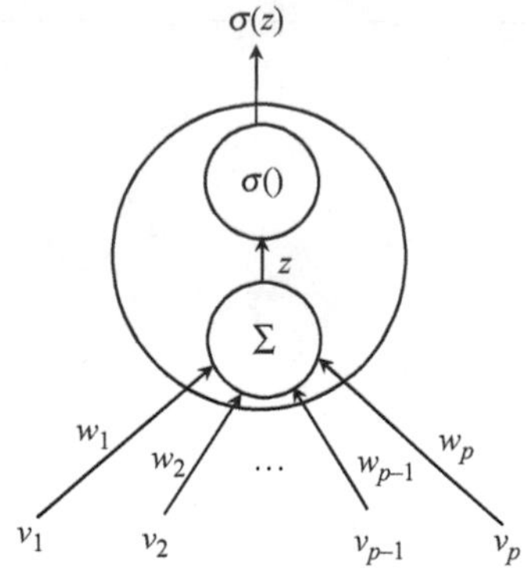

Figure 7.11 An illustration of the mathematical neuron; all inputs are given different weights and summed. The sum of the weighted inputs is then processed by the neuron nonlinear firing function

modeling any nonlinear mapping using the correct number of hidden neurons and optimal weights. In this configuration, the ANN maps the input parameters $\boldsymbol{x} \in \Re^n$ to the output responses $\boldsymbol{y} \in \Re^m$. We assume that we have p hidden neurons in this case.

The target of the training process is to determine the optimal weights that implement a certain mapping $\boldsymbol{y}=\boldsymbol{y}(\boldsymbol{x})$. This mapping is usually not known analytically. It is represented as a finite number of input-output pairs $\{\boldsymbol{x}, \boldsymbol{y}\}$ that are used in the training process. For example, the operational function of a MOSFET transistor may be given by the input-output pairs $\{[v_{GS} \quad v_{DS}]^T, i_{DS}\}$ where, in this case, $n=2$ and $m=1$.

The weights of the neural network are denoted by $\boldsymbol{w} = [\boldsymbol{w}^{hT} \quad \boldsymbol{w}^{oT}]^T$, where $\boldsymbol{w}^h \in \Re^{np}$ is the set of weights connecting the inputs to the hidden neurons and $\boldsymbol{w}^o \in \Re^{mp}$ is the set of weights connecting the hidden layer neurons to the output layer neurons. The weight w_{ik}^h is assigned to the connection between the kth parameter and the ith neuron of the hidden layer, $k=1, 2, \ldots, n$ and $i=1, 2, \ldots, p$. The weight w_{ji}^o is the weight assigned to the connection between the output of the ith neuron of the hidden layer and the jth neuron of the output layer, $i=1, 2, \ldots, p$ and $j=1, 2, \ldots, m$.

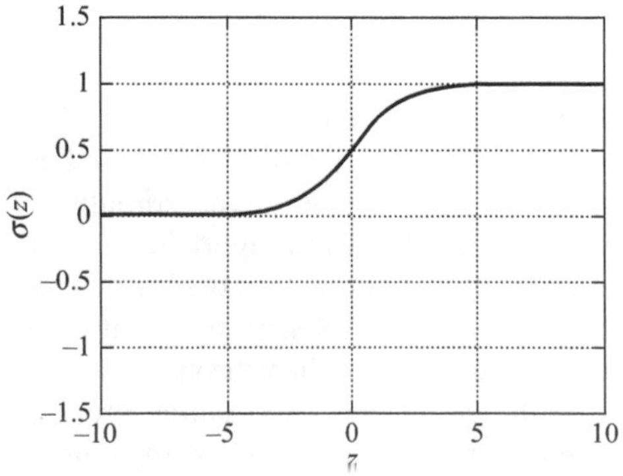

Figure 7.12 The Sigmoid activation functions utilized in ANNs

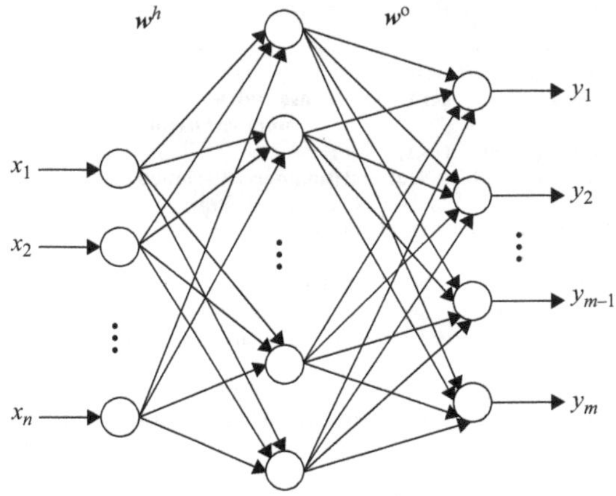

Figure 7.13 A single layer feed forward ANN with n inputs and m outputs

The output of the *j*th neuron of the output layer is used to approximate the *j*th response. Using the ANN configuration in Figure 7.13, the output of this neuron is given by:

$$R_j(\boldsymbol{w}, \boldsymbol{x}) = \sigma(z_j) = \sigma\left(\sum_{i=1}^{p} w_{ji}^{o}\sigma(v_i)\right) = \sigma\left(\sum_{i=1}^{p} w_{ji}^{o}\sigma\left(\sum_{k=1}^{n} w_{ik}^{h}x_k\right)\right) \tag{7.44}$$

where $\sigma(v_i)$ is the output of the *i*th neuron of the hidden layer. The value v_i, $i=1, 2, \ldots, p$, is the total input to the *i*th neuron of the hidden layer and is a weighted sum of the input parameters.

The training problem of the ANN involves determining the optimal set of weights $\boldsymbol{w}^*$ that makes the outputs of the neural network match the desired outputs for the given training set. A possible objective function for this problem is given by:

$$\underset{w}{\text{minimize}}\ f(\boldsymbol{w}) = \sum_{l=1}^{M} (\|\boldsymbol{R}(\boldsymbol{w}, \boldsymbol{x}_l) - \boldsymbol{y}(\boldsymbol{x}_l)\|_2)^2 \tag{7.45}$$

Here, the training data is given by the pairs $\{\boldsymbol{x}_l, \boldsymbol{y}(\boldsymbol{x}_l)\}$, $l=1, 2, \ldots, M$. The error vector $\boldsymbol{E}_l = \boldsymbol{R}(\boldsymbol{w}, \boldsymbol{x}_l) - \boldsymbol{y}(\boldsymbol{x}_l)$ represents the difference between the outputs of the ANN and the corresponding desired outputs for the *l*th training sample. The problem (7.45) aims at minimizing the sum of the square of the L_2 norms of the errors at all training pairs. If the ANN is properly trained, it should be able to reasonably predict the output for inputs not in the training set.

Any optimization technique can be used for the solution of (7.45). If a gradient-based optimization algorithm is utilized, then the gradient of the outputs with respect to the weights should be determined. Luckily, (7.44) offers an analytical relationship between the output of the ANN and all the weights of the ANN. Using the chain rule of differentiation, we can calculate all these derivatives. For example, the derivative of the *j*th output with respect to the weight w_{ji}^{o} is given by:

$$\frac{\partial R_j}{\partial w_{ji}^{o}} = \sigma'(z_j)\sigma(v_i), \quad j = 1, 2, \ldots, m \quad \text{and} \quad i = 1, 2, \ldots, p \tag{7.46}$$

```
%M7.6
%This function returns the response and Jacobian of the response of an artificial neural network
%The vector weight contains all the weights and it has size np+pm
function [ANNResponse, Jacobian]=getANNResponseAndDerivative(Parameters, Weights,
                                    numberOfHiddenNeurons, numberOfOutputs)
numberOfParameters=size(Parameters,1); %this is n
numberOfWeights=numberOfHiddenNeurons*(numberOfParameters+numberOfOutputs);
%First we estimate the inputs to the hidden neurons, their outputs, and their derivatives
hiddenNeuronsInput=zeros(numberOfHiddenNeurons,1); %storage for v
hiddenNeuronsResponse=zeros(numberOfHiddenNeurons,1); %storage for Sigma(v)
hiddenNeuronsDerivatives=zeros(numberOfHiddenNeurons,1); %storage for Sigma'(v)
for i=1:numberOfHiddenNeurons %calculate for each hidden neuron
 hiddenNeuronsInput(i,1)=Weights(((i-1)*numberOfParameters+1):(i*numberOfParameters))'
                                                        *Parameters; %Vi
 hiddenNeuronsResponse(i,1)=getNeuronResponse(hiddenNeuronsInput(i,1)); %Sigma(Vi)
 hiddenNeuronsDerivatives(i,1)=getNeuronDerivative(hiddenNeuronsInput(i,1)); % Sigma'(Vi)
end
%we then get the inputs to the output layer, the ANN output, and their derivatives
outputNeuronsInput=zeros(numberOfOutputs,1); %storage for Z
outputNeuronsResponse=zeros(numberOfOutputs,1); %storage y=Sigma(z)
outputNeuronsDerivatives=zeros(numberOfOutputs,1); %storage for Sigma'(z)
Shift=numberOfParameters*numberOfHiddenNeurons; %start of weights of output layer
for j=1:numberOfOutputs
 outputNeuronsInput(j,1)=Weights((Shift+(j-1)*numberOfHiddenNeurons+1):
                  (Shift+j*numberOfHiddenNeurons))'*hiddenNeuronsResponse; %zj
 outputNeuronsResponse(j,1)=getNeuronResponse(outputNeuronsInput(j,1)); %yj=Sigma(Zj)
 outputNeuronsDerivatives(j,1)=getNeuronDerivative(outputNeuronsInput(j,1)); % Sigma'(Zj)
end
ANNResponse=outputNeuronsResponse; %This is the vector of outputs
%We then evaluate the Jacobian matrix of the output responses
Jacobian=zeros(numberOfOutputs, numberOfWeights); %Jacobian matrix
for j=1: numberOfOutputs %repeat for all rows of the Jacobian matrix
  %we first calculate the sensitivities relative to the np hidden neuron weights
  outputNeuronDerivative=outputNeuronsDerivatives(j); %jth output derivative
  for i=1:numberOfHiddenNeurons
    hiddenNeuronDerivative=hiddenNeuronsDerivatives(i,1); % Sigma'(Vi)
    Weight=Weights(Shift+(j-1)*numberOfHiddenNeurons+i); %wji of the output layer
    for k=1:numberOfParameters
      Input=Parameters(k,1); % kth input parameter
      %evaluate sensitivity relative to Wik of the hidden layer
      Jacobian(j,(i-1)*numberOfParameters+k)=
              outputNeuronDerivative*Weight*hiddenNeuronDerivative*Input;
    end
  end
  % we then evaluate the derivatives relative to the p weights wji of the output layer
  % i=1,2,..,p
  for i=1:numberOfHiddenNeurons
    hiddenNeuronResponse=hiddenNeuronsResponse(i,1);
    Jacobian(j,Shift+(j-1)*numberOfHiddenNeurons+i)=
         outputNeuronDerivative*hiddenNeuronResponse; %derivative w.r.t. wji output
  end
end
```

The derivative of the neuron activation function is used in (7.46). Similarly, using chain rule, the derivative of the *j*th ANN output with respect to the weight w_{ik}^h of the hidden layer is given by:

$$\frac{\partial R_j}{\partial w_{ik}^h} = \sigma'(z_j) w_{ji}^o \sigma'(v_i) x_k, \quad i = 1, 2, \ldots, p \quad \text{and} \quad k = 1, 2, \ldots, n \qquad (7.47)$$

Using (7.46) and (7.47), the Jacobian of all the outputs $\boldsymbol{J} \in \Re^{m\times(np+mp)}$ can be calculated. The *j*th row of this matrix is the transpose of the gradient of the *j*th response with respect to all weights. Notice that this Jacobian depends also on the

```
%M7.7
%This function formulates the L2 objective function for the ANN and its gradient.
function [Objective, Gradient]=getANNObjectiveAndDerivative(Weights)
load TrainingData.mat ;% load trainingPoints, trainingResponses, and numberOfHiddenNeurons
numberOfParameters=size(trainingPoints,1); %number of parameters
numberOfTrainingPoints=size(trainingPoints,2); %number of training points
numberOfOutputs=size(trainingResponses,1); %number of outputs
numberOfWeights=size(Weights,1); %number of weights
Objective=0; %initialize objective
Gradient=0; %initialize Gradient
for i=1:numberOfTrainingPoints %repeat for all training point
   %get the ANN response and Jacobian at the ith training point
   [ANNResponse, Jacobian]=getANNResponseAndDerivative(trainingPoints(:,i),
                                    Weights, numberOfHiddenNeurons, numberOfOutputs)
   Error=(ANNResponse-trainingResponses(:,i)); %errror vector between ANN response and
                                                  %ith response
   Objective=Objective+(norm(Error))^2; %update objective function
   Gradient=Gradient+2*Jacobian'*Error; %update gradient with ith training point contribution
end
```

```
%M7.8
%This program implements the DFB for minimization of functions for training
%Artificial Neural networks (ANNs). Only single hidden layer feed forward
%ANNs are considered
numberOfOutputs=1; %this is m
numberOfInputs=2; %this is n
numberOfHiddenNeurons=2; %this is p
numberOfWeights=numberOfHiddenNeurons*(numberOfOutputs+numberOfInputs);%np+mp
Epsilon=1.0e-3; %perturbation used in finite differences
% initialize weights
oldWeights=-3*ones(numberOfWeights,1)+6*rand(numberOfWeights,1); % random values [-3.0, 3.0]
[oldValue oldGradient]=getANNObjectiveAndDerivative(oldWeights) %Get the objective function and
                                                   %its gradient at the starting point
BOld=eye(numberOfWeights); %initialize the approximation to the Hessian inverse by the identity
Step=0.5; % initial step=0.5 in all iterations
while(oldValue>Epsilon) %repeat until the objective function is small enough
   QuasiNewtonDirection=-1.0*BOld*oldGradient; %get the current search direction
   % we will reduce the step until a successful step is reached
   newValue=1.5*oldValue; %initialization before logical condition
   while (newValue>oldValue) %repeat until a successful step is made
      newWeights=oldWeights+Step*QuasiNewtonDirection %get new point
      [newValue, newGradient]=getANNObjectiveAndDerivative(newWeights) %get new value
      if(newValue>oldValue) %unsuccessful step
         Step=Step*0.8; %reduce step size and repeat
      end
   end
   %now we update the inverse of the Hessian for the next iteration
   deltaGradient=newGradient-oldGradient; %this is the gradient difference
   deltaWeights=newWeights-oldWeights; %this is the weights difference
   tempVector=BOld*deltaGradient; %one of the formula's terms
   BNew=BOld+((deltaWeights*deltaWeights')/(deltaWeights'*deltaGradient))-
           ((tempVector*tempVector')/(tempVector'*deltaGradient)) %This is the DFP formula
   %Now we make the new values the current values
   BOld=BNew
   oldWeights=newWeights;
   oldValue=newValue
   oldGradient=newGradient
   Step=0.5; % reset step to ideal value
end
oldValue
oldWeights
```

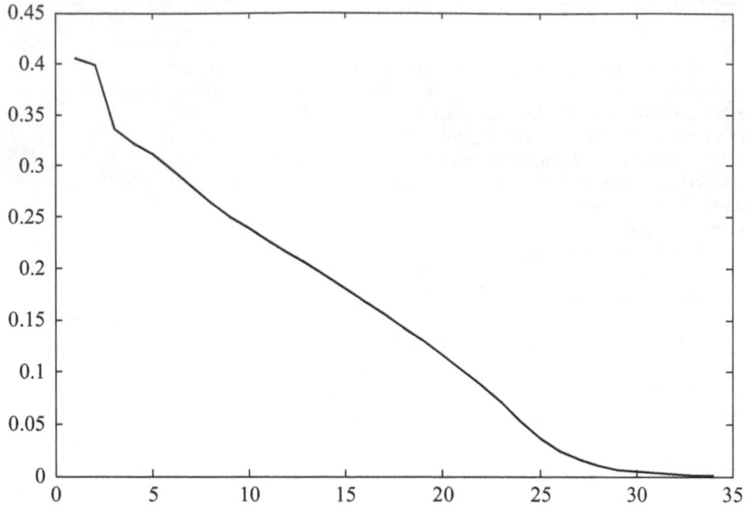

Figure 7.14 *The objective function at every iteration during the training of the ANN of Example 7.6*

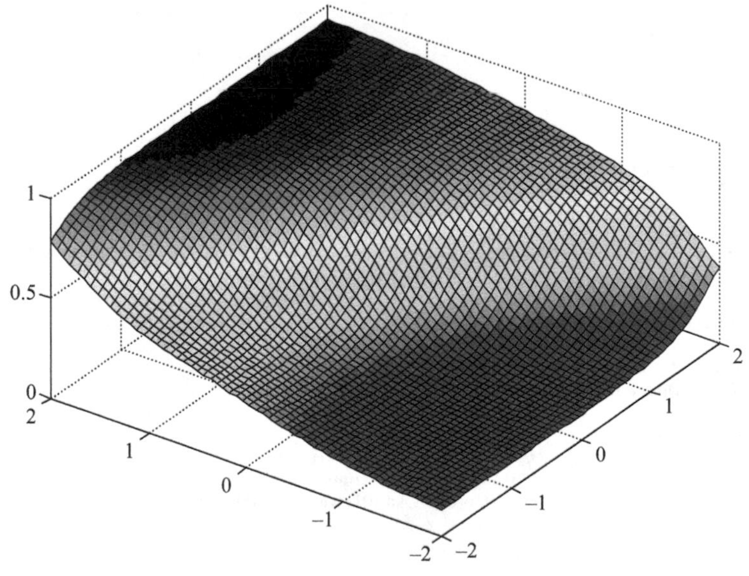

Figure 7.15 *The response surface of the trained ANN of Example 7.6. The response surface matches well the given logic table at all four training points*

values of the input vector $\boldsymbol{x}$. Using this Jacobian, the gradient of the objective function (7.45) is given by:

$$\frac{\partial f}{\partial \boldsymbol{w}} = 2\sum_{l=1}^{M} \boldsymbol{J}^T(\boldsymbol{w}, \boldsymbol{x}_l)\boldsymbol{E}_l \tag{7.48}$$

It follows that each training point contributes to the gradient of the objective function. The MATLAB listing M7.6 shows an implementation of a function that calculates the ANN output $\boldsymbol{R}(\boldsymbol{x}, \boldsymbol{w})$ for given input parameters and weights. This function also evaluates the Jacobian of the response by evaluating (7.46) and (7.47).

The MATLAB listing M7.7 shows an implementation of the function that calculates the training objective function (7.45) and its gradient (7.48). This function calls the neural network evaluating function for every training point to determine the ANN response and Jacobian at this point.

The following example illustrates the applications of the training process to ANN.

Example 7.6: Utilize the MATLAB listings M7.6 and M7.7 and a quasi-Newton approach to train an ANN to model the OR logic function. The four training points of the data are given by:

$$\left\{\boldsymbol{x}_1 = \begin{bmatrix} 0 \\ 0 \end{bmatrix}, y_1 = 0\right\}, \left\{\boldsymbol{x}_2 = \begin{bmatrix} 0 \\ 1 \end{bmatrix}, y_2 = 1\right\}$$

$$\left\{\boldsymbol{x}_3 = \begin{bmatrix} 1 \\ 0 \end{bmatrix}, y_3 = 1\right\}, \left\{\boldsymbol{x}_4 = \begin{bmatrix} 1 \\ 1 \end{bmatrix}, y_4 = 1\right\}$$

Solution: This problem has $n=2$ and $m=1$. We utilize an ANN with a single hidden layer of neurons with only two neurons. The total number of weights is thus 6. The MATLAB listing M7.8 shows a modified DFP code for the training of the ANN. The initial weights are started randomly within a given range. The algorithm terminates when the error drops below 1.0e-3. The objective function value at every iteration is shown in Figure 7.14. The surface response of the ANN is shown at the final set of weights in Figure 7.15. It is obvious that the ANN has been trained to correctly model the OR function. This problem is an under-determined problem because the number of given data pieces is less than the number of unknowns. The problem thus has more than one unique solution.

References

1. Singiresu S. Rao, *Engineering Optimization Theory and Practice*, Third Edition, John Wiley & Sons Inc, New York, 1996
2. Edwin K.P. Chong and Stanislaw H. Zak, *An Introduction to Optimization*, John Wiley & Sons Inc, Second Edition, New Jersey, 2001
3. Andreas Antoniou and Wu-Sheng Lu, *Practical Optimization Algorithms and Engineering Applications*, Springer, New York, 2007
4. John W. Bandler and R.M. Biernacki, *ECE3KB3 Courseware on Optimization Theory*, McMaster University, Canada, 1997
5. Simon Haykin, *Neural Networks: A Comprehensive Foundation*, 2nd Edition, Prentice-Hall, New Jersey, 1999

Problems

7.1 Apply Newton's method to the minimization of the function $f(x_1, x_2) = x_1^2 + 4x_2^2 + x_1x_2 + 2x_2 + 1$. Start from the starting point $\boldsymbol{x} = [0 \quad 2]^T$. Confirm your answer analytically by enforcing first-order necessary conditions.

7.2 Utilize the Levenberg–Marquardt method to minimize the three-dimensional function $f(\boldsymbol{x}) = \frac{1}{1-(x_1-x_2)^2} + \sin\left(\frac{\pi}{2}x_2x_3\right) + \exp\left(-\left(\frac{x_1+x_3}{x_2} - 2\right)^2\right)$ starting with the initial point $\boldsymbol{x}^{(0)} = [0 \quad 1 \quad 2]^T$. Utilize the MATLAB listing M7.1 after making the necessary changes.

7.3 Minimize the function $f(\boldsymbol{x}) = \frac{x_1^4}{3} + \frac{x_2^2}{2} - x_1x_2 + x_1 - x_2$ using the DFP quasi-Newton method starting from the two points $\boldsymbol{x}^{(0)} = [1.0 \quad 1.0]^T$ and $\boldsymbol{x}^{(0)} = [1.5 \quad 1.5]^T$. Utilize the MATLAB listing M7.3. Do you get the same answer in both cases? Verify your answers by obtaining a contour plot of this function.

7.4 Adapt the MATLAB listing M7.3 to implement the BFGS formula. Apply this code to minimize the function $f(\boldsymbol{x}) = (x_1^2 + x_2^2 - 9)^2 + (x_1 + x_2 - 2)^2$ starting from the point $\boldsymbol{x}^{(0)} = [1.0 \quad 1.0]^T$.

7.5 Measurements of a sinusoidal signal $s(t) = A\sin(\omega t + \theta)$ of unknown amplitude, frequency, and phase at different time instants are given by the table below:

t (msec)	0	0.1	0.2	0.3	0.4	0.5	0.6	0.7	0.8	0.9	1.0
Value	2.83	3.99	3.00	0.37	−2.45	−3.95	−3.31	−0.87	2.04	3.84	3.56

Utilize the MATLAB listing M7.5 to solve the corresponding nonlinear least squares problems for the unknowns $\boldsymbol{x} = [A \quad \omega \quad \theta]^T$ after making any necessary adjustments. Notice that the parameters have different orders of magnitude and scaling may be needed.

7.6 Solve the system of nonlinear equations $x^2 + y^2 - 9 = 0$ and $y - x^2 + 1 = 0$. Utilize the Gauss–Newton code M7.5 with the starting point $[x^{(0)} \quad y^{(0)}]^T = [1.0 \quad 1.0]^T$. Verify your answer through a graphical solution.

7.7 Consider the following data:

x	1.0	1.5	2.0	2.5	3.0	3.5	4.0	4.5	5
y	5	6.4	8.05	9.5	11.1	12.7	13.95	15.62	17.2

Fit this data to the line $y = ax + b$. Utilize linear least squares to solve for the unknowns a and b.

7.8 Train a three-layer feed forward ANN with three hidden neurons to model the nonlinear function $f(\boldsymbol{x}) = \frac{x_1^2}{3} + \frac{x_2^2}{2} + 0.2$ over the interval $-1.0 \le x_1 \le 1.0$ and $-1.0 \le x_2 \le 1.0$. Utilize a training set of 16 points over this domain. The MATLAB listings M7.6 and M7.7 should be used after making the necessary changes. Check the accuracy of your trained ANN by plotting its response surface and comparing it to the original function.

Chapter 8

Constrained optimization techniques

8.1 Introduction

In the last three chapters, we addressed the unconstrained optimization problem. In this problem, the solution can be any point in the parameter space. Practical problems, however, always impose some form of constraints. For example, the current that can be drawn from a battery must be below a certain value to guarantee a constant voltage. The price of a product cannot be reduced below a certain limit to cover the production costs and a small profit margin. The distance between electric components on a high-frequency chip cannot be reduced below a certain value to avoid coupling that may seriously reduce the performance. There are always some constraints that limit the minimization of the considered objective function.

Many of the concepts utilized in unconstrained optimization techniques can be extended to the constrained case. This is why the study of unconstrained optimization is useful as a start. The optimality conditions for the constrained case, however, are different. In some cases, the constraints are redundant and the solution obtained using unconstrained optimization techniques is the same one obtained using techniques for constrained optimization. In other cases, the constraints limit the feasibility region giving rise to a solution completely different from the unconstrained one.

Similar to the unconstrained optimization case, some of the addressed techniques are derivative free while other techniques utilize gradients of the objective functions and the constraints. I show in all these algorithms how optimization is utilized to locate the optimal solution over the feasible region as defined by all the constraints.

8.2 Problem definition

The problem we aim at solving is to minimize an objective function over all points in the feasible region defined by the constraints. This problem may be given by the following formulation:

$$\begin{aligned} & \boldsymbol{x}^* = \min_{\boldsymbol{x}} f(\boldsymbol{x}) \\ \text{subject to} \quad & y_j(\boldsymbol{x}) \le 0, \quad j = 1, 2, \ldots, m \\ & h_k(\boldsymbol{x}) = 0, \quad k = 1, 2, \ldots, p \end{aligned} \tag{8.1}$$

where m is the number of inequality constraints and p is the number of equality constraints. The functions $y_j(\boldsymbol{x})$ and $h_k(\boldsymbol{x})$ are, in general, nonlinear functions of the optimization variables. Most design problems in the area of electrical engineering give rise to constrained optimization problems with only inequality constraints. Equality constraints are more stringent and rarely used in reality. An equality constraint can be substituted by two inequality constraints. We will thus mainly focus on design problems with only inequality constraints.

8.3 Possible optimization scenarios

The inequality constraints $y_j(\boldsymbol{x}) \leq 0, j = 1, 2, \ldots, m$ define a subset of the parameter space where all the constraints are negative (satisfied). This region interacts with the constant value surfaces of the objective function to determine the possible solution of the optimization problem. A number of possible scenarios may take place. These scenarios are illustrated in Figures 8.1–8.3 for simple two-dimensional problems. In Figure 8.1, the unconstrained minimum of the objective functions is inside the feasible region. This means that the constraints are redundant in this case. This minimum could have been reached using unconstrained optimization techniques. A valid first approach to a design problem is to first ignore the constraints and use the much simpler unconstrained optimization techniques. If the achieved minimum is not feasible, constrained optimization techniques may be then used.

Figure 8.2 illustrates the case where the unconstrained minimum of the objective function is outside the feasible region defined by the constraints. Applying unconstrained optimization techniques would give a non-feasible solution. The minimum of the constrained problem lies on one of the constraints or on the intersection of a number of them. This minimum point satisfies the KKT conditions discussed in Chapter 3.

Figure 8.3 shows the case where the objective function has two unconstrained minima. One of these minima lies inside the feasible region and thus is a possible solution to the constrained optimization problem. The interaction between the contours of the objective functions and the boundaries of the feasible region creates another local minimum at the boundary. The constrained optimization problem has thus two minima.

Most of the techniques we discuss in this chapter are local constrained optimization techniques. They focus on finding only one local solution of the constrained optimization problem. Global optimization techniques are discussed in Chapter 9.

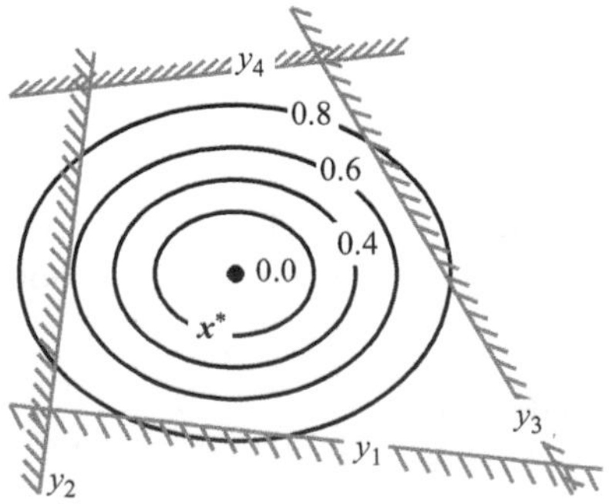

Figure 8.1 An illustration of a possible scenario for a two-dimensional problem with four constraints; the unconstrained minimum of the objective function is inside the feasible region. Unconstrained optimization would give the same result as constrained optimization because the constraints are redundant

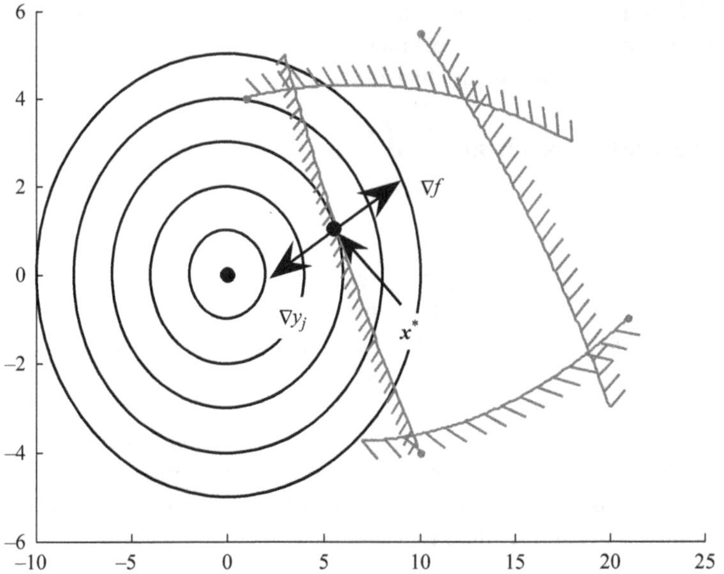

Figure 8.2 *An illustration of a possible scenario for a two-dimensional constrained problem; the unconstrained objective function minimum is outside the feasible region. The minimum of the constrained problem is on the boundary where the objective function contour touches the jth constraint y_j*

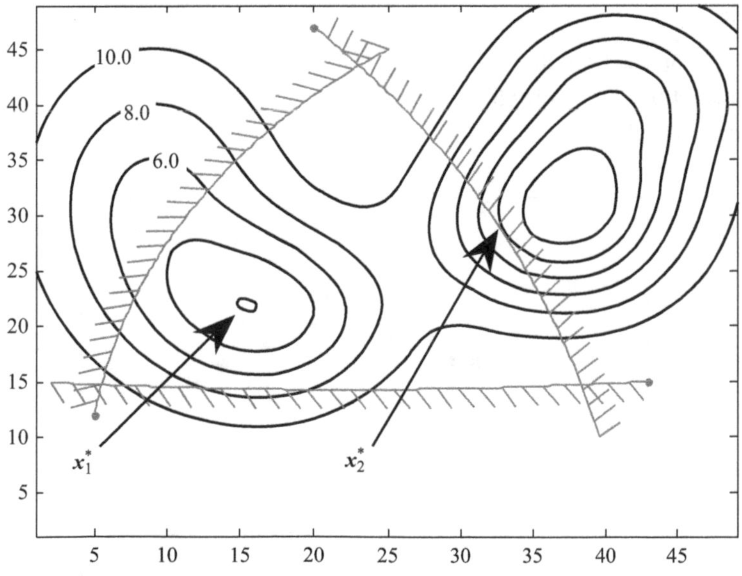

Figure 8.3 *An illustration of a possible scenario for a two-dimensional constrained problem; the objective function has two unconstrained minima with only one of them feasible (x_1^*). The interaction between the contours of the objective function and the boundary creates a second local minimum for the constrained problem (x_2^*)*

We discuss in this chapter both derivative-free and gradient-based constrained optimization techniques. Some of these techniques are a direct extension to unconstrained optimization techniques discussed earlier.

8.4 A random search method

The basic concept in random search approaches is to randomly generate points in the parameter space. Only feasible points satisfying $y_j(\boldsymbol{x}) \leq 0, j = 1, 2, \ldots, m$ are considered, while non-feasible points with at least one $y_j(\boldsymbol{x}) > 0$ for some j are rejected. The algorithm keeps track of the feasible random point with the least value of the objective function. This requires checking, at every iteration, if the newly generated feasible point has a better objective function than the best value achieved so far. Figure 8.4 illustrates the random search approach for the two-dimensional case. The MATLAB® listing M8.1 shows a possible implementation of this algorithm. The main disadvantage of this algorithm is that a large number of objective function calculations may be required especially for problems with large n.

The following example illustrates this technique.

Example 8.1: Find a solution for the two-dimensional constrained optimization problem:

$$\begin{aligned} &\min_{\boldsymbol{x}} (x_1^2 + x_2^2) \\ &x_1 - x_2^2 - 4 \geq 0 \\ \text{subject to} \quad &x_1 - 10 \leq 0 \end{aligned} \tag{8.2}$$

using the MATLAB listing M8.1. Verify your answer by applying the KKT conditions.

Solution: The output from the MATLAB code M8.1 for Problem 8.2, after 100,000 iterations, is:

IterationCounter = 100000
OldPoint = [4.0584 −0.1856]′
OldValue = 16.5051

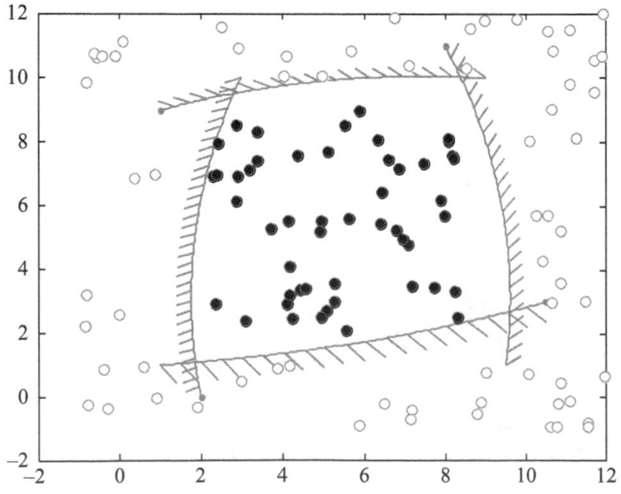

Figure 8.4 The generation of random points for a constrained optimization problem. Only the feasible points (•) are considered while non-feasible points (○) are discarded

```
%M8.1
%The Random Search Approach for constraint problems
NumberOfParameters=2; %This is n for this problem
NumberOfContsraints=2; %This is m for this problem
UpperValues=[10  10]'; %upper values
LowerValues=[-10  -10]'; %lower values
OldValue=1.0e9; %select a large initial value for the minimum
MaximumNumberOfIterations=100000; %maximum number of allowed iterations
IterationCounter=0; %iteration counter
while(IterationCounter<MaximumNumberOfIterations) %repeat until maximum number of iteration
  RandomVector=rand(NumberOfParameters,1); %get a vector of random variables
  NewPoint=LowerValues+RandomVector.*(UpperValues-LowerValues); %Get new random point
  NewValue=getObjective(NewPoint); %get new objective function value
  ConstraintsValues=getConstratints(NewPoint); %get the value of ALL constraints at the new point
  if((NewValue<OldValue)&&(max(ConstraintsValues)<0)) %is there an improvement and the new
                                                      %point is feasible?.
    OldPoint=NewPoint; %adjust best value
    OldValue=NewValue;
  end
  IterationCounter=IterationCounter+1; %increment the iteration counter
end
IterationCounter
OldPoint
OldValue
```

The point returned by the random optimization algorithm is $\boldsymbol{x}^* = [4.0584 \quad -0.1856]^T$. To apply the KKT conditions, the first constraint must be put in the standard form by multiplying both sides by -1. The gradients of the objective function and the constraints are given by:

$$\nabla f = \begin{bmatrix} 2x_1 \\ 2x_2 \end{bmatrix}, \quad \nabla y_1 = \begin{bmatrix} -1 \\ 2x_2 \end{bmatrix}, \quad \nabla y_2 = \begin{bmatrix} 1 \\ 0 \end{bmatrix} \tag{8.3}$$

Because we have two constraints, there are four possibilities: none of the constraints is active at the optimal point, only the first one is active, only the second one is active, or both are active. The KKT conditions should be checked for all these cases. First, assuming that none of the constraints is active at the optimal point, we then obtain the optimal solution:

$$\nabla f = \begin{bmatrix} 2x_1 \\ 2x_2 \end{bmatrix} = \begin{bmatrix} 0 \\ 0 \end{bmatrix} \Rightarrow x_1 = 0 \text{ and } x_2 = 0 \tag{8.4}$$

This point, however, violates the first constraint and is thus a non-feasible point. This eliminates the unconstrained minimum of the objective function. If we assume that only the first constraint is active, we have the optimality conditions:

$$\nabla f + \lambda_1 \nabla y_1 = 0 \Rightarrow \begin{bmatrix} 2x_1 \\ 2x_2 \end{bmatrix} + \lambda_1 \begin{bmatrix} -1 \\ 2x_2 \end{bmatrix} = \begin{bmatrix} 0 \\ 0 \end{bmatrix} \tag{8.5}$$

Using (8.5), we can express the parameters in terms of the Lagrange multiplier to get $x_1 = 0.5\lambda_1$ and $x_2 = 0$. Substituting into the active constraint $x_1 - x_2^2 - 4 = 0$, we get $\lambda_1 = 8$ and $x_1 = 4$. The point $\boldsymbol{x} = [4.0 \quad 0]^T$ is a feasible point as the second constraint ($x_1 \leq 10$) is satisfied. It follows that this point is indeed a solution of the constrained optimization problem (8.2). It is obvious that the solution obtained by the random solver is close to this optimal point.

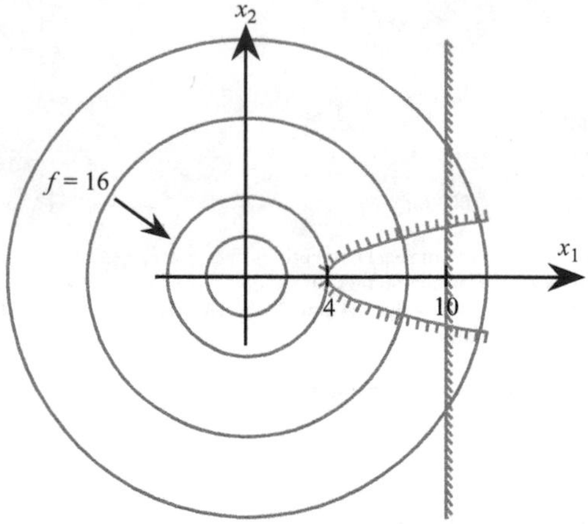

Figure 8.5 The contours of the objective function and the constraints of Example 8.1

It is instructive to check the other two possibilities even though a solution was already found. If we assume that only the second constraint is active, the optimality conditions become:

$$\nabla f + \lambda_2 \nabla y_2 = 0 \Rightarrow \begin{bmatrix} 2x_1 \\ 2x_2 \end{bmatrix} + \lambda_2 \begin{bmatrix} 1 \\ 0 \end{bmatrix} = \begin{bmatrix} 0 \\ 0 \end{bmatrix} \tag{8.6}$$

which implies that $x_1 = -0.5\lambda_2$ and $x_2 = 0$. Substituting in the active second constraint, we get $\lambda_1 = -20$ and $x_1 = 10$. This point is not a local minimizer as the corresponding multiplier has a negative value.

The last case is to assume that both constraints are active. The optimality conditions are thus given by:

$$\nabla f + \lambda_1 \nabla y_1 + \lambda_2 \nabla y_2 = 0 \Rightarrow \begin{bmatrix} 2x_1 \\ 2x_2 \end{bmatrix} + \lambda_1 \begin{bmatrix} -1 \\ 2x_1 \end{bmatrix} + \lambda_2 \begin{bmatrix} 1 \\ 0 \end{bmatrix} = \begin{bmatrix} 0 \\ 0 \end{bmatrix} \tag{8.7}$$

which gives:

$$x_1 = 0.5\lambda_1 - 0.5\lambda_2 \text{ and } x_2 = -\lambda_1 x_1 \tag{8.8}$$

Substituting into the two active constraints, we get $x_1 = 10$ and $x_2 = \pm\sqrt{6}$. It follows that the two constraints intersect at two possible points. Substituting in (8.8), we get at least one negative Lagrange multiplier for both points. It follows that these two points are not local minima of the problem (8.2). A graphical solution of the problem (8.2) is shown in Figure 8.5.

8.5 Finding a feasible starting point

Many of the constrained optimization algorithms adopt an iterative approach. Starting from an initial feasible point $\boldsymbol{x}^{(0)}$, the algorithm proceeds to generate other feasible points with better values of the objective function. The iterative approach is illustrated in Figure 8.6. These approaches require a feasible starting point. Such a point may be easily guessed in some problems. In other problems, it is not obvious how to get an initial feasible point.

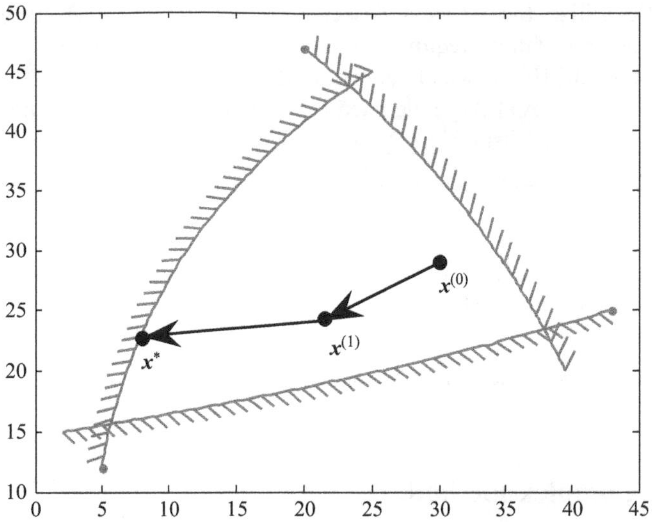

Figure 8.6 *An illustration of an iterative constrained optimization algorithm; starting with a feasible point* $\boldsymbol{x}^{(0)}$*, the algorithm generates other feasible points with better values of the objective function*

A two-phase approach is thus adopted by some algorithms. These algorithms first solve the unconstrained optimization problem:

$$\min_{x} \left(\sum_{j=1}^{m} Y_j^2 + \sum_{k=1}^{p} H_k^2 \right) \tag{8.9}$$

where the functions $Y_j(\boldsymbol{x})$ and $H_k(\boldsymbol{x})$ are given by:

$$\begin{aligned} Y_j(\boldsymbol{x}) &= \max\{0, y_j(\boldsymbol{x})\}, \quad j = 1, 2, \ldots, m \\ H_k(\boldsymbol{x}) &= h_k(\boldsymbol{x}), \quad k = 1, 2, \ldots, p \end{aligned} \tag{8.10}$$

Notice that (8.9) is a sum of squares. Its lowest possible value should be zero and is achieved when all of its terms are zeros. Each term Y_j would assume a value of zero only if the corresponding constraint $y_j(\boldsymbol{x})$ is negative or zero. This means that the corresponding constraint is satisfied. Similarly, each term $H_k(\boldsymbol{x})$ reaches the value of zero if the corresponding equality constraint is satisfied ($h_k(\boldsymbol{x}) = 0$). It follows that the minimum of the objective function (8.9) is zero and is achieved when all inequality constraints are satisfied ($y_j(\boldsymbol{x}) \leq 0, \forall j$) and all equality constraints are satisfied ($h_k(\boldsymbol{x}) = 0, \forall k$). This minimum is then used as a starting point for solving the constrained optimization problem (8.1).

The following example illustrates this approach for starting a constrained optimization problem.

Example 8.2: Find a feasible starting point for a constrained optimization problem with the following constraints:

$$\begin{aligned} x_2 &\leq x_1^2 + 4 \\ x_2 &\leq (x_1 - 2)^2 + 3 \end{aligned} \tag{8.11}$$

starting with the non-feasible point $\boldsymbol{x} = [1.0 \quad 5.0]^T$.

Solution: We utilize the steepest descent method listing M6.1 in solving this problem. The algorithm required only one iteration to reach the point $x = [-2.5768 \quad 3.2107]^T$ which corresponds to a value of 0 of the objective function (8.9). The MATLAB code used to evaluate the objective function (8.9) is given in the MATLAB listing M8.2.

```
%M8.2
%This function evaluates an auxiliary objective function for finding a feasible starting point
% of a constrained optimization problem
function Objective=getObjective(Point)
ConstraintsValues=getConstraints(Point); %get the vector of constraint values
Indices=find(ConstraintsValues>0); %determine the indices of violated constraints
%sum the square of the values of violated constraints
Objective=ConstraintsValues(Indices)'*ConstraintsValues(Indices);
```

8.6 The Complex method

The Complex method is a derivative-free constrained optimization technique. It extends the Simplex method discussed in Chapter 5 to constrained optimization. This algorithm starts with $(n+1)$ feasible points forming an initial simplex. These initial points can be formed by using small perturbations from a feasible starting point. At every iteration, the simplex point x_h with the highest value for the objective function is determined. The centroid of all other simplex points is then given by:

$$x_o = \frac{1}{N+1} \sum_{j, j \neq h} x_j \tag{8.12}$$

Similar to the Simplex method, the point x_h is then reflected along the point x_o to obtain the reflected point x_r (see Figure 8.7). The reflected point is given by the formula:

$$x_r = (1+\alpha)x_o - \alpha x_h \tag{8.13}$$

A main difference between the complex and the simplex approach is that the reflection ratio is adjusted to guarantee that the reflected point x_r is feasible, i.e., $y_j(x_r) \leq 0$, $\forall j$ and that a reduction in the objective function is achieved ($f(x_r) < f(x_h)$). A possible implementation of the Complex method is shown in the MATLAB listing M8.3.

Example 8.3: Solve the following constrained optimization problem:

$$\begin{aligned} &\min_x x_1^2 + x_2^2 - 6x_1 - 8x_2 + 10 \\ &\text{subject to} \begin{cases} 4x_1^2 + x_2^2 \leq 16 \\ 3x_1 + 5x_2 \leq 15 \\ x_1 \geq 0, x_2 \geq 0 \end{cases} \end{aligned} \tag{8.14}$$

starting with $x^{(0)} = [1.0 \quad 1.0]^T$. Utilize the complex implementation in the listing M8.3.

Solution: The point $x^{(0)}$ is obviously a feasible point as all constraints are satisfied. The initial simplex is created by making small perturbations around the initial points in all coordinate directions. The points forming the initial simplex are given by:

$$\left\{ \begin{bmatrix} 1.0 \\ 1.0 \end{bmatrix}, \begin{bmatrix} 1.5 \\ 1.0 \end{bmatrix}, \begin{bmatrix} 1.0 \\ 1.5 \end{bmatrix} \right\} \tag{8.15}$$

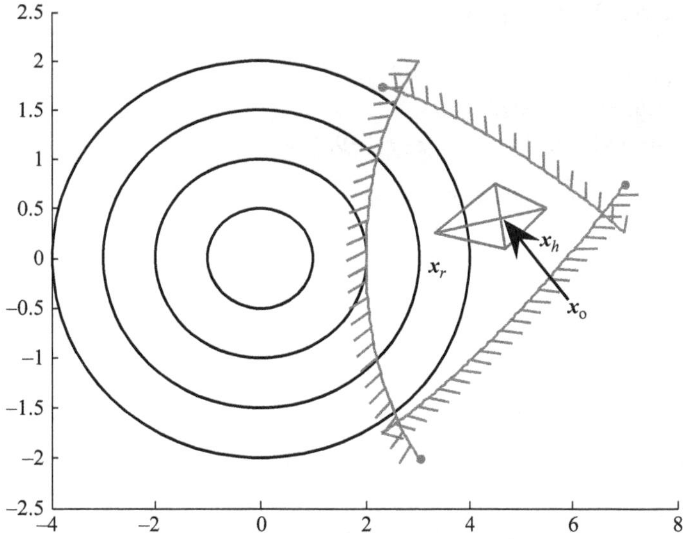

Figure 8.7 An illustration of the Complex method; the algorithm expands and contracts the simplex at every iteration within the feasible region

```
%M8.3
% Implementation of the Complex method
NumberOfParameters=2; %This is n for this problem
NumberOfConstraints=4; %This is m for this problem
[Simplex, SimplexValues]=getInitialFeasibleSimplex(); %get initial feasible simplex
IterationCounter=0;
while(IterationCounter<100) %repeat
  [MaxValue MaxIndex]=max(SimplexValues); %Get the maximum value and its index
  CenterLower=GetCenter(Simplex,MaxIndex);%Get center of points with lower function values
  ReflectionAccepted=0;
  ReflectionCoefficient=1.0;
  while(ReflectionAccepted==0) %repeat until an accepted reflected point is found
    ReflectedPoint=(1.0+ReflectionCoefficient)*CenterLower-ReflectionCoefficient*Simplex
                                                    (:,MaxIndex);
    ReflectedValue=feval('getObjective',ReflectedPoint); %Get the reflected value
    ReflectedConstraintsValues=getConstraints(ReflectedPoint); %get values of constrants
    if((ReflectedValue<MaxValue)&&(max(ReflectedConstraintsValues)<0)) %is there an
                                                    %improvement?
      Simplex(:,MaxIndex)=ReflectedPoint; %now we store the reflected point
      SimplexValues(1,MaxIndex)=ReflectedValue;
      ReflectionAccepted=1.0;
    else %no improvment. use a smaller reflection coefficient
      ReflectionCoefficient=ReflectionCoefficient*0.8;
    end
  end
  IterationCounter=IterationCounter+1; %increase the iteration counter
end
```

After seven iterations, the code gives the output:

$$\text{Simplex} = [1.3445 \quad 1.5812 \quad 1.4378$$
$$2.1868 \quad 2.0440 \quad 2.1368]$$
$$\text{SimplexValues} = [-8.9718 \quad -9.1611 \quad -9.0879]'$$
$$\text{ReflectedConstraintsValues} = [-3.1653 \quad -0.0027 \quad -1.4378 \quad -2.1368]'$$

The final output from the program is:

Simplex = [1.7447   1.7447   1.7447
          1.9532   1.9532   1.9532]
SimplexValues = [−9.2347 −9.2347 −9.2347]′

It is obvious that the simplex has shrunk to virtually one point at $\boldsymbol{x}^* = [\,1.7447 \quad 1.9532\,]^T$ with a corresponding objective −9.2347.

The result obtained through the Complex method can be verified analytically by applying the KKT conditions. Because we have 4 constraints, there are 16 different combinations of active/inactive constraints. We will consider here only the case when the second constraint is active. In this case, the KKT conditions become:

$$\nabla f + \lambda_2 \nabla y_2(\boldsymbol{x}) = 0 \tag{8.16}$$

which gives:

$$\begin{bmatrix} 2x_1 - 6 \\ 2x_2 - 8 \end{bmatrix} + \lambda_2 \begin{bmatrix} 3 \\ 5 \end{bmatrix} = 0 \Rightarrow x_1 = -\frac{3}{2}\lambda + 3, x_2 = -\frac{5}{2}\lambda + 4 \tag{8.17}$$

Substituting into the assumed active second constraint, we get:

$$3\left(-\frac{3}{2}\lambda + 3\right) + 5\left(-\frac{5}{2}\lambda + 4\right) = 15 \Rightarrow \lambda = 0.8235 \tag{8.18}$$

Using (8.18), this positive Lagrange multiplier gives the values $x_1 = 1.7647$ and $x_2 = 1.9500$. These values are very close to the ones obtained using the Complex method.

8.7 Sequential linear programming

The sequential linear programming (SLP) aims at utilizing linear programming techniques (discussed in Chapter 2) to solve nonlinear constrained optimization problems. The basic idea is illustrated in Figure 8.8. Starting with a feasible point $\boldsymbol{x}^{(0)}$, both nonlinear constraints and objective function are linearized at this point to

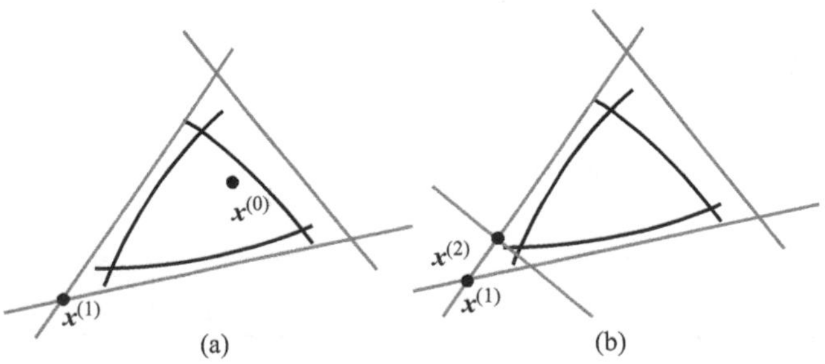

Figure 8.8 An illustration of two iterations of the Sequential Linear Programming (SLP) approach; (a) in the first iteration, all constraints are linearized at the feasible point $\boldsymbol{x}^{(0)}$. $\boldsymbol{x}^{(1)}$ is the solution of the corresponding linear program, and (b) the most violated constraint is linearized at the point $\boldsymbol{x}^{(1)}$ to obtain a better approximation of the feasible region. The point $\boldsymbol{x}^{(2)}$ is the solution of the corresponding linear programming problem

obtain a linear program (LP). This LP is then solved to obtain the point $\boldsymbol{x}^{(1)}$ that may not be feasible. The most violated constraint at the point $\boldsymbol{x}^{(1)}$ is then linearized again and then added to the set of linearized constraint. The LP is then resolved again to obtain the point $\boldsymbol{x}^{(2)}$. The process of linearizing the most violated constraint and adding it to the set of linear constraints is repeated until the nonlinear region has been sufficiently approximated by these linearizations. This will be the case if the original feasible region is convex and can thus be approximated by a sufficient number of linear hyperplanes. The solution of the LP should be close enough to the solution of the original nonlinear problem.

Mathematically, the original nonlinear problem (8.1) is converted in the first iteration to the equivalent LP:

$$\begin{aligned} &\min_{x} f(\boldsymbol{x}^{(0)}) + \nabla f(\boldsymbol{x}^{(0)})^T(\boldsymbol{x} - \boldsymbol{x}^{(0)}) \\ \text{subject to}\quad & y_j(\boldsymbol{x}^{(0)}) + \nabla y_j(\boldsymbol{x}^{(0)})^T(\boldsymbol{x} - \boldsymbol{x}^{(0)}) \le 0, \quad j = 1, 2, \ldots, m \end{aligned} \tag{8.19}$$

The LP (8.18) is then solved. At each of the following iterations the most violated constraint is linearized at the new point and added to the initial set of m linearizations in (8.19). This process is repeated until a termination condition has been met. The resulting LP at the kth iteration is given by:

$$\begin{aligned} &\min_{x} \boldsymbol{c}^T\boldsymbol{x} \\ &\boldsymbol{A}\boldsymbol{x} \le \boldsymbol{b} \end{aligned} \tag{8.20}$$

where $\boldsymbol{c} = \nabla f$. The matrix $\boldsymbol{A}$ and vector $\boldsymbol{b}$ are given by:

$$\boldsymbol{A} = \begin{bmatrix} \nabla y_1(\boldsymbol{x}^{(0)})^T \\ \vdots \\ \nabla y_m(\boldsymbol{x}^{(0)})^T \\ \nabla y_{I_1}(\boldsymbol{x}^{(1)})^T \\ \vdots \\ \nabla y_{I_K}(\boldsymbol{x}^{(k)})^T \end{bmatrix} \quad \text{and} \quad \boldsymbol{b} = \begin{bmatrix} \nabla y_1(\boldsymbol{x}^{(0)})^T\boldsymbol{x}^{(0)} - y_1(\boldsymbol{x}^{(0)}) \\ \vdots \\ \nabla y_m(\boldsymbol{x}^{(0)})^T\boldsymbol{x}^{(0)} - y_m(\boldsymbol{x}^{(0)}) \\ \nabla y_{I_1}(\boldsymbol{x}^{(1)})^T\boldsymbol{x}^{(1)} - y_1(\boldsymbol{x}^{(0)}) \\ \vdots \\ \nabla y_{I_K}(\boldsymbol{x}^{(k)})^T\boldsymbol{x}^{(k)} - y_{I_K}(\boldsymbol{x}^{(k)}) \end{bmatrix} \tag{8.21}$$

where $y_{Ij}(\boldsymbol{x}^{(j)})$ is the value of the linearized constraint at the jth iteration. A possible MATLAB implementation of the SLP is given in the MATLAB listing M8.4. The following example illustrates the application of this method to nonlinear optimization.

Example 8.4: Solve the following variation of Example 8.3:

$$\begin{aligned} &\min_{x} x_1^2 + x_2^2 - 6x_1 - 8x_2 + 10 \\ \text{subject to}\quad & 4x_1^2 + x_2^2 \le 16 \\ & 3x_1 + 5x_2 \le 15 \end{aligned} \tag{8.22}$$

starting from the point $\boldsymbol{x}^{(0)} = [1.0 \quad 1.0]^T$. Utilize the SLP implementation in the listing M8.4.

Solution: The output from the MATLAB code for this problem is as follows:
$x(0) = [1 \quad 1]'$

A(1:2,:) = [8 2
3 5]
b(1:2,:) = [21
15]

```
%M8.4
%Implementation of the Sequential Linear Programming
NumberOfParameters=2; %This is n for this problem
NumberOfConstraints=2; %This is m for this problem
MaximumLinearizedConstraints=20;
MaximumIterations=MaximumLinearizedConstraints-NumberOfConstraints;
UpperValues=[3  3]'; %upper values
LowerValues=[0  0]'; %lower values
A=zeros(MaximumLinearizedConstraints, NumberOfParameters); % linear program matrix
b=zeros(MaximumLinearizedConstraints,1); %right hand side
c=zeros(MaximumLinearizedConstraints,1); % linear program cost vectr
StartingPoint=[1.0  1.0]'; %This is x(0)
ConstraintsValues=getConstraints(StartingPoint); % get constraints values at x(0)
ConstraintsGradients=getConstraintsGradients(StartingPoint); %get the gradient of the constraints
ObjectiveValue=getObjective(StartingPoint); %get objective value at x(0)
ObjectiveGradient=getObjectiveGradient(StartingPoint); % get gradient of f at x(0)
% We now build the initial linear program given by (8.19)
A(1:NumberOfConstraints,:)=ConstraintsGradients; % fill the first m rows of A
b(1:NumberOfConstraints,:)=ConstraintsGradients*StartingPoint-ConstraintsValues;
c=ObjectiveGradient; %copy gradient
IterationCounter=0;
while(IterationCounter<MaximumIterations)
%Solve the LP problem to get x(k+1)
  NewPoint=linprog(c,A(1:NumberOfConstraints+IterationCounter,:),
                   b(1:NumberOfConstraints+IterationCounter),[],[],LowerValues,UpperValues);
  ConstraintsValues=getConstraints(NewPoint); %get constraints values at x(k+1)
  [MaxValue  MaxIndex]=max(ConstraintsValues); %determine the most violated constraint
  ConstraintsGradients=getConstraintsGradients(NewPoint); %constraints gradients at x(k+1)
  ObjectiveGradient=getObjectiveGradient(NewPoint); %gradient of f at x(k+1)
  IterationCounter=IterationCounter+1; %increment iteration counter
 %augment the linear program with the linearization of the most violated constraint
  A(NumberOfConstraints+IterationCounter,:) =ConstraintsGradients(MaxIndex,:);
  b(NumberOfConstraints+IterationCounter,:)=ConstraintsGradients(MaxIndex,:)*NewPoint-
                                      ConstraintsValues(MaxIndex,1); %update b
end
NewPoint %print final solution
```

NewPoint = [2.2059 1.6765]′

$$A(1{:}3,:) = \begin{bmatrix} 8.0000 & 2.0000 \\ 3.0000 & 5.0000 \\ 17.6471 & 3.3529 \end{bmatrix}$$

$$b(1{:}3,:) = \begin{bmatrix} 21.0000 \\ 15.0000 \\ 38.2742 \end{bmatrix}$$

NewPoint = [1.8046 1.9172]′

$$A(1{:}4,:) = \begin{bmatrix} 8.0000 & 2.0000 \\ 3.0000 & 5.0000 \\ 17.6471 & 3.3529 \\ 14.4368 & 3.8345 \end{bmatrix}$$

$$b(1{:}4,:) = \begin{bmatrix} 21.0000 \\ 15.0000 \\ 38.2742 \\ 32.7021 \end{bmatrix}$$

NewPoint = [1.7467 1.9520]′

The rows of the matrix A are the gradients of the linearized constraints. Four linearizations are needed to achieve the final solution. This solution

$\bar{x} = [1.7467 \quad 1.9520]^T$ is very close to the analytical solution obtained earlier ($x^* = [1.7647 \quad 1.9500]^T$).

8.8 Method of feasible directions

This method aims at generating a sequence of feasible points with continuously decreasing values of the objective function. It is different from the SLP method discussed earlier where the solution of LP may be non-feasible. At the kth iteration, the algorithm assumes that we have a feasible point $x^{(k)}$. A descent direction $s^{(k)}$ for the objective function is then determined. The algorithm then carries out a line search to obtain a new point $x^{(k+1)} = x^{(k)} + \lambda_k s^{(k)}$ such that this new point is also a feasible point with a better value of the objective function.

Two possible scenarios arise for the generation of a descent direction $s^{(k)}$ at the kth iteration. These two scenarios are illustrated in Figure 8.9. In Figure 8.9(a), the point $x^{(k)}$ is an interior point of the feasible region, i.e., no constraint is active at this point. In this case, the boundaries do not limit our choice of the descent direction $s^{(k)}$. It can then be chosen as the negative of the gradient of the objective function at the current point ($s^{(k)} = -\nabla f(x^{(k)})$). A line search is then carried out to obtain a feasible point with a better value of the objective function.

The second scenario is illustrated in Figure 8.9(b). In this case, the point $x^{(k)}$ is a boundary point, i.e., at least one of the constraints is active at this point. A new search direction must be generated such that it is a descent direction for the objective function and it moves away from the boundary, i.e., in the direction of decrease of the active constraints. Mathematically, the sought-after search direction should satisfy:

$$s^{(k)T} \nabla f(x^{(k)}) < 0, \quad s^{(k)T} \nabla y_j(x^{(k)}) \leq 0, \quad j \in I_{A_k} \tag{8.23}$$

where I_{A_k} is the index set of active constraints at the kth iteration. The first inequality of (8.23) imposes the condition that the search direction should have an obtuse angle with the gradient of the objective function. This implies that it is a direction of objective function descent. The second set of constraints imposes the condition that this direction should make an obtuse angle with the gradients of all active constraints. This means that the value of these constraints will not increase along that direction, thus keeping the new point feasible. As mentioned earlier, the constraints are violated if they become positive.

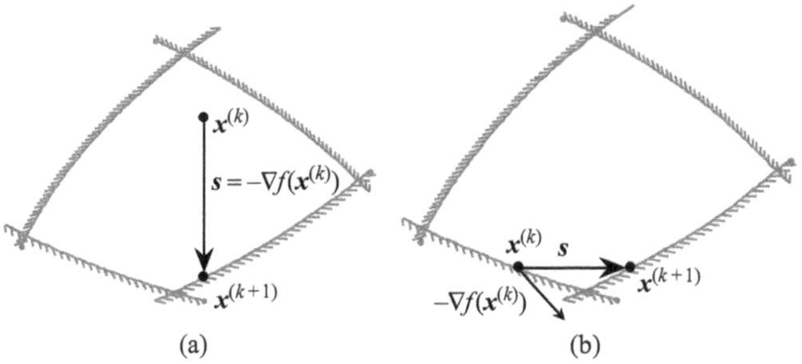

Figure 8.9 An illustration of the method of feasible directions; (a) for an interior point, the search direction is the steepest descent direction and (b) for a boundary point, a feasible direction is found that reduces the objective function but maintains feasibility

The method of feasible directions obtains a direction s_k by solving the auxiliary LP:

$$\begin{aligned} &\min\ (-\alpha) \\ \text{such that}\quad & \boldsymbol{s}^{(k)T}\nabla f(\boldsymbol{x}^{(k)}) + \alpha \le 0, \\ & \boldsymbol{s}^{(k)T}\nabla y_i(\boldsymbol{x}^{(k)}) + \alpha \le 0, \quad i \in I_{A_k} \\ & -1 \le s_j^{(k)} \le 1, \quad j = 1, 2, \ldots, n \end{aligned} \tag{8.24}$$

The LP (8.24) solves for both the slack variable α and the search direction $\boldsymbol{s}^{(k)}$. There are thus $(n+1)$ unknowns in this problem. This LP attempts to maximize the slack variable α by minimizing its negative. This should result in the largest possible value possible of this variable. This value, because of the imposed constraints, pushes the values $\boldsymbol{s}^{(k)T}\nabla f(\boldsymbol{x}^{(k)})$ and $\boldsymbol{s}^{(k)T}\nabla y_i(\boldsymbol{x}^{(k)})$ to be as negative as possible. This guarantees that $\boldsymbol{s}^{(k)}$ is a feasible and descent direction. The last set of conditions in (8.24) limit the norm of each component in the vector $\boldsymbol{s}^{(k)}$ to guarantee a unique solution.

Few points to notice about the LP (8.24): First, if there are no active constraints, i.e., the set I_{A_k} is empty, there is no need to solve (8.24). Its solution is known in advance ($\boldsymbol{s}^{(k)} = -\nabla f(\boldsymbol{x}^{(k)})$). Second, the ith constraint is considered active at a point $\boldsymbol{x}^{(k)}$ if it satisfies:

$$\left| y_i(\boldsymbol{x}^{(k)}) \right| \le \varepsilon \tag{8.25}$$

where ε is a small enough positive threshold. Several termination conditions may be used with the method of feasible directions. They may impose conditions on the change in the objective function or the change in the solution from one iteration to another. A possible MATLAB implementation of the method of feasible directions is given in the MATLAB listing M8.5. It should be noted that the line search within this code limits its search to only feasible points.

The following example illustrates the application of method of feasible directions.

Example 8.5: Solve the following problem:

$$\begin{aligned} &\min_{x}\ (x_1 - 1)^2 + (x_2 - 5)^2 \\ \text{subject to}\quad & -x_1^2 + x_2 - 4 \le 0 \\ & -(x_1 - 2)^2 + x_2 - 3 \le 0 \end{aligned} \tag{8.26}$$

starting with the feasible point $\boldsymbol{x}^{(0)} = \begin{bmatrix} -1.0 & 1.0 \end{bmatrix}^T$.

Solution: Using the MATLAB listing M8.5, the program generates the following output:

OldPoint = [−1.0 1.0]′
ObjectiveValue = 20
ConstraintsValues = [−4 −11]′
NewPoint = [0.7639 4.5279]′
ObjectiveValue = 0.2786
ConstraintsValues = [−0.0557 0.0000]′
NewPoint = [0.7455 4.5558]′
ObjectiveValue = 0.2621
ConstraintsValues = [−0.0000 −0.0180]′

The solution obtained using the method of feasible directions in two iterations is $\boldsymbol{x}^* = \begin{bmatrix} 0.7455 & 4.556 \end{bmatrix}^T$. This solution can be verified by applying the KKT

```
%M8.5
%An implementation of the method of Feasible Directions
NumberOfParameters=2; %This is n for this problem
NumberOfConstraints=2; %This is m for this problem
MaximumIterations=20; %maximum number of iterations
MaximumNumberOfEquations=1+2*NumberOfParameters+NumberOfConstraints;
A=zeros(MaximumNumberOfEquations, NumberOfParameters+1);
b=zeros(MaximumNumberOfEquations,1); %right hand side
c=zeros(NumberOfParameters+1,1); %objective function
Identity=eye(NumberOfParameters); %This is the identity
%Store constraints on the length of the steps taken  -1<=Sj <=1
A(2:(NumberOfParameters+1),1:NumberOfParameters)=Identity;
b(2:(NumberOfParameters+1),1)=1.0; %
A((NumberOfParameters+2):(2*NumberOfParameters+1),1:NumberOfParameters)=-1.0*Identity;
b((NumberOfParameters+2):(2*NumberOfParameters+1),1)=1.0;
c(NumberOfParameters+1)=-1.0; %cost vector multiplying the slack variable alpha
OldPoint=[-1.0  1.0]'; %This is the starting point
ConstraintsValues=getConstraints(OldPoint); %get values of constraints at the current point
ObjectiveValue=getObjective(OldPoint); %get objective function value at starting point
ObjectiveGradient=getObjectiveGradient(OldPoint); %get gradient of objective function
Epsilon=1.0e-3; %feasibility threshold
LambdaMax=3.0; %maximum lambda for a line search
Tolerance=1.0e-5; %termination tolerance for a line search
IterationCounter=0;
while(IterationCounter<MaximumIterations) %Repeat for a number of iterations
  if(max(ConstraintsValues)<(-1.0*Epsilon)) %are all constraints satisfied
    SearchDirection=-1.0*ObjectiveGradient; %use negative of objective function gradient
  else
    [NumberOfActiveConstraints, ActiveIndexes]
                =getActiveConstraints(ConstraintsValues,Epsilon); %determine active constraints
    %number of new equations
    TotalNumberOfEquations=1+2*NumberOfParameters+NumberOfActiveConstraints;
    %Fill matrix for new linear program. Notice that RHS is zero except for norm constraints
    A(1,1:NumberOfParameters)=ObjectiveGradient';
    A(1,NumberOfParameters+1)=1;
    ConstraintsGradients=getConstraintsGradients(OldPoint); %Get gradient of all constraints
    %Store gradients of active constraints in matrix
    A((2*NumberOfParameters+2):(TotalNumberOfEquations),1:NumberOfParameters)
                                      =ConstraintsGradients(ActiveIndexes,:);
    %All coefficients of the slack variable alpha in matrix are 1s
    A((2*NumberOfParameters+2):(TotalNumberOfEquations),(NumberOfParameters+1))=1;
    %Solve linear program
    SearchDirection=linprog(c,A(1:TotalNumberOfEquations,:),b(1:TotalNumberOfEquations));
  end
  LambdaOptimal =GoldenSection('getModifedObjective',Tolerance,
                          OldPoint,SearchDirection(1:NumberOfParameters),LambdaMax);
  NewPoint=OldPoint+LambdaOptimal*SearchDirection(1:NumberOfParameters); %get new point
  OldPoint=NewPoint; %make new point old point
  ConstraintsValues=getConstraints(OldPoint);
  ObjectiveValue=getObjective(OldPoint);
  ObjectiveGradient=getObjectiveGradient(OldPoint);
  IterationCounter=IterationCounter+1;
end
```

conditions. We have two constraints for this example and thus there are four possible cases. We first check the case where both constraints are simultaneously active. In this case, the KKT conditions are given by:

$$\nabla f + \lambda_1 \nabla y_1 + \lambda_2 \nabla y_2 = 0$$

$$\Rightarrow \begin{bmatrix} 2(x_1 - 1) \\ 2(x_2 - 5) \end{bmatrix} + \lambda_1 \begin{bmatrix} -2x_1 \\ 1 \end{bmatrix} + \lambda_2 \begin{bmatrix} -2(x_1 - 2) \\ 1 \end{bmatrix} = \begin{bmatrix} 0 \\ 0 \end{bmatrix} \tag{8.27}$$

Solving the two constraints (8.26), we obtain the solution $x_1^* = 0.75$ and $x_2^* = 4 + (0.75)^2 = 4.5625$. Substituting these values in (8.27), we get $\lambda_1 = 0.42$ and $\lambda_2 = 0.46$. These positive values verify that this point is indeed a minimizer of the problem (8.26). The analytical solution is very close to the one obtained using the method of feasible directions.

8.9 Rosen's projection method

Rosen's method utilizes essentially the same concept of the method of feasible directions but applies it differently. It is suited for problems with linear constraints and arbitrary objective function. In the kth iteration, this method aims at finding a suitable search direction $\boldsymbol{s}^{(k)}$ that would result in the maximum possible reduction in the objective function but maintains the same active constraints at the current point $\boldsymbol{x}^{(k)}$. Finding this direction is formulated as the auxiliary nonlinear constrained optimization problem:

$$\begin{aligned} &\min_{\boldsymbol{s}} \ \boldsymbol{s}^{(k)T}\nabla f(\boldsymbol{x}^{(k)}) \\ \text{such that} \quad &\boldsymbol{s}^{(k)T}\ \nabla y_j(\boldsymbol{x}^{(k)}) = 0, \quad j \in I_{A_k} \\ &\boldsymbol{s}^{(k)T}\boldsymbol{s}^{(k)} = 1 \end{aligned} \tag{8.28}$$

The problem (8.28) can be put in the more compact matrix form:

$$\begin{aligned} &\min_{\boldsymbol{s}} \ \boldsymbol{s}^{(k)T}\nabla f(\boldsymbol{x}^{(k)}) \\ \text{such that} \quad &\boldsymbol{N}^{(k)T}\boldsymbol{s}^{(k)} = 0 \\ &\boldsymbol{s}^{(k)T}\boldsymbol{s}^{(k)} = 1 \end{aligned} \tag{8.29}$$

where $\boldsymbol{N}^{(k)}$ is the matrix whose columns are the gradients of the active constraints at the point $\boldsymbol{x}^{(k)}$. We consider in the matrix $\boldsymbol{N}^{(k)}$ only the gradients that make an acute angle with the negative of the gradient of the objective function. The reason is that only these constraints may be violated by moving in a descent direction of the objective function. This is illustrated in Figure 8.10.

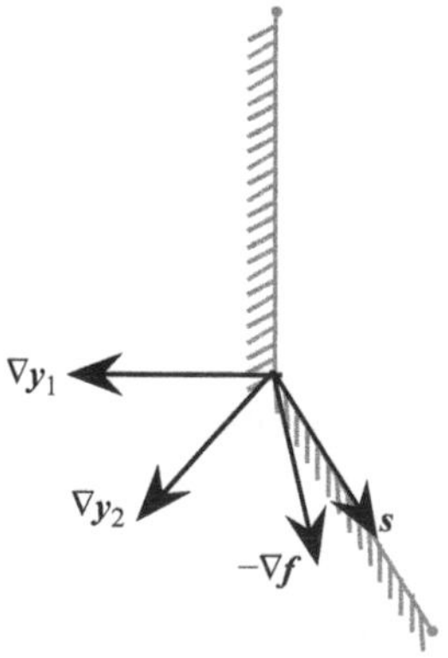

Figure 8.10 An illustration of the selection of active constraint gradient in Rosen's projection method; the active constraint gradient ∇y_1 is ignored as it makes an obtuse angle with the steepest descent direction. Only the gradient ∇y_2 of the second active constraint is used for projecting the gradient

The direction $\boldsymbol{s}^{(k)}$ is found by forming the Lagrangian function:

$$L(\boldsymbol{s}^{(k)T}, \boldsymbol{\lambda}, \beta) = \boldsymbol{s}^{(k)T}\nabla f(\boldsymbol{x}^{(k)}) + \boldsymbol{\lambda}^T \boldsymbol{N}^{(k)T}\boldsymbol{s}^{(k)} + \beta(\boldsymbol{s}^{(k)T}\boldsymbol{s}^{(k)} - 1) \tag{8.30}$$

Applying the KKT condition $\partial L/\partial \boldsymbol{s}^{(k)} = 0$, we get:

$$\nabla f(\boldsymbol{x}^{(k)}) + \boldsymbol{N}^{(k)}\boldsymbol{\lambda} + 2\beta \boldsymbol{s}^{(k)} = \boldsymbol{0} \Rightarrow \boldsymbol{s}^{(k)} = -\frac{1}{2\beta}(\nabla f(\boldsymbol{x}^{(k)}) + \boldsymbol{N}^{(k)}\boldsymbol{\lambda}) \tag{8.31}$$

Substituting from (8.31) into the first set of constraints $\boldsymbol{N}^{(k)T}\boldsymbol{s}^{(k)} = 0$, we get:

$$\begin{aligned} \boldsymbol{N}^{(k)T}\boldsymbol{s}^{(k)} &= -\frac{1}{2\beta}\left(\boldsymbol{N}^{(k)T}\nabla f(\boldsymbol{x}^{(k)}) + \boldsymbol{N}^{(k)T}\boldsymbol{N}^{(k)}\boldsymbol{\lambda}\right) = 0 \\ \Rightarrow \boldsymbol{\lambda} &= -\left(\boldsymbol{N}^{(k)T}\boldsymbol{N}^{(k)}\right)^{-1}\boldsymbol{N}^{(k)T}\nabla f(\boldsymbol{x}^{(k)}) \end{aligned} \tag{8.32}$$

Using (8.32) to eliminate the Lagrange multipliers $\boldsymbol{\lambda}$ in (8.31), we get:

$$\begin{aligned} \boldsymbol{s}^{(k)} &= -\frac{1}{2\beta}\left(\nabla f(\boldsymbol{x}^{(k)}) - \boldsymbol{N}^{(k)}\left(\boldsymbol{N}^{(k)T}\boldsymbol{N}^{(k)}\right)^{-1}\boldsymbol{N}^{(k)T}\nabla f(\boldsymbol{x}^{(k)})\right) \\ &= -\frac{1}{2\beta}\boldsymbol{P}\nabla f(\boldsymbol{x}^{(k)}) \end{aligned} \tag{8.33}$$

The multiplier β affects only the norm of the vector $\boldsymbol{s}^{(k)}$ but not its direction. It follows that the direction $\boldsymbol{s}^{(k)}$ is obtained by projecting the negative of the gradient in the null space of the matrix $\boldsymbol{N}^{(k)T}$. The projection matrix $\boldsymbol{P}$ is determined once the matrix $\boldsymbol{N}^{(k)}$ is known. Similar projection matrices have been explained earlier in Chapter 1. The step taken in the direction $\boldsymbol{s}^{(k)}$ is determined such that none of the constraints is violated. Figure 8.11 illustrates the projection of the gradient in the null space of the matrix $\boldsymbol{N}^{(k)}$.

Rosen's projection approach can be implemented as in the MATLAB listing M8.6. The following example illustrates the application of Rosen's projection method.

Example 8.6: Solve the constrained optimization problem:

$$\min_x \; (x_1 - 1)^2 + (x_2 - 2)^2 - 4$$

$$\text{subject to} \; \begin{cases} x_1 + 2x_2 \le 5 \\ 4x_1 + 3x_2 \le 10 \\ 6x_1 + x_2 \le 7 \\ 0 \le x_1, 0 \le x_2 \end{cases} \tag{8.34}$$

starting with the initial point $\boldsymbol{x}^{(0)} = [0.5 \quad 1.0]^T$. Utilize Rosen's projection method.

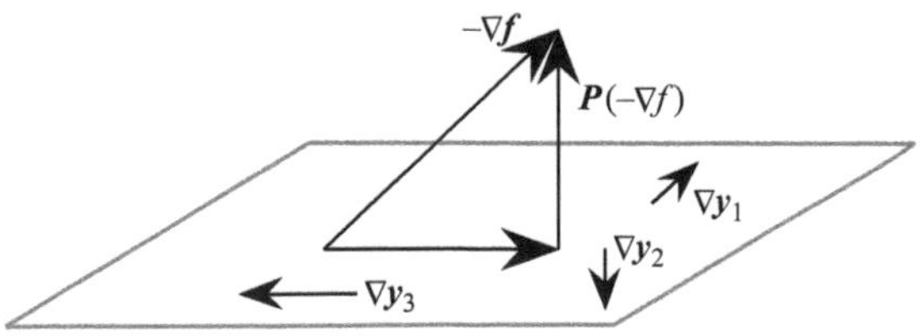

Figure 8.11 An illustration of Rosen's projection method; the vector $-\boldsymbol{P}\nabla f$ is normal to the subspace spanned by the gradients of the considered active constraints

```
%M8.6
%An Implementation of Rosen's Projection Method
NumberOfParameters=2; %This is n for this problem
NumberOfConstraints=5; %This is m for this problem
MaximumIterations=20; %maximum number of iterations
OldPoint=[0.5  1.0]'; %This is the feasible point
ConstraintsValues=getConstraints(OldPoint);
ObjectiveValue=getObjective(OldPoint); %get objective
ObjectiveGradient=getObjectiveGradient(OldPoint);
Epsilon=1.0e-3; %feasibility threshold
LambdaMax=3.0; %maximum lambda for a line search
Tolerance=1.0e-5; %termination for a line search
IterationCounter=0;
Identity=eye(NumberOfParameters); %identity matrix
TotalProjectionMatrix=zeros(NumberOfParameters, NumberOfConstraints); %this is the projection
                                                                      % matrix
while(IterationCounter<MaximumIterations) %Repeat for a number of iterations
  if(max(ConstraintsValues)<(-1.0*Epsilon)) %Is the current point an interior point?
    SearchDirection=-1.0*ObjectiveGradient  %search direction is the negative of objective gradient
  else %at least one constraint is active
    %determine indexes of active constraints and their number
    [NumberOfActiveConstraints, ActiveIndexes]
                    =getActiveConstraints(ConstraintsValues,Epsilon);
    %get gradients of all active constraints at x(k)
    ConstraintsGradients=getConstraintsGradients(OldPoint);
    NumberOfProjectionConstraints=0;  %initialize number of included projection matrices
    for j=1:NumberOfActiveConstraints  %check all active constraints
      CurrentConstraint=ActiveIndexes(j);
      CurrentGradient=ConstraintsGradients(CurrentConstraint,:);
      if((CurrentGradient*ObjectiveGradient)<0) %Does it make an obtuse angle with objective
                                                %gradient?
        NumberOfProjectionConstraints=NumberOfProjectionConstraints+1; %add to projection
        TotalProjectionMatrix (:,NumberOfProjectionConstraints) =CurrentGradient';
      end
    end
    %Get the projection matrix from the total storage matrix
    ProjectionMatrix=TotalProjectionMatrix(:,1:NumberOfProjectionConstraints);
    %Project the -ve of the gradient
    SearchDirection=(Identity-(ProjectionMatrix*inv(ProjectionMatrix'*ProjectionMatrix)
                              *ProjectionMatrix'))*-1.0*ObjectiveGradient;
  end
  %carry out a line search while keeping the search points feasible
  LambdaOptimal = GoldenSection('getModifedObjective',Tolerance,OldPoint,
                                              SearchDirection,LambdaMax);
  NewPoint=OldPoint+LambdaOptimal*SearchDirection; %get x(k+1)
  %make new values the current values
  OldPoint=NewPoint;
  ConstraintsValues=getConstraints(OldPoint);
  ObjectiveValue=getObjective(OldPoint);
  ObjectiveGradient=getObjectiveGradient(OldPoint);
  IterationCounter=IterationCounter+1; %set k=k+1
end
```

Solution: Using the MATLAB listing M8.6, we obtain the following sequence of outputs:

$$\text{OldPoint} = [0.5000 \quad 1.0000]'$$

$$\text{ConstraintsValues} = [-2.5000 \quad -5.0000 \quad -3.0000 \quad -0.5000 \quad -1.0000]$$

$$\text{SearchDirection} = [1.0 \quad 2.0]'$$

NewPoint = [0.8750 1.7500]′

ConstraintsValues = [−0.6250
−1.2500
−0.0000
−0.8750
−1.7500]

SearchDirection = [−0.0743 0.4460]′

NewPoint = [0.8378 1.9730]′

ConstraintsValues = [−0.2162
−0.7297
−0.0000
−0.8378
−1.9730]

SearchDirection = 1.0e−006 *[−0.0993 0.5956]′

Notice that at the initial point all constraints have negative values, implying that this is an interior point. In this case, the matrix $\boldsymbol{N}^{(0)}$ is empty and the search direction is equal to negative of gradient at the starting point. In the second iteration, only the third constraint is active and thus the matrix $\boldsymbol{N}^{(1)}$ has only one column. In the final iteration the search direction has effectively a zero value, indicating that there is no possible feasible search direction that can reduce the objective function any further.

8.10 Barrier and penalty methods

A possible way for solving a constrained optimization problem is to convert it into unconstrained problem by integrating the constraints within the objective function. Two approaches for such an approach are possible: barrier approaches and penalty approaches. As the name implies, barrier approaches do not allow the iterations to leave the feasible region by enforcing a strong increase in the new objective function value near the boundaries. A common barrier method is given by:

$$\varphi(\boldsymbol{x}, r) = f(\boldsymbol{x}) - r\sum_{j} \frac{1}{y_j(\boldsymbol{x})} \tag{8.35}$$

where r is the barrier parameter. During the optimization iterations, if a point $\boldsymbol{x}$ approaches the boundary, the barrier term in (8.35) becomes high in value thus creating an objective function barrier. Decreasing the value of the barrier coefficient r reduces the effect of the barrier. Ideally, we should allow the iterations to get as close as possible to the boundary if the solution is on the boundary. It follows that the problem (8.35) is solved several times for different values of the parameter r and the solution is extrapolated for the case $r = 0$. Figure 8.12 illustrates the barrier method for a simple one-dimensional problem.

Another approach for solving problems with inequality constraints is to penalize iterations outside the feasible region. The further the iteration points drift outside the feasible region, the more the penalty term increases. This approach is

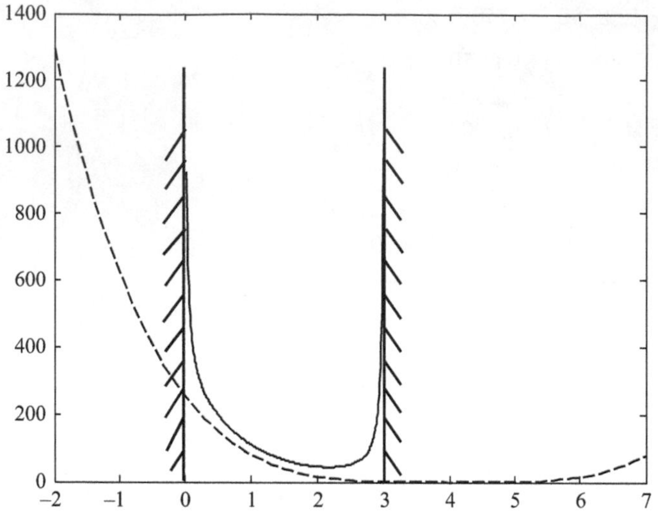

Figure 8.12 An illustration of the internal barrier method for a one-dimensional example; the original objective function is $f(x) = (x - 4)^2$ (shown in dashed line). The constraints are $x \geq 0$ and $x \leq 3$. The new objective function (shown as a solid line) incorporates the constraints with a barrier coefficient of $r = 20$. The minimum of the constrained problem is close to $x = 3$ as expected. Making r smaller gets a better estimate of the minimum of the constrained problem

different from the barrier method as the iterations are allowed to become non-feasible. A possible external barrier is given by:

$$\varphi(\boldsymbol{x}, r) = f(\boldsymbol{x}) + r\sum_{j} (Y_j(\boldsymbol{x}))^q \tag{8.36}$$

where $Y_j(\boldsymbol{x}) = \max(y_j(\boldsymbol{x}), 0)$. r is the external penalty parameter. The higher the value of r, the more a non-feasible point is penalized. q is a suitable positive exponent. The formulation (8.36) guarantees that once a constraint is violated, the objective function sees an increase in its value, thus keeping the iterations close to the feasible region. To get the actual solution of the constrained problem, the problem (8.36) is solved several times for increased values of the external penalty parameter. The solution is then extrapolated to the case $r = \infty$. The external penalty approach is illustrated in Figure 8.13.

The following example illustrates these approaches.

Example 8.7: Solve the constrained optimization problem:

$$\begin{aligned} &\min_{x} f(\boldsymbol{x}) = x_1^2 + x_2^2 - 6x_1 - 8x_2 + 15 \\ \text{subject to} \quad &4x_1^2 + x_2^2 \leq 16 \\ &3x_1 + 5x_2 \leq 15 \end{aligned} \tag{8.37}$$

using an external penalty method.

Solution: Any of the unconstrained optimization techniques discussed earlier may be used to solve the external penalty objective functions (8.36). The only change required is in the calculation of the objective function. The MATLAB listing M8.7

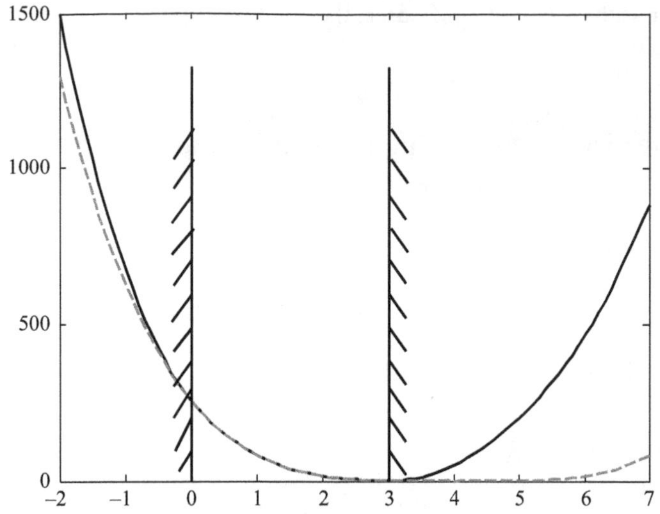

Figure 8.13 *An illustration of the external penalty method for a one-dimensional example; the original objective function is $f(x) = (x - 4)^2$ (shown in dashed line). The constraints are $x \geq 0$ and $x \leq 3$. The new objective function (shown as a solid line) incorporates the constraints with a penalty parameter $r = 50$. The minimum of the constrained problem is achieved at $x = 3$ as expected*

```
%M8.7
%Objective function modification to include external penalty term
function value=getObjective(x)
x1=x(1); %get first parameter
x2=x(2); %get second parameter
r=50.0; %penalty paramater
FirstConstraintTerm=(max(0,(4*x1*x1+x2*x2-16)))^2;
SecondConstraintTerm=(max(0,(3*x1+5*x2-15)))^2;
OriginalObjectiveFunction=x1*x1+x2*x2-6*x1-8*x2+15;
value=OriginalObjectiveFunction+r*(FirstConstraintTerm+SecondConstraintTerm);
```

illustrates the objective function calculations of the problem (8.37) using an external penalty approach.

The problem (8.37) was solved several times for increasing values of the penalty parameter r. Unconstrained optimization gives the following output:

r = 1.0
OldPoint = [1.7191   2.0458]′

r = 5
OldPoint = [1.7396   1.9723]′

r = 10
OldPoint = [1.7416   1.9624]′

r = 20
OldPoint = [1.7434   1.9579]′

r = 50
OldPoint = [1.7437   1.9546]′

We see that as the value of r increases, the solution approaches the actual solution of this problem.

A8.1 Electrical engineering application: analog filter design

To illustrate the application of constrained optimization to analog filter design, we consider the active filter shown in Figure 8.14. The parameters of this filter are $\boldsymbol{x} = [R_1 \quad R_2 \quad R_3 \quad C]^T$. Assuming an ideal Op-Amp, a closed form expression of the response of this filter is obtained. Making the typical ideal assumptions for Op-Amps ($V^+ = V^-$) and assuming that no current is flowing into the Op-Amp input terminals, we get the relationship:

$$\frac{V_o}{V_s} = \frac{-s}{\left(CR_1 s^2 + \dfrac{2R_1}{R_3}s + \left(\dfrac{1}{R_3 C} + \dfrac{R_1}{R_2 R_3 C}\right)\right)} \tag{8.38}$$

which is a typical bandpass response. A more general approach for analyzing active filters with linearized active elements is to replace the active element by its equivalent circuit as shown in Figure 8.15. This figure takes into account the high but finite input resistance of the Op-Amp, its small output resistance, and its high

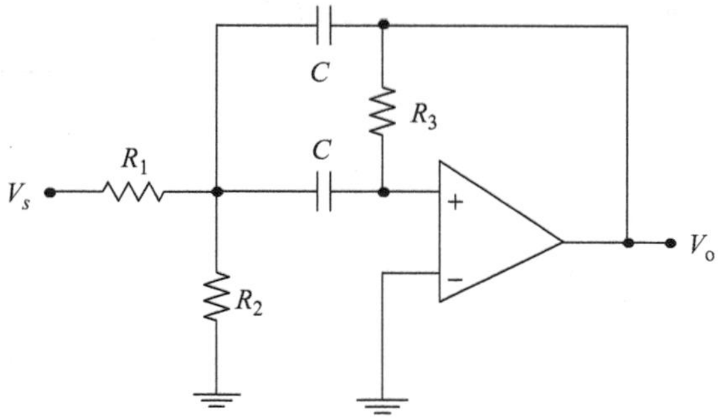

Figure 8.14 The active bandpass filter of application A8.1

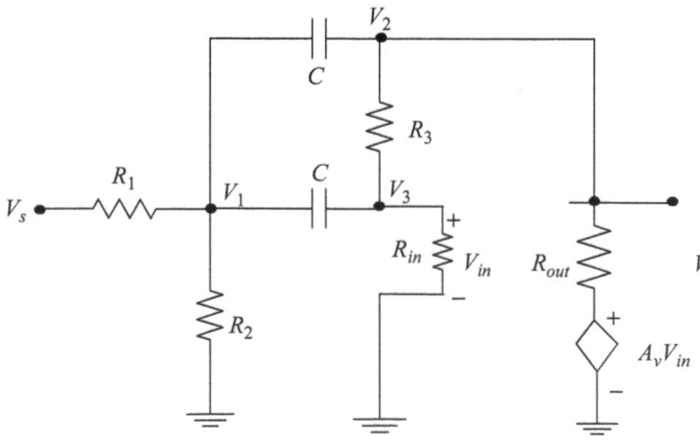

Figure 8.15 The active bandpass filter with the operational amplifier replaced by an equivalent circuit

but finite gain. Writing the nodal equations at the three nodes, we get the three equations:

$$\begin{aligned}
&\frac{(V_1 - V_s)}{R_1} + (V_1 - V_3)Cs + (V_1 - V_2)Cs + \frac{V_1}{R_2} = 0 \\
&(V_2 - V_1)Cs + \frac{(V_2 - V_3)}{R_3} + \frac{(V_2 - A_v V_3)}{R_{out}} = 0 \\
&(V_3 - V_1)Cs + \frac{V_3}{R_{in}} + \frac{(V_3 - V_2)}{R_3} = 0
\end{aligned} \tag{8.39}$$

where A_v is the Op-Amp voltage gain, R_{in} is its input resistance, and R_{out} is its output resistance. Putting (8.39) in the form of a system of linear equations we get:

$$\begin{bmatrix} (2Cs + 1/R_1 + 1/R_2) & -Cs & -Cs \\ -Cs & (Cs + 1/R_3 + 1/R_{out}) & -(1/R_3 + A_v/R_{out}) \\ -Cs & -1/R_3 & (Cs + 1/R_3 + 1/R_{in}) \end{bmatrix} \begin{bmatrix} V_1 \\ V_2 \\ V_3 \end{bmatrix} = \begin{bmatrix} \dfrac{V_s}{R_1} \\ 0 \\ 0 \end{bmatrix} \tag{8.40}$$

For a given source voltage V_s, the system (8.40) can be solved to determine the output voltage at any frequency. The output voltage V_o is given by $V_o = V_2$. The gain at a specific frequency is then given by $V_o \cdot V_s$ is the source excitation which is selected to have a unity value. The MATLAB listing M8.8 shows a possible code for building and solving the linear system of equations governing the active filter.

Two things to notice about the listing M8.8: First, the values sent to this function are scaled back to their physical values. The optimization code deals with parameters of the same order of magnitude to avoid ill-conditioning. Second, to get the response at any specific frequency *freq*, the Laplace parameter s is replaced by $J\omega = J2\pi freq$, where $J = \sqrt{-1}$. A comparison between the response obtained through the closed formula (8.38) and the solution obtained through this MATLAB listing is shown in Figure 8.16 for a sweep of frequencies at a specific set of parameters. Good match between both approaches is achieved.

It is required to utilize the simulator given by the listing M8.8 to solve the constrained optimization problem:

$$\begin{aligned}
&\min_{x} \; -|V_o(freq^*)| \\
\text{subject to} \quad &R_1 \le 3\ K\Omega
\end{aligned} \tag{8.41}$$

where $x = [R_1 \quad R_2 \quad R_3 \quad C]^T$ as explained earlier. $freq^* = 400$ Hz is the target resonant frequency of the active filter. By minimizing the negative of the gain at this frequency, we are actually maximizing the gain and forcing the center frequency to be shifted to $freq^*$. The starting point for optimization is $x^{(0)} = [0.5\ K\Omega \quad 0.5\ K\Omega \quad 0.5\ K\Omega \quad 3.0\ \mu F]^T$

The penalty method is used to integrate the constraint in (8.41) in the objective function. The steepest descent method is then used to solve the unconstrained

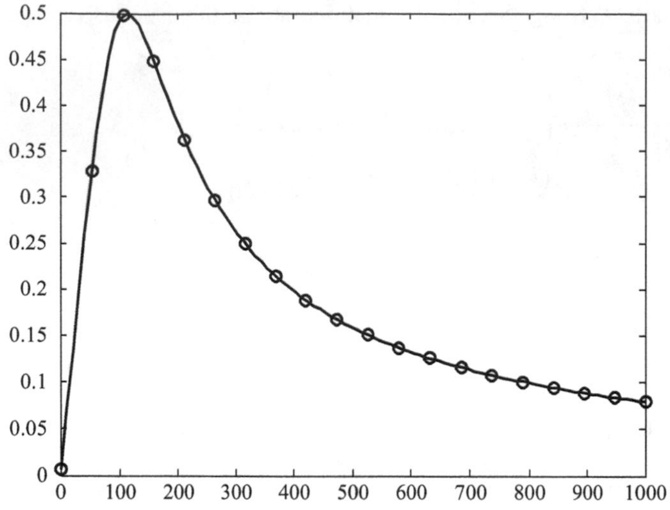

Figure 8.16 The response of the active bandpass filter obtained using a linearized operational amplifier model (—) and using ideal assumptions (○) for the point $\boldsymbol{x} = [1.0\ K\Omega \quad 1.0\ K\Omega \quad 1.0\ K\Omega \quad 2.0\ \mu\text{F}]^T$

optimization problem. The utilized code for calculating the objective function is shown in the listing M8.9.

The steepest descent approach carries out only two iterations. The achieved optimal point $\boldsymbol{x}^* = [1.1107\ \text{K}\Omega \quad 0.5301\ \text{K}\Omega \quad 1.1399\ \text{K}\Omega \quad 0.6214\ \mu\text{F}]^T$. Figure 8.17 shows the response at the initial and final point. The center frequency of the filter was indeed shifted to the desired one.

```
%M8.8
%constructing and solving the linear system of equations for the application A8.1
function Response=getFilterResponse(R1value, R2value, R3value, Cvalue, Frequency)
i=sqrt(-1)
C=Cvalue*1.0e-6; %recover the scale used for optimization.
R1=R1value*1.0e3; %value in Kohm
R2=R2value*1.0e3; %value in KOhm
R3=R3value*1.0e3; %value in KOhm
OpAmpGain=2.0e6; %gain of op amp
Rin=1.0e6; %input resistance of op amp
Rout=1; %output resistance of Op amp
Vs=1.0; %unity excitation with zero angle assumed
Omega=2*pi*Frequency;
% Build matrix values
A11=(2*C*i*Omega+(1/R2)+(1/R1));
A12=-C*i*Omega;
A13=-C*i*Omega;
A21=-C*i*Omega;
A22=(C*i*Omega+(1/R3)+(1/Rout));
A23=-((1/R3)+(OpAmpGain/Rout));
A31=-i*C*Omega;
A32=-1/R3;
A33=(i*C*Omega+(1/Rin)+(1/R3));
A=[A11 A12 A13
  A21 A22 A23
  A31 A32 A33]; %build matrix
b=[(Vs/R1) 0 0]'; %right hand side vector
NodeVoltages=inv(A)*b; %solve system of equations
Vout=NodeVoltages(2); %get the output voltage
Response=abs(Vout); %return the absolute value of the output voltage
```

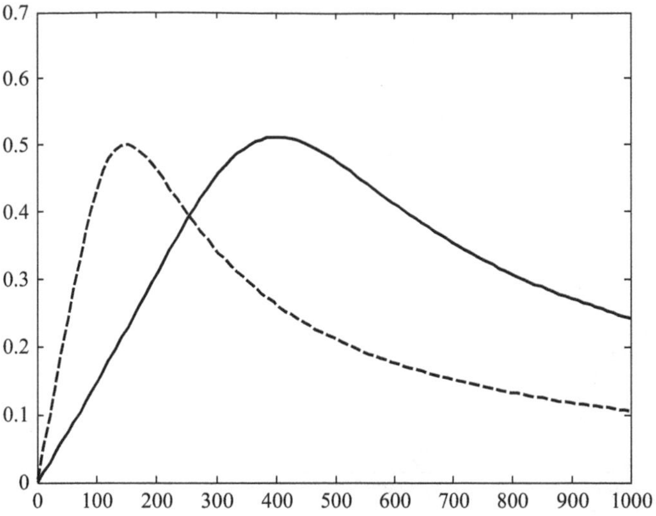

Figure 8.17 *The response of the active bandpass filter at the starting point (–) ($\mathbf{x}^{(0)} = [0.5\ K\Omega \quad 0.5\ K\Omega \quad 0.5\ K\Omega \quad 3.0\ \mu F]^T$) and at the optimal point (—) ($[1.0624\ K\Omega \quad 0.5277\ K\Omega \quad 1.0893\ K\Omega \quad 0.8097\ \mu F]^T$). The filter's center frequency has been successfully moved to $f^* = 400$ Hz*

```
% M8.9
%This file evaluates the penalized objective function for the opAmp active filter
function objective=getActiveFilterObjective2(Point);
CurrentFrequency=400; % target is at 400 Hz
%Get scaled parameter values
R1=Point(1);
R2=Point(2);
R3=Point(3);
C=Point(4);
Response=getFilterResponse(R1, R2, R3, C, CurrentFrequency); %modulus of gain
OriginalObjective=-1.0*(Response); %minimizing the negative is equivalent to maximization
if(R1<3) % Is R1 less than 3 KOhm
   PenaltyTerm1=0;
else
   PenaltyTerm1=(R1-3)^2;
end
objective=OriginalObjective+10*PenaltyTerm1; %objective function with penalty term
```

If the constraint in (8.41) is more stringent ($R_1 \leq 1.0$ KΩ), the unconstrained minimum becomes outside the feasible region. The optimal design obtained in this case is $[1.0624\ K\Omega \quad 0.5277\ K\Omega \quad 1.0893\ K\Omega \quad 0.8097\ \mu F]^T$ which slightly violates the constraint and gives a solution that is not as good as the one shown in Figure 8.18. Using a higher penalty parameter would bring the solution closer to the feasible region.

A8.2 Spectroscopy

The area of spectroscopy aims at identifying the frequency-dependent characteristics of a sample by illuminating it by high-frequency radiation. Most materials in nature exhibit unique finger prints at specific frequencies. We can thus identify the

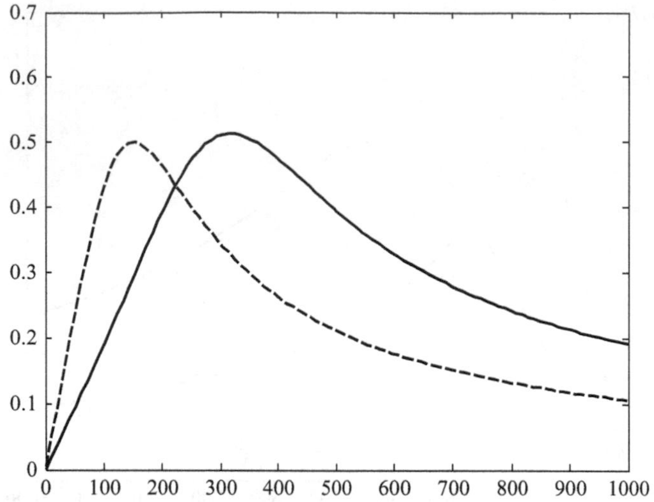

Figure 8.18 The response of the active bandpass filter at the starting point (–) $\mathbf{x}^{(0)} = [0.5\ \text{K}\Omega \quad 0.5\ \text{K}\Omega \quad 0.5\ \text{K}\Omega \quad 3.0\ \mu\text{F}]^T$ and at the optimal point (—) ($[1.0624\ \text{K}\Omega \quad 0.5277\ \text{K}\Omega \quad 1.0893\ \text{K}\Omega \quad 0.8097\ \mu\text{F}]^T$) obtained using the constraint $R_1 \leq 1.0\ \text{K}\Omega$. The filter's center frequency is only moved close to $f^* = 400$ Hz

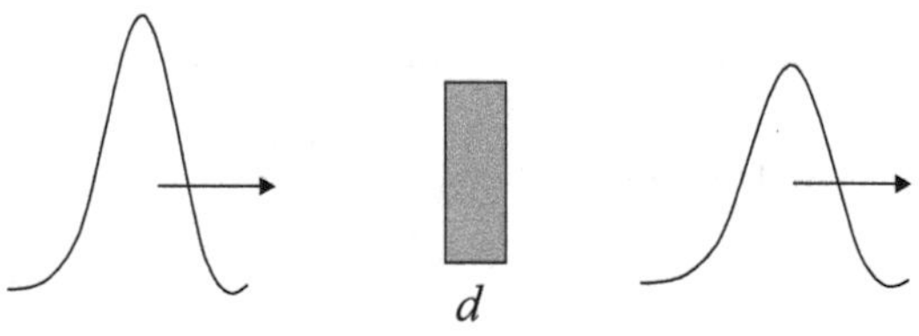

Figure 8.19 An illustration of high-frequency spectroscopy; the sample is illuminated with an electromagnetic wave with certain time-domain profile and the transmitted field is recorded. The changes in amplitude and phase at every frequency are used to extract the sample properties

material by illuminating it with the proper type of radiation and observing the reflected or transmitted fields.

Figure 8.19 shows a spectroscopy system. A narrow electromagnetic (EM) pulse is sent and the transmitted pulse is observed. The ratio between the incident pulse and the measured transmitted pulse at any frequency ω is given by $T_{\text{exp}}(\omega)$. The amplitude and phase of the transmitted pulse at each frequency are used to determine the material properties at that frequency.

We focus here on THz spectroscopy. There are many variations of this approach. We focus here on only one approach. Considering the structure in Figure 8.19, we assume that the sample thickness d is known. In this case, the transfer function describing how the transmitted pulse is changed in amplitude and phase is known analytically and is given by [7]:

$$T_a(\omega) = \frac{4\bar{n}}{(\bar{n}+1)^2} \times \exp(-j(\bar{n}-1)\omega d/c) \tag{8.42}$$

where $\bar{n} = \bar{n}(\omega) = n(\omega) - J\kappa(\omega)$, is the unknown complex refractive index. $n(\omega)$ is the real part of the refractive index and it represents the relative dielectric constant of the material. This parameter must be greater than one at all frequencies for physical reasons. The imaginary part $\kappa(\omega)$ is called the extinction rate of the material. It represents the electric losses in the medium which causes reduction of the field energy. The extinction rate must be non-negative for all frequencies for physical reasons. The parameter c is the velocity of light. The analytical transfer function (8.42) results in a complex number given a certain complex refractive index, sample thickness, and frequency. This complex number describes how the amplitude and phase of the incident pulse change to result in the transmitted pulse as illustrated in Figure 8.19. The target of spectroscopy is to determine the real refractive indexes $\boldsymbol{n} = [n(w_1) \quad n(w_2) \quad \ldots n(w_m)]^T$ and the extinction rates $\kappa = [\kappa(w_1) \quad \kappa(w_2) \quad \ldots \kappa(w_m)]^T$ at a number of frequencies $\{\omega_1, \omega_2, \ldots, \omega_m\}$ by fitting the measured transfer function $T_{\text{exp}}(\omega)$ to the analytical transfer function. We thus have $2m$ unknowns and $2m$ pieces of information (the real and imaginary part of the transfer function at every frequency).

In Reference 8, the $2m$ unknowns are solved for through the constrained optimization problem:

$$\begin{aligned} &\min_{n,k} \ \delta \boldsymbol{M}^T \delta \boldsymbol{M} + \zeta \delta \boldsymbol{\phi}^T \delta \boldsymbol{\phi} \\ \text{subject to} \quad &n(\omega_i) \geq 1, \ i = 1, 2, \ldots, m \\ &\kappa(\omega_i) \geq 0, \ i = 1, 2, \ldots, m \end{aligned} \tag{8.43}$$

where δM_i and $\delta \phi_i$ represent the differences between the amplitudes and phases of the experimental transfer function and analytical transfer function. In Reference 8, these errors were given by:

$$\begin{aligned} \delta M_i &= \ln\Big(|T_a(\omega_i)|/|T_{\text{exp}}(\omega_i)|\Big) \\ \delta \phi_i &= \angle T_a(\omega_i) - \angle T_{\text{exp}}(\omega_i) \end{aligned} \tag{8.44}$$

Notice how the natural logarithm function was used to penalize more the difference in amplitudes between measurements and the analytical transfer function.

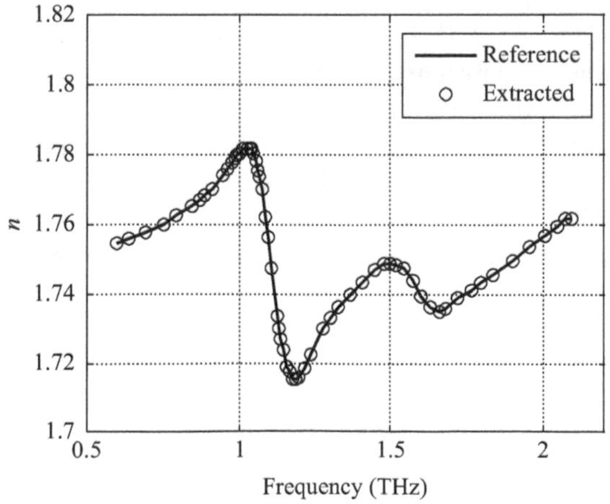

Figure 8.20 The extracted real part of the complex refractive index as a function of frequency

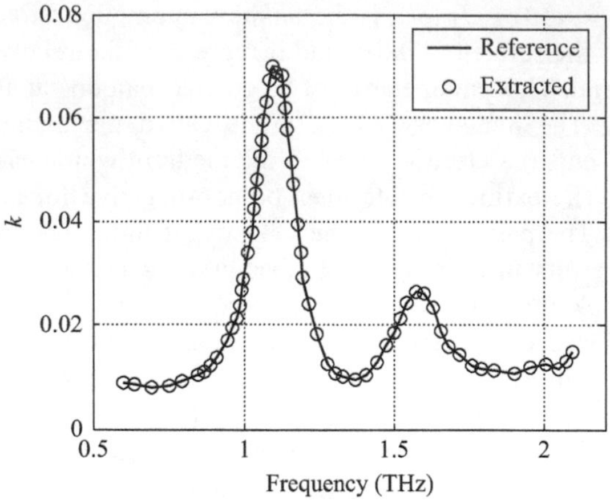

Figure 8.21 The extracted imaginary part of the complex refractive index as a function of frequency

The parameter ζ scales the errors in the angles to the same order of magnitude as the magnitude errors.

The problem (8.43) is a constrained optimization problem with linear constraints. It can be solved using any of the techniques discussed in this chapter. In Reference 8, a Sequential Quadratic Programming (SQP) algorithm was used.

Figures 8.20 and 8.21 show the results obtained for α-naphthol for a sample thickness of 10.0 μm. The real and imaginary parts of the complex refractive index are extracted using $m = 300$. The results match very well with those reported in the literature [9].

References

1. Singiresu S. Rao, *Engineering Optimization Theory and Practice*, Third Edition, John Wiley & Sons Inc, New York, 1996
2. Andreas Antoniou and Wu-Sheng Lu, *Practical Optimization Algorithms and Engineering Applications*, Springer, New York, 2007
3. John W. Bandler and R.M. Biernacki, *ECE3KB3 Courseware on Optimization Theory*, McMaster University, Canada, 1997
4. Jasbir S. Arora, *Introduction to Optimum Design*, Third Edition, Elsevier Inc., Oxford, UK, 2012
5. Richard C. Jaeger, *Microelectronic Circuit Design*, WCB/McGraw-Hill, New York, 1997
6. John O. Attia, *Electronics and Circuit Analysis Using MATLAB*, CRC Press, Florida, 1999
7. L. Duvillaret, F. Garet, and J. L. Coutaz, 'Highly precise determination of optical constants and sample thickness in terahertz time-domain spectroscopy', *Applied Optics*, vol. 38, no. 2, pp. 409–415, January 1999
8. O.S. Ahmed, M.A. Swillam, M.H. Bakr, and X. Li, 'An efficient optimization approach for accurate parameter extraction with terahertz time domain spectroscopy', *IEEE Journal of Lightwave Technology*, vol. 28, pp. 1685–1692, 2010
9. J. Han, H. Xu, Z. Zhu, X. Yu, and W. Li, 'Terahertz spectroscopy of naphthalene, α-naphthol, β-naphthol, biphenyl and anthracene', *Chemical Physics Letters*, vol. 392, no. 4–6, pp. 348–351, July 2004

Problems

8.1 Utilize the random search method to solve the constrained three-dimensional optimization problem:

$$\begin{array}{ll} \text{minimize} & f(\boldsymbol{x}) = x_1^2 + x_2^2 + x_3^2 + x_1x_2 - x_1 - 2x_2 \\ \text{subject to} & x_1 + x_2 + x_3 \leq 4 \\ & x_1, x_2, x_3 \geq 0 \end{array}$$

Utilize the MATLAB listing M8.1 after making the necessary changes. Use 10,000 points over the three-dimensional cube $0 \leq x_1, x_2, x_3 \leq 5.0$.

8.2 Obtain a feasible point satisfying the constraints:

$$\begin{array}{l} x_3 \leq 3 - x_1^2 - x_2^2 \\ x_1, x_2, x_3 \geq 0 \end{array}$$

Utilize the auxiliary objective function (8.9) and the steepest descent listing M6.1 starting from the non-feasible point $\boldsymbol{x} = [1.0 \quad 1.0 \quad 5.0]^T$.

8.3 Utilize the Complex method in solving the constrained optimization problem:

$$\text{minimize} \quad f(\boldsymbol{x}) = x_1^2 + 3x_2^2 + x_3^2$$

$$\text{subject to} \quad \begin{cases} 5 \leq x_1 \leq 20 \\ 3 \leq x_2 \leq 11 \\ 10 \leq x_3 \leq 40 \\ x_1 + x_2 + x_3 \leq 40 \end{cases}$$

Utilize the MATLAB listing M8.3. Construct the initial feasible simplex around the feasible point $\boldsymbol{x} = [8 \quad 8 \quad 20]^T$. Verify your answer by applying the KKT optimality conditions.

8.4 Solve the constrained optimization problem:

$$\text{minimize} \quad f(\boldsymbol{x}) = x_1^2 + 3x_2^2$$

$$\text{subject to} \quad \begin{cases} x_1^2 + x_2^2 \leq 4 \\ (x_1 - 2)^2 + (x_2 - 2)^2 \leq 4 \end{cases}$$

using sequential linear programming. Utilize the MATLAB listing M8.4 with the feasible starting point $\boldsymbol{x}^{(0)} = [1.0 \quad 1.0]^T$.

8.5 Repeat Problem 8.4 using the method of feasible directions. Exploit the MATLAB listing M8.5 after making any necessary changes.

8.6 Solve the quadratic program:

$$\begin{array}{l} \text{minimize } f(\boldsymbol{x}) = 3x_1^2 + 2x_2^2 - 10x_1 - 20x_2 \\ \text{subject to } x_1 + x_2 \leq 3, \quad x_1, x_2 \geq 0 \end{array}$$

Using Rosen's projection method. Utilize the MATLAB code M8.6 after making any necessary changes starting with the feasible point $\boldsymbol{x}^{(0)} = [1.0 \quad 1.0]^T$.

8.7 Solve the optimization problem:

$$\begin{array}{ll} \text{minimize} & f(x_1, x_2) = x_1^2 + 3x_2^2 \\ \text{subject to} & x_2 \geq 1.0, \quad x_1 - x_2 \geq 0 \end{array}$$

using the external penalty method with penalty coefficient of $r = 1$, 10, and 50. Utilize an unconstrained optimization code and objective function similar to M8.7 starting from the point $\boldsymbol{x}^{(0)} = [4.0 \quad 2.0]^T$. Verify your answer by applying the KKT conditions.

Chapter 9

Introduction to global optimization techniques

9.1 Introduction

In previous chapters, we addressed a number of optimization techniques for solving unconstrained and constrained optimization problems. All these techniques obtain a local minimum of the problem. This minimum may not be the best possible solution. The optimization problem may have a better minimum with an improved value of the objective function.

To illustrate this case, consider the objective function:

$$f(x) = -\frac{\sin(x)}{x} \tag{9.1}$$

This objective function has an infinite number of local minima. A number of these local minima are shown in Figure 9.1. This problem has only one global minimum at the point $x = 0$. The value of the objective function at this optimal point is $f^* = -1.0$, which is lowest value over all other values of the parameter x. While in some problems, finding a local minimum with a reasonable value of the objective function is acceptable, in other applications it is mandatory to find the global minimum of the problem.

Over the years, many techniques were developed for finding the global minimum of a nonlinear optimization problem. These techniques include Statistical Optimization [3], Simulated Annealing [4], Genetic Algorithms [5], Particle Swarm Optimization (PSO) [6], Weed Invasive Optimization [7], Wind Optimization [8], and Ant Eolony Optimization [9], just to mention a few. All of these techniques introduce an element of randomness in the iterations to escape local minima. Some of these techniques are inspired by nature, which is always able to find the global minimum of its optimization problems.

We review in this chapter some of these techniques. We illustrate the basic concepts of these algorithms and illustrate them with MATLAB® implementations.

9.2 Statistical optimization

The basic idea of statistical optimization is to convert a global optimization problem into a number of local optimization problems with random starting points. Because the starting points are different, the algorithm may converge to different local minima. The minimum with the best value of the objective function is then used as the global minimum.

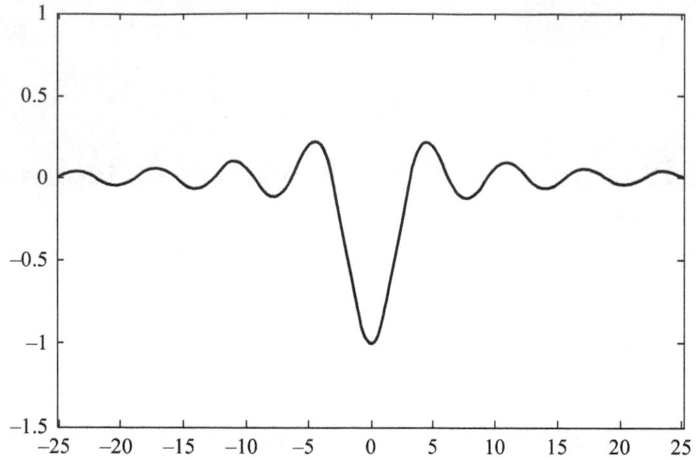

Figure 9.1 A one-dimensional function with one global minimum and infinite number of local minima

Assuming that we want to get the best possible minimum over a region in the parameter space given by:

$$l_i \le x_i \le u_i, \quad i = 1, 2, \ldots, n \tag{9.2}$$

where l_i and u_i are lower and upper bounds on the ith parameter. Statistical optimization then carries out a sequence of local optimization techniques. The starting point of the jth local optimization problem is given by:

$$\boldsymbol{x}_j^{(0)} = \begin{bmatrix} x_1 \\ x_2 \\ \vdots \\ x_n \end{bmatrix} = \begin{bmatrix} l_1 + r_{1,j}(u_1 - l_1) \\ l_2 + r_{2,j}(u_2 - l_2) \\ \vdots \\ l_n + r_{n,j}(u_n - l_n) \end{bmatrix} \tag{9.3}$$

where $r_{i,j} \in [0, 1]$ are random numbers of the jth optimization problem. Utilizing this starting point, a solution $\boldsymbol{x}_j^*$ is reached with a corresponding objective function value of f_j^*. The global minimum $\boldsymbol{x}_G$ is then taken as the minimum with best (lowest) value of the objective function. It is thus given by:

$$\boldsymbol{x}_G = \arg \min_j f_j^* \tag{9.4}$$

Figure 9.2 illustrates the statistical optimization approach for a function with more than one local minimum. The MATLAB listing M9.1 shows a possible implementation of this approach. The following example illustrates the statistical analysis approach.

Example 9.1: Utilize the MATLAB listing 9.1 to find the global minimum of the function:

$$f(x_1, x_2) = 2x_1^2 - 1.05x_1^4 + 0.16666x_1^6 - x_1x_2 + x_2^2 \tag{9.5}$$

within the interval $-6.0 < x_1 < 6.0$, $-6.0 \le x_2 \le 6.0$.

Solution: The MATLAB listing M9.1 applies the steepest descent method starting from random starting points within the interval of interest. We allowed for only

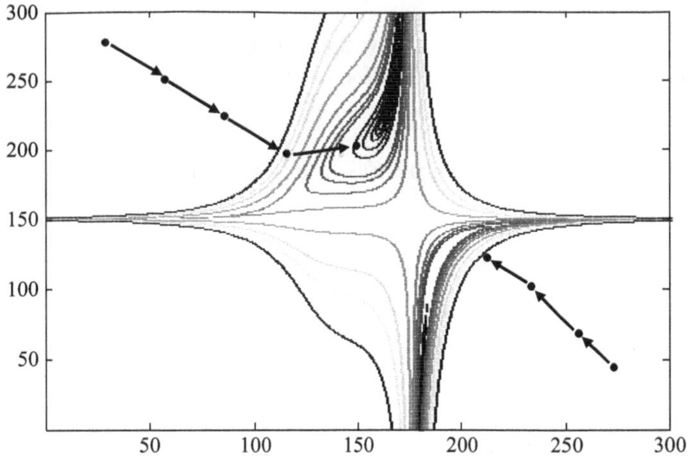

Figure 9.2 An illustration of the statistical optimization approach; starting from different starting points, the local optimization algorithms may converge to different local minima

```
%M9.1
%This program carries out global optimization using steepest
%descent approach with random starting points
NumberOfParameters=2; %This is n for this problem
numberOfRandomPoints=10;
numberOfIterations=25;
lowerValues=[-6  -6]'; %lower values of parameters
upperValues=[6   6]'; %upper values of parameters
OptimalPoints=zeros(2,numberOfRandomPoints); %allocate memory for optimal points
OptimalValues=zeros(numberOfRandomPoints,1); %memory for optimal values
Tolerance=0.001; %terminating tolerance for line search
Epsilon=0.001; %exploration step
LambdaMax=4.0; %maximum value of lambda for line search
for i=1:numberOfRandomPoints
  RandomVector=rand(NumberOfParameters,1); %get a vector of random variables
  %create a random starting point
  OldPoint= lowerValues+RandomVector.*(upperValues-lowerValues);
  OldValue=getObjective(OldPoint) %Get the objective function at the starting point
  k=1; %counter for local optimization iterations
  while(k<25) %repeat until maximum number of iterations is achieved
   Gradient=getGradient('getObjecive',OldPoint, Epsilon); %get the gradient at the old point
   NormalizedNegativeGradient=-1.0*Gradient/norm(Gradient); %normalize the gradient to avoid
                                                            %large or small values of lambda
   LambdaOptimal = GoldenSection('getObjective',Tolerance,OldPoint,
                              NormalizedNegativeGradient,LambdaMax); %get the optimal value
                                                     % in the direction of u starting from YOld
   NewPoint=OldPoint+LambdaOptimal*NormalizedNegativeGradient; %Get new point
   NewValue=feval('getObjective',NewPoint); %Get the New Value
   StepNorm=norm(NewPoint-OldPoint); %get the norm of the step
   OldPoint=NewPoint %update the current point
   OldValue=NewValue %update the current value
   k=k+1;
  end
  %store optimal solution and corresponding objective function
  OptimalPoints(:,i)=OldPoint;
  OptimalValues(i)=OldValue;
end
[GlobalValue, Index]=min(OptimalValues);      %get index of global minimum
GlobalMinimum=OptimalPoints(:,Index);
```

25 steps within the steepest descent algorithm. We utilized 10 random starting points for this function. The 10 achieved optimal points and their corresponding objective function values are given by the output:

Optimal Points =

−0.0020	−0.0029	−0.0156	−0.0006	−0.0758	1.7468	−1.7474	1.7480
−1.7365	0.1478						
−0.0057	−0.0016	−0.0036	−0.0009	−0.1258	0.8727	−0.8737	0.8730
−0.8649	0.1267						

Optimal Values =

0.0000
0.0000
0.0004
0.0000
0.0177
0.2985
0.2984
0.2985
0.2992
0.0405

The steepest descent algorithm terminated at a number of points. The global minimum of this objective function is approximated by local minimum with the lowest objective function and is given by the output:

Global Minimum =

1.0e-03*
−0.5630
−0.9198

Global Value =

9.6211e-07

The global minimum of this function is at the point $\boldsymbol{x} = [0 \quad 0]^T$. The contour plot of the objective function (9.5) is shown in Figure 9.3.

9.3 Nature-inspired global techniques

Humans have always been amazed at the ability of nature to find global minima of optimization problems. For example, the concept of "survival for the fittest" applies to all living creatures. Only elements that have enough fitness can survive and pass their genes to the next generation. Here, the concept of fitness is equivalent to the negative of the objective function value we aim at minimizing. Another example is how bee swarms manage through their collective intelligence to find areas with high nectar content. The level of nectar in a certain area is very similar to the concept of an objective function. The examples from nature are so many.

Nature inspired mathematicians to formula optimization techniques that imitate these natural phenomena. These techniques aim at finding the global minimum of an

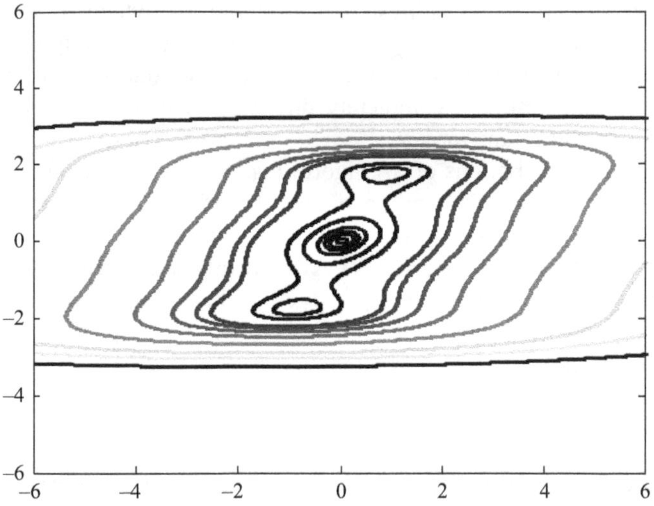

Figure 9.3 The contours of the objective function in Example 9.1; this objective function has three local minima with the global minimum at the point $\boldsymbol{x} = [0 \quad 0]^T$

optimization problem rather than just local minima. There is already a wealth of such techniques. They include genetic algorithms, simulated annealing, particle swarm optimization (PSO), weed invasive optimization, ant colony optimization, and wind optimization. There is actually a wealth of published research papers in this area that applies these techniques to many areas in engineering design problems.

Most of these techniques do not have robust convergence proofs like local optimization techniques. They are shown through examples to reach the global minima. Most of them also require extensive number of objective function evaluations making them very costly as compared to local optimization techniques. As will be seen, the computational cost can be extensive thus limiting the application of some of these techniques to only problems with small function evaluation time.

We will focus in this chapter on only three global optimization approaches. These are simulated annealing, genetic algorithms, and PSO.

9.3.1 Simulated annealing

Simulated annealing (SA) is an optimization technique that finds its roots in statistical mechanics. It was noticed that in some crystalline materials when they are heated up and then allowed to cool down, these materials form different crystalline states. These crystalline states correspond to different organizations of the atoms and to different energy levels of the system.

This concept inspired a similar approach in optimization theory. Every possible crystalline state is actually a local minimum of the energy of the system. Then, by the same concept, we can allow an optimization problem to "heat" and then cool down to move from the current local minimum to another local minimum with a better objective function value. Of course, we cannot heat an optimization problem using thermal energy as in the case of physical systems. Rather, we allow moving to a higher value of the objective function with probability that is decreasing with every iteration. The algorithm can thus escape a local minimum and converge to a completely different minimum with a better objective function value.

The basic concept of SA is illustrated in Figure 9.4, which shows a bouncing ball on a surface with many local minima. The ball can completely rest in any one of these local minima. However, if we allow the ball to bounce up, it may escape a local minimum and rest in a completely different local minimum with a longer depth (better objective function).

There are many variations of SA algorithms. We focus here on the basic steps of the algorithm. At the kth iteration of the algorithm, we assume that the best solution we have so far is $\boldsymbol{x}^{(k)}$. The algorithm then creates a random step $\Delta\boldsymbol{x}^{(k)}$ resulting in a suggested point $\boldsymbol{x}_s = \boldsymbol{x}^{(k)} + \Delta\boldsymbol{x}^{(k)}$. If this suggested point has a better value of the objective function ($f(\boldsymbol{x}_s) < f(\boldsymbol{x}^{(k)})$) then it is accepted resulting in $\boldsymbol{x}^{(k+1)} = \boldsymbol{x}_s$. Otherwise, we calculate the probability of accepting this point. This probability is given by:

$$\text{probability} = e^{-\beta \Delta f} \tag{9.6}$$

where β is the positive annealing parameter. This parameter is increased with every iteration, thus reducing the probability of accepting a point with a higher objective function. This parameter corresponds to the inverse of temperature in annealed physical systems. The quantity $\Delta f = f(\boldsymbol{x}_s) - f(\boldsymbol{x}^{(k)})$ is the increase in the objective function from the value at the current point. The probability of making an ascending step decreases as Δf increases. If the point with a higher objective function is accepted, we have $\boldsymbol{x}^{(k+1)} = \boldsymbol{x}^{(k)} + \Delta\boldsymbol{x}^{(k)}$. This random "jump" in the objective function allows the algorithm to escape local minimum. The algorithm proceeds until a termination condition is satisfied.

The algorithm steps of an SA algorithm are summarized by the following steps:

Step 0: Initialization: Set $k = 0$ and initialize β. Given a starting point $\boldsymbol{x}^{(0)}$. Evaluate $f(\boldsymbol{x}^{(0)})$.

Step 1: Generate a random direction $\boldsymbol{s}^{(k)}$ with a unity norm ($\|\boldsymbol{s}^{(k)}\| = 1.0$).

Step 2: Evaluate the suggested point $\boldsymbol{x}_s = \boldsymbol{x}^{(0)} + \alpha\, \boldsymbol{s}^{(k)}$. Evaluate $f(\boldsymbol{x}_s)$ and $\Delta f = f(\boldsymbol{x}_s) - f(\boldsymbol{x}^{(0)})$.

Comment: The line search parameter α is determined through line search or through taking a constant step.

Step 3: If $\Delta f < 0$, set the probability of acceptance $p = 1$. Otherwise, $p = e^{-\beta\Delta f}$.

Comment: The parameter p is the probability of accepting the suggested point. If $p = 1$, the point is accepted. If $p < 1$, the point may or may not be accepted.

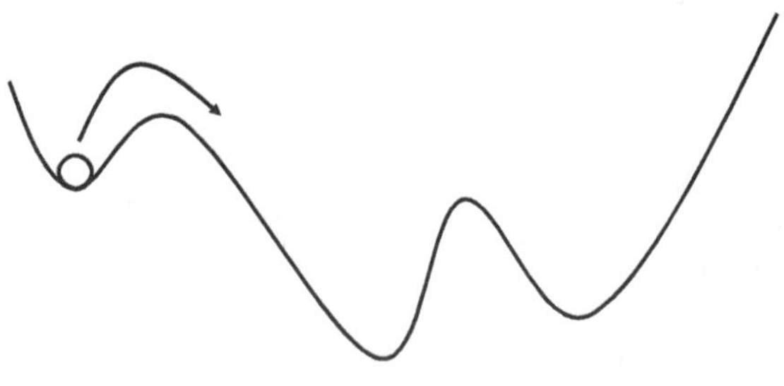

Figure 9.4 An illustration of simulated annealing (SA) optimization; a ball can have several resting positions (local minima) on a curvy surface. Allowing the ball to bounce up may enable it to escape from a resting position to another one with a longer depth (better objective function)

Step 4: Generate a random number r, with $0 \leq r \leq 1$. If $r \leq p$, accept the new point and set $\boldsymbol{x}^{(k+1)} = \boldsymbol{x}_s$; otherwise keep current point $\boldsymbol{x}^{(k+1)} = \boldsymbol{x}^{(k)}$.

Comment: The random parameter r allows for accepting a suggested point with a probability p.

Step 5: If termination condition is satisfied, stop.

Step 6: Set $k = k + 1$ and go to Step 2.

The steps given above do not take into account the objective function sensitivities that may be available. These sensitivities can, however, be integrated in determining the search direction $\boldsymbol{s}^{(k)}$. The MATLAB listing M9.2 shows a possible implementation of the SA approach.

```
%M9.2
%An implementation of Simulated Annealing Approach
NumberOfParameters=2; %This is the number of parameters
CurrentPoint=[-5 -5]'; %This is the current point
Perturbation=0.001; %perturbation used in calculating sensitivities
CurrentValue=getObjective(CurrentPoint); %get current value
NumberOfIterations=1000; %this is the number of iterations
LambdaOptimal=1.0; %initial value of line search parameter
LambdaMax=10.0; %maximum step in line search
Tolerance=1.0e-3; %this is the tolerance
%Beta is initialized so that a 50% change in objective function gives a 70% acceptance probability
DeltaF=0.5*CurrentValue; %This is 50% change from starting point
AcceptanceProbability = 0.7; % 70% probability
Beta = -log(AcceptanceProbability)/DeltaF; %get Beta
if(abs(Beta)<1) %we do not start below 1
  Beta = 1;
end
if (Beta > 0) % a positive Beta is of course not allowed
  Beta = -1*Beta;
end
for IterationCounter=1:NumberOfIterations %Repeat for a number of iterations
  Direction= rand(NumberOfParameters,1); % Get a random search direction
  Sign = rand(NumberOfParameters,1); % used to randomly change signs for s
  for j = 1:NumberOfParameters
    if (Sign(j,1)> 0.5)
      Direction(j,1) = -1*Direction(j,1);
    end
  end
  PerturbedPoint=CurrentPoint+Perturbation*Direction; %get a slightly perturbed point
  PerturbedValue=getObjective(PerturbedPoint); %get the perturbed value
  Slope=(PerturbedValue-CurrentValue)/Perturbation; %get slope along this direction
  if(Slope < 0) %does function decrease?
    LambdaOptimal = GoldenSection('getObjective',Tolerance,CurrentPoint,
                                               Direction,LambdaMax);%line search
    NewPoint=CurrentPoint+LambdaOptimal*Direction; %this is the new point
    NewValue=getObjective(NewPoint); %get the new value
    CurrentPoint=NewPoint; %update parameters
    CurrentValue=NewValue;
  else
    NewPoint = CurrentPoint + LambdaOptimal*Direction; %get new point
    NewValue=getObjective(NewPoint); %get the new value
    DeltaF=NewValue-CurrentValue; %get change in objective function
    AcceptanceProbability = exp(Beta*DeltaF);
    random1 = rand(1,1); % get a random number
    if (random1 <= AcceptanceProbability) % Jump
      CurrentPoint=NewPoint;
      CurrentValue=NewValue;
    end
    Beta=Beta*1.03; %increase Beta
  end
end
```

The MATLAB listing M9.2 tests the slope of the function in the suggested random direction. If this is a descending direction, a line search is carried out along that direction. If it is an ascending direction, the code uses the same value of the line search parameter determined in the previous step to create the suggested point $\boldsymbol{x}_s$. This point may or may not be accepted depending on the acceptance probability. The following example illustrates this method.

Example 9.2: Find the global minimum of the function:

$$f(x_1,x_2) = -20\frac{\sin\left(\sqrt{0.1+(x_1-4)^2+(x_2-4)^2}\right)}{\sqrt{0.1+(x_1-4)^2+(x_2-4)^2}} \tag{9.7}$$

starting from the initial points $\boldsymbol{x}^{(0)} = [5.0 \quad 5.0]^T$. Utilize the MATLAB listing M9.2.

Solution: The function values obtained during the optimization iterations are shown in Figure 9.5. This figure shows that the iterations approach a number of minima during the optimization iterations. However, in some of these iterations, the algorithm is able to break and reach the global minimum with an objective function value of $f^* = -19.6683$. The point with the best objective function reached by the algorithm is $\boldsymbol{x}^* = [3.9932 \quad 3.9935]^T$. The global minimum of this problem is at $\boldsymbol{x}_G = [4.0 \quad 4.0]^T$.

9.3.2 Genetic algorithms

Another class of global optimization problems that utilize a nature-inspired approach is genetic algorithms (GAs). These algorithms apply the concept of survival for the fittest to optimization. Most of the terms utilized in these algorithms are borrowed from the area of biology as will become evident. There are a large number of research papers in this area. We aim here at only covering the basic concepts.

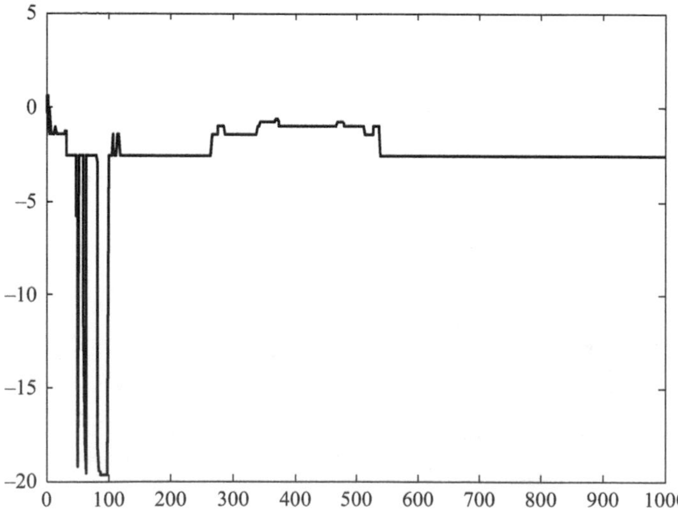

Figure 9.5 The values of the objective function reached during the iterations of Example 9.2. It is obvious that the iterations were able to escape the local minimum and reached the actual global minimum a number of times

Different from the algorithm discussed so far, these algorithms use at every iteration a set of points in the parameter space. This set of points is called the "population". At the kth iteration, the current population is denoted by $P^{(k)}$. At every iteration, a new population $P^{(k+1)}$ is created from the previous population through the processes of crossover, mutation, and immigration. A new "generation" of points is thus created from the old one. Similar to biology, two points (parents) in one generation mix their genes to give rise to their children. In digital GAs, every point in the parameter space is first encoded by a string that includes a number of symbols. This string is usually referred to as a chromosome. For example, a point in a two-dimensional parameter space may be given by the following string:

$$\underbrace{01011100}_{x_1}\underbrace{11110101}_{x_2} \tag{9.8}$$

In (9.8), every parameter is assigned eight symbols and the length of the string is 16 symbols. The number of bits (symbols) assigned to each parameter determines the value of that parameter. It follows that the string determines the value of all the parameters of this point. Every point has a fitness value associated with it. The target is to maximize the fitness. The fitness is thus taken as the negative of the objective function to be minimized.

At the kth iteration, the population $P^{(k)}$ of points are first encoded. For example, for a two-dimensional problem, assuming only a population of three members, the encoded chromosomes are given by:

$$P^{(k)} = \{1000101011110001, 1010110101110010, 1100101000110011\} \tag{9.9}$$

where every parameter is encoded by eight bits. The algorithm then determines the fitness value of each chromosome. As the algorithm utilizes a survival for the fittest approach, only chromosomes with good fitness values are allowed to give rise to children. A mating pool $M^{(k)}$ is formed using the population $P^{(k)}$ from the points with good fitness. It has though the same number of members as the population. Elements of $P^{(k)}$ with good fitness appear at higher probability. Algorithms such as the tournament scheme or the roulette scheme are used to generate $M^{(k)}$ from $P^{(k)}$. Once the mating pool has been constructed, the next generation is formed from this set by mating elements. Operations such as crossover, mutation, and migration are utilized to generate $P^{(k+1)}$ using $M^{(k)}$. In the crossover operation, each two chromosomes give rise to two offsprings by sharing their genes. Consider, for example, the following two chromosome parents, each with eight symbols:

$$\text{Parent1} = \{11111111\}, \quad \text{Parent2} = \{00000000\} \tag{9.10}$$

By sharing half of their genes with their offsprings, they give rise to the following children:

$$\text{Child1} = \{11110000\}, \quad \text{Child2} = \{00001111\} \tag{9.11}$$

It is very important to grasp the meaning of the crossover process. The two parents are members of the population with a good fitness. Their position in the parameter space is encoded into their genes. By sharing genes, as in (9.13), we are creating offsprings that are not very far in the parameter space from the parents but they still explore new domains where better minima may exist.

While the process of crossover gives rise to most of the elements in $P^{(k+1)}$, other imitations of biological processes can also take place. For example, the process of mutation in biology implies a random change in some of the genes. This may give rise to members of the population with special characteristics. The same can be applied to GA. For example, a member of $P^{(k+1)}$ can be mutated by randomly picking one of its symbols and changing it as follows:

$$\begin{aligned}&\text{Child before mutation} = \{11110000\}\\&\text{Child after mutation} = \{11110010\}\end{aligned} \tag{9.12}$$

This mutation implies that we introduce a random change in the position in the parameter space of some of the members of the new generation. This allows for exploring new domains in the parameter space with possibly better fitness. Similar to mutations in biology, this process happens with only very little probability.

Another procedure which is used in GA is the process of immigration. Many countries allow some of the highly educated and industrious members of other countries to migrate to them and become members of the population. This process raises the fitness of the population as a whole and enriches its diversity. The same concept is applied in GA. Random chromosomes are created and added to the population to improve its diversity and improve the possibility of finding better minima. The steps of crossover, mutation, and immigration satisfy two of the main targets of GA: exploration and exploitation.

A possible algorithm for a GA is given by the following steps:

Step 0: Given N_{max}. Set $k=0$. Create the initial population $P^{(0)}$.

Comment: N_{max} is the number of generations (iterations) allowed. The initial population is generated in a random way with upper and lower bounds on each parameter.

Step 1: Evaluate the fitness for every chromosome in $P^{(k)}$.

Comment: The fitness is the negative of the objective function. The higher the fitness, the better the objective function is.

Step 2: If $k > N_{max}$, stop. Otherwise, select $M^{(k)}$ using $P^{(k)}$.

Comment: $M^{(k)}$ has the same number of elements as $P^{(k)}$. Chromosomes with higher fitness appear with higher probability in $M^{(k)}$.

Step 3: Obtain $P^{(k+1)}$ from $M^{(k)}$ through evolution.

Comment: Evolution includes the steps crossover, mutation, and immigration.

Step 4: Set $k=k+1$ and go to Step 1.

Equations (9.8)–(9.12) illustrate the encoded GA approach. A variation of GAs that is easier to implement is real GA. In this approach, no encoding is needed. Each chromosome is simply a point in the parameter space. The operations of crossover, mutation, and immigration are also utilized in this approach. For example, given two parents $\boldsymbol{x} = [x_1 \quad x_2 \quad \ldots \quad x_n]^T$ and $\boldsymbol{y} = [y_1 \quad y_2 \quad \ldots \quad y_n]^T$, two offsprings may be given by:

$$\begin{aligned}&\text{first child} = [x_1 \quad x_2 \ldots x_{r-1} \quad y_r \quad y_{r+1} \ldots y_n]^T\\&\text{second child} = [y_1 \quad y_2 \ldots y_{r-1} \quad x_r \quad x_{r+1} \ldots x_n]^T\end{aligned} \tag{9.13}$$

The two offsprings in (9.13) each include mixed components from both parents. The index r is called the truncation index. Another possible crossover formulation for real GA utilizes convex linear combination of the two parents of the form:

$$\begin{aligned} &\text{first child} = \lambda_1 \boldsymbol{x} + \lambda_2 \boldsymbol{y} \\ &\text{second child} = \lambda_2 \boldsymbol{x} + \lambda_1 \boldsymbol{y} \end{aligned} \tag{9.14}$$

where $\lambda_1 + \lambda_2 = 1$, $0 \leq \lambda_1, \lambda_2 \leq 1.0$ are random numbers. The two children are thus located in the parameter space along the line connecting both parents. The process of mutation can also be implemented by introducing a random change to a randomly selected component of a chromosome.

The MATLAB listing M9.3 shows a possible implementation of a real GA.

```
%M9.3
%This code carries out a real Genetic Algorithm
PopulationSize=100; %number of chromosomes
NumberOfParameters=2; %number of designable parameters
NumberOfGenerations=5; %number of generations
LowerBounds=[-10 -10]'; %lower bounds on parameters
UpperBounds=[10 10]'; %lower bounds on parameters
MutationProbability=0.01; %probability of mutation
Fitness=zeros(PopulationSize,1); %fitness of all chromosomes
BestFitness=zeros(NumberOfGenerations,1); %storage for best fit in every iteration
BestFitnessPoints=zeros(NumberOfParameters,NumberOfGenerations); %storage for best fit
                                                                  %point in every iteration
%now we do initialization
Population=Initialize(NumberOfParameters, PopulationSize, LowerBounds, UpperBounds);
%This is the main loop
for GenerationCounter=1:NumberOfGenerations %repeat for all generations
  for i=1:PopulationSize %repeat for all chromosomes
    Fitness(i,1)=getFitness(Population(:,i)); %get fitness of the ith element
  end
  [MaxValue MaxValueIndex]= max(Fitness); %get best fitness and its index
  BestFitness(GenerationCounter,1)= MaxValue%store the best fit in this generation
  BestFitnessPoints(:,GenerationCounter)=Population(:,MaxValueIndex); %store best fit point
  %now we create the mating pool using the population
  MatingPool=Select(Population,Fitness); %select the fittest from inside this population
  %now we do simply crossover
  OffSprings=CrossOver(MatingPool); %do simple crossover
  Population=Mutation(OffSprings, MutationProbability); %do mutation
end
BestFitness
BestFitnessPoints
```

The MATLAB listing M9.3 initializes the population in a random way. The function **Select** creates the mating pool using the current population. The evolution to the next generation is implemented in the functions **CrossOver** and **Mutation**. The implementation of all these functions is shown in the MATLAB listings M9.4 and M9.5.

The function **Select** picks randomly two chromosomes from the population and compares their fitness values. The one with the higher fitness is allowed to join the mating pool. This is repeated until the mating pool is full. This results in mating pool where the chromosomes with higher fitness have a higher probability. The mating pool will of course have some repeated members as a result of the selection process.

The function **CrossOver** implements the crossover approach explained by (9.13). Two points are picked randomly from the mating pool. The truncation index r is then determined in a random way as well. The two offsprings are then given by copying the first r components from a parent and the following $(n-r)$ components from the other parent.

The function **Mutation** implements a simple mutation process. A random number is generated and if its value is below or equal to the mutation probability, mutation takes place. This involves randomly picking a component and changing its value.

```
%M9.4
%This function carries out the selection part of the real genetic algorithm
function MatingPool=Select(Population, Fitness)
NumberOfParameters=size(Population,1);
PopulationSize=size(Population,2);
MatingPool=zeros(NumberOfParameters,PopulationSize); %allocate memory for mating pool
for i=1:PopulationSize %we have to fill all locations
  random1=rand(1,1); %get a random number
  Index1=ceil(random1*PopulationSize); %get index of 1st chromosomes
  random2=rand(1,1); %get a second random number
  Index2=ceil(random2*PopulationSize); %get index of 2nd chromosome
  if(Fitness(Index1)>Fitness(Index2)) %which chromosome is better
    MatingPool(:,i)=Population(:,Index1); %store the fittest chromosome
  else
    MatingPool(:,i)=Population(:,Index2);
  end
end

%This function carries out a simple crossover procedure
function OffSprings=CrossOver(MatingPool)
NumberOfParameters=size(MatingPool,1);
PopulationSize=size(MatingPool,2);
OffSprings=zeros(NumberOfParameters,PopulationSize); %allocate memory for mating pool
OffSpring1=zeros(NumberOfParameters,1); %first offspring
OffSpring2=zeros(NumberOfParameters,1); %second offspring
for i=1:2:PopulationSize %we have to fill all locations
  random1=rand(1,1); %get a random number
  Index1=ceil(random1*PopulationSize); %get index of 1st parent
  random2=rand(1,1); %get a second random number
  Index2=ceil(random2*PopulationSize); %get index of 2nd parent
  %now we do crossover
  random3 = rand(1,1); %get a random number between 0 and 1
  %determine the index of the crossover
  if random3 > 0.5
    random4 = floor(NumberOfParameters*random3);
  else
    random4 = ceil(NumberOfParameters*random3);
  end
  %now we do crossover at location random4
  for j = 1:NumberOfParameters
   if j <= random4
     OffSpring1(j,1) = MatingPool(j,Index1);
     OffSpring2(j,1) = MatingPool(j,Index2);
   else
     OffSpring2(j,1) = MatingPool(j,Index1);
     OffSpring1(j,1) = MatingPool(j,Index2);
   end
  end
  OffSprings(:,i)=OffSpring1; %store first offspring
  OffSprings(:,i+1)=OffSpring2; %store 2nd offspring
end
```

```
%M9.5
%This function carries out the mutation  part of the real genetic algorithm
function Population=Mutation(OffSprings, MutationProbability)
NumberOfParameters=size(OffSprings,1); %get number of parameters
PopulationSize=size(OffSprings,2); %get size of population
Population=zeros(NumberOfParameters, PopulationSize); %allocate memory for population
for i=1:PopulationSize %repeat for all offsprings
  random1=rand(1,1); %get a random number between 0 and 1
  Population(:,i)=OffSprings(:,i); %copy ith offspring
  if(random1<=MutationProbability)
    random2=rand(1,1); %get another random number
    Index=ceil(random2*NumberOfParameters); %get index of mutation
    %change the value of that component
    Population(Index,i)=random1*Population(Index,i);
  end
end
```

The following two examples illustrate the application of GAs to optimization problems:

Example 9.3: Find the global minimum of the objective function $f(x_1, x_2) = x_1^2 + x_2^2$ using genetic algorithms. Use the MATLAB listing M9.3.

Solution: This listing utilizes a population of 100 chromosomes. It allows for five generations (iterations). It follows that a total of 500 function evaluations are needed. The code keeps track of the best chromosome obtained in every generation. The output of this code is given by:

Best Fitness =

−0.0646
−0.4947
−0.4133
−0.0002
−0.0315

Best Fitness Points =

−0.0379	0.6858	−0.6180	0.0136	0.0136
−0.2514	0.1560	−0.1771	−0.0011	−0.1771

It follows that the best point was achieved in the fourth generation with the highest fitness of −0.0002. This corresponds to an objective function value of 0.0002. The corresponding solution is thus $[0.0136 \quad -0.0011]^T$. The actual minimum of this function is $\boldsymbol{x}^* = [0 \quad 0]^T$.

Example 9.4: Find the global minimum of the function:

$$f(x_1, x_2) = (1.5 - x_1(1 - x_2))^2 + (2.25 - x_1(1 - x_2^2))^2 + (2.625 - x_1(1 - x_2^3))^2$$

Utilize the MATLAB listing M9.3 with 1000 chromosomes and five generations.

Solution: This function exhibits a higher degree of nonlinearity than the one in the previous example. The results obtained for the best solution in every generation are given by the output:

Best Fitness =

−0.1118
−0.0307
−0.0339
−0.0071
−0.0385

Best Fitness Points =

3.4184	3.0358	3.0705	3.2202	3.4819
0.6344	0.5432	0.5521	0.5443	0.5787

It follows that the best solution was obtained in the fourth generation with a corresponding fitness of -0.0071. The corresponding optimal point is $[3.2202 \quad 0.5443]^T$. The contours of the objective function are shown in Figure 9.6. This figure shows that the result obtained by the GA is very close to the actual global minimum.

9.3.3 Particle swarm optimization

Another nature-based class of optimization problems is PSO. This approach utilizes concepts borrowed from the field of social psychology. It was noticed in this field, for example, that humans have a better chance of solving a problem when communicating to each other their individual experiences. Another example is bee swarms that communicate with each other to determine the spots with abundant nectar. This social learning and social influence enables a member of a group to maintain cognitive consistency by not being limited to the personal experience but by tapping into the collective knowledge of the group.

The best way to explain the basic idea behind the PSO technique is to imagine a swarm of particles (points) traveling together in the parameter space. At every iteration, every particle moves in a certain way in search of better local minima. Each individual particle remembers the position in the parameter space where this particle achieved the best value of the objective function. This is called the individual best position. In addition, the whole swarm keeps track of the position where the best value of the whole swarm was achieved. Each member of the swarm moves according to a relationship that is influenced by its individual best value and the swarm best value. This approach integrates the collective cognitive experience of the swarm into the optimization process.

Mathematically, we assume that we have N particles (points) within the swarm moving together in the parameter space. The position and velocity of the ith particle in the kth iteration are denoted by $\boldsymbol{x}_i^{(k)}$ and $\boldsymbol{v}_i^{(k)}$, respectively. At every iteration,

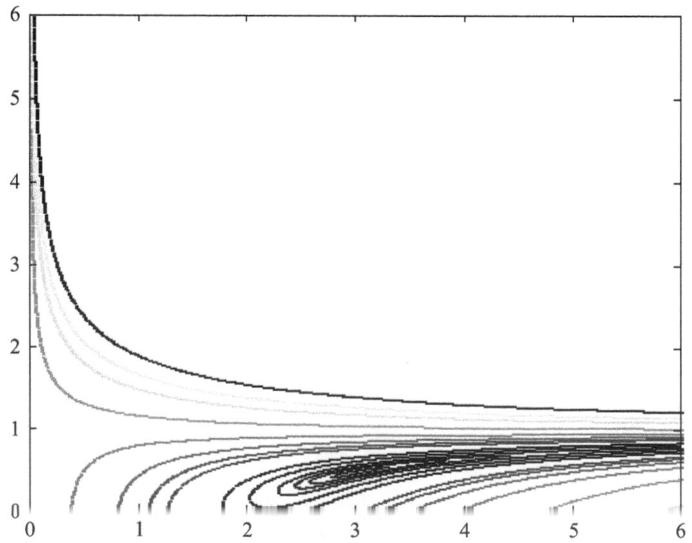

Figure 9.6 The contours of the function of Example 9.4

these parameters are updated based on the individual and collective knowledge of the swarm. A possible update formula is given by [6]:

$$\begin{aligned} \boldsymbol{v}_i^{(k+1)} &= \omega \boldsymbol{v}_i^{(k)} + C_1 r_1 (\boldsymbol{x}_i^* - \boldsymbol{x}_i^{(k)}) + C_2 r_2 (\boldsymbol{x}_G - \boldsymbol{x}_i^{(k)}) \\ \boldsymbol{x}_i^{(k+1)} &= \boldsymbol{x}_i^{(k)} + \boldsymbol{v}_i^{(k+1)} \end{aligned} \tag{9.15}$$

Notice that the velocity vector represents the change in the position of the particle at every iteration. The parameter ω is called the inertia parameter and is usually given values 0.95–0.99. The parameter C_1 is called the cognitive parameter. Its value is chosen heuristically to be around 2.0. The parameter C_2 is called the social parameter. Its value is usually chosen equal to C_1. The parameters r_1 and r_2 are random numbers satisfying $0 \le r_i \le 1$, $i = 1, 2$. The position $\boldsymbol{x}_i^*$ is the position of the point with the best value of the objective function reached so far by the ith particle. $\boldsymbol{x}_G$ is the position of the best point reached by the swarm as a whole.

It can be seen that in (9.15), the change applied to the position of the ith particle depends on the change applied in the previous iteration and the locations of its individual best and the global best. It is being biased randomly between its own experience of the objective function and the collective swarm experience.

A possible algorithm for PSO optimization is given by the following steps:

Step 0: Initialize $\boldsymbol{x}_i^{(0)}$ and $\boldsymbol{v}_i^{(0)}, \forall i$. Set $k = 0$.

Step 1: Evaluate $f_i^{(k)} = f(\boldsymbol{x}_i^{(k)}), \forall i$. Update the local minimum $\boldsymbol{x}_i^* = \arg\min\limits_k f_i^{(k)}$ and the global minimum $\boldsymbol{x}_G = \arg\min\limits_{i,k} f_i^{(k)}$.

Comment: The global minimum position is the best point achieved over all points over all iterations. The individual best minimum is the best value of the particle over all iterations.

Step 2: Update $\boldsymbol{v}_i^{(k)}$ and $\boldsymbol{x}_i^{(k)}$ using (9.15).

Step 3: If the termination condition is satisfied, stop. Otherwise, set $k = k + 1$. Go to Step 1.

The MATLAB listing M9.6 shows a possible implementation of the PSO.

Example 9.5: Resolve Example 9.2 using PSO. Utilize the MATLAB listing M9.6.

Solution: The MATLAB listing M9.6 utilizes 50 particles in the solution. These parameters are initialized randomly within the given upper and lower limits for each parameter. The final optimal point obtained by the algorithm is $\boldsymbol{x} = [4.0151 \quad 4.3410]^T$ with a corresponding objective function value of -19.2861.

A9.1 Least *p*th optimization of filters

In Chapter 2, we addressed the minimax optimization problem. This problem is given by:

$$\min_{\boldsymbol{x}} \{\max_j q_j\} \tag{9.16}$$

where q_j is the jth error function. The target of this problem is to make the errors as negative as possible. This implies, for example, pushing the response of the filter as low as possible below an upper bound or pushing the response as high as possible above a lower bound.

A variation of this problem tries to push all the errors to match a given target response exactly. The aim here is not to make the errors as negative as possible but

```
%M9.6
%This program executes a simple canonical PSO algorithm
NumberOfParticles=50; %number of particles in the swarm
NumberOfParameters=2; %number of parameters
LowerValues=[-8 -8]'; %lower bounds on parameters
UpperValues=[8 8]'; %upper bounds on parameters
Positions=zeros(NumberOfParameters, NumberOfParticles); %allocate memory for positions
LocalBestPositions=zeros(NumberOfParameters, NumberOfParticles); %this is local best position
LocalBest=zeros(NumberOfParticles,1); %This is best value of each paricle
LocalBest(:,1)=-1.0e20; %put a very small fitness as initialization
GlobalBest=-1.0e9; %initialize value of global best
GlobalBestPosition=zeros(NumberOfParameters,1); %this is the position of the global best
Velocities=zeros(NumberOfParameters, NumberOfParticles); %allocate memory for velocities
InertiaCoefficient=0.99; %this is omega
CognitiveCoefficient=1.99; %This is the parameter c1
SocialCoefficient=1.99; %This is the parameter c2
NumberOfIterations=1000; %this is the number of iterations
%now we initialize the positions. Velocities are already zeros
for i=1:NumberOfParticles
  for j=1:NumberOfParameters
    Positions(j,i)=LowerValues(j,1)+rand(1,1)*(UpperValues(j,1)-LowerValues(j,1))
  end
end
%repeat for all iterations
for IterationCounter=1:NumberOfIterations
  CurrentFitnessValues=getFitness(Positions); %get the vector of fitness for the current positions
  %now we update the local best
  for i=1:NumberOfParticles
    if(CurrentFitnessValues(i,1)>LocalBest(i,1)) %is the this the local best
      LocalBest(i,1)=CurrentFitnessValues(i,1); %update local best
      LocalBestPositions(:,i)=Positions(:,i); %update location of local best
    end
  end
  [BestFitness BestFitnessIndex]=max(CurrentFitnessValues); %get the best particle and its index
  if(BestFitness>GlobalBest) %There is an improvement over the previous global best
    GlobalBest=BestFitness; %update the global best value
    GlobalBestPosition=Positions(:,BestFitnessIndex); %update the global best value position
  end
  %now we update positions
  for i=1:NumberOfParticles
    for j=1:NumberOfParameters
      r1=rand(1,1); %first random number
      r2=rand(1,1); %second random number
      Velocities(j,i)=InertiaCoefficient*Velocities(j,i)+CognitiveCoefficient*
       r1*(LocalBestPositions(j,i)-Positions(j,i))+SocialCoefficient*r2*(GlobalBestPosition(j,1)-
                                                                          Positions(j,i));
      Positions(j,i)=Positions(j,i)+Velocities(j,i); %update position
    end
  end
end
GlobalBest
GlobalBestPosition
```

rather trying to reduce the deviation of the response from the target response. Mathematically, this problem is given by:

$$\min_{x} \{\max_{j} |q_j|\} \tag{9.17}$$

The absolute minimax problem (9.17) minimizes the error function with the maximum modulus. Figure 9.7 shows the possible response of a filter and the design specifications. The optimal design of the filter parameters $\boldsymbol{x}^*$ should ideally drive the modulus of all errors to 0. This, however, may not be possible in some cases.

In previous chapters, we utilized the L_2 norm as an objective function. The L_2 norm of the error vector $\boldsymbol{q} = [q_1\ q_2, \ldots, q_m]^T$ is given by:

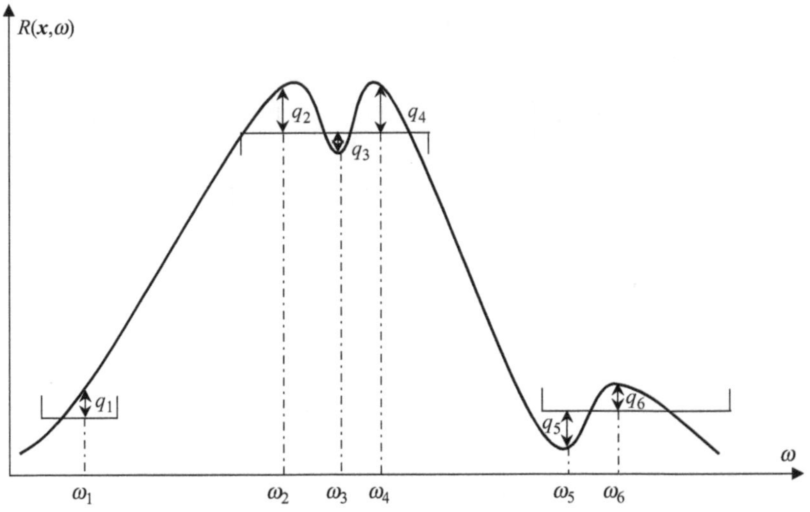

Figure 9.7 An illustration of the absolute minimax problem; six frequencies are used to impose the specifications on the response. The target is to minimize the modulus of the deviations from the specifications q_i, $i = 1, 2, \ldots, 6$ by minimizing their maximum value

$$U = \left(\sum_{j=1}^{m} |q_j|^2 \right)^{1/2} \tag{9.18}$$

The objective function (9.18) puts more emphasis on larger errors but it does not ignore small errors as well. The norm used in (9.18) is generalized to the L_p norm by using an arbitrary exponent p to get:

$$U = \left(\sum_{j=1}^{m} |q_j|^p \right)^{1/p} \tag{9.19}$$

By using a large positive exponent p, more emphasis is put on minimizing the larger errors. Actually, if we move the error with the largest modulus outside the summation, we have:

$$U = |q_{j_{\max}}| \left(1 + \sum_{\substack{j=1, \\ j \neq j_{\max}}}^{m} \left(\frac{|q_j|}{|q_{j_{\max}}|} \right)^p \right)^{1/p} \tag{9.20}$$

where $j_{\max}$ is the index of the error with largest modulus. If the exponent p is made very large, we have:

$$U = \lim_{p \to \infty} |q_{j_{\max}}| \left(1 + \sum_{\substack{j=1, \\ j \neq j_{\max}}}^{m} \left(\frac{|q_j|}{|q_{j_{\max}}|} \right)^p \right)^{1/p} = |q_{j_{\max}}| = \max_j \ |q_j| \tag{9.21}$$

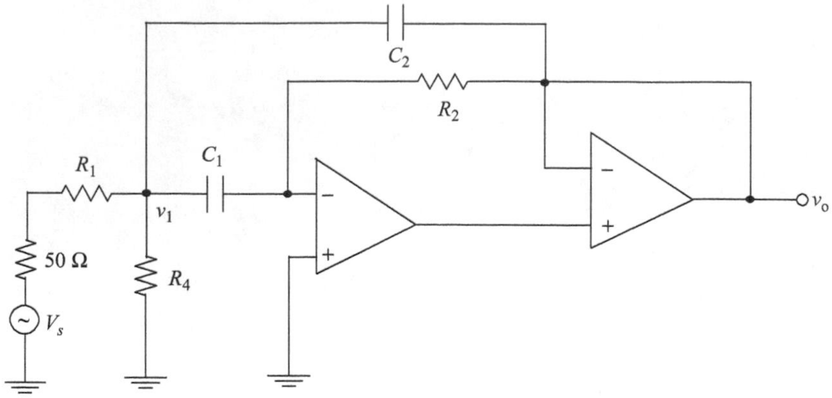

Figure 9.8 The Op-Amp bandpass filter circuit of application A9.1

In (9.21), we utilized the fact that raising a less-than-unity positive quantity to an infinite power gives zero. It follows from (9.21) that solving the absolute minimax problem (9.17) is equivalent to minimizing the pth norm of the errors for sufficiently large values of p. This approach is called least pth optimization. Any unconstrained optimization problem can be used for minimizing (9.19). Global optimization techniques may also be used to avoid getting trapped in a local minima of the objective function.

To illustrate this approach, we consider the active bandpass filter shown in Figure 9.8. The design constraints of this problem are:

$$\begin{aligned} &|V_o/V_s| = 0, \quad \text{freq} \leq 200\ \text{Hz} \\ &|V_o/V_s| = 0, \quad \text{freq} \geq 500\ \text{Hz} \\ &|V_o/V_s| = 50, \quad 375\ \text{Hz} \leq \text{freq} \leq 425\ \text{Hz} \end{aligned} \tag{9.22}$$

We thus want to move the passband to the range [375 Hz, 425 Hz]. Two stop bands are desired below 200 Hz and above 500 Hz. We select four frequencies in the first stop band {50 Hz, 100 Hz, 150 Hz, 200 Hz}, three points in the desired passband {375 Hz, 400 Hz, 425 Hz}, and only one point in the second stop band {500 Hz}. It follows that in this problem we have eight errors. By using $V_s = 1.0\angle 0$, the value of each error is given by:

$$q_i(\boldsymbol{x}) = |V_o(\boldsymbol{x}, freq_i)| - c_i, \quad i = 1, 2, \ldots, 8 \tag{9.23}$$

where $freq_i$ is the ith frequency and c_i is the corresponding constraint with $c = [0 \quad 0 \quad 0 \quad 0 \quad 50 \quad 50 \quad 50 \quad 0]^T$. We solve this problem by using the pth norm with $p = 5$. The parameters for this problem are $\boldsymbol{x} = [R_1 \quad C_1 \quad R_2 \quad C_2 \quad R_4]^T$. The source impedance has the fixed value $R_s = 50$. The error functions and objective function calculations are shown in the MATLAB listing M9.7.

The function **getFilterResponse** returns the value of $V_o(\boldsymbol{x}$, freq) at the given frequency for the given parameters. Using ideal Op-Amp assumptions, it can be shown that V_o is given by:

$$V_o(\boldsymbol{x}, s) = -\frac{sC_1}{s^2 C_1 C_2 R_1 + s\left(\dfrac{R_1}{R_2}\right)(C_1 + C_2) + \left(\dfrac{R_1}{R_2}\right)\left(\dfrac{1}{R_4} + \dfrac{1}{R_1}\right)} \tag{9.24}$$

```
%M9.7
%This function returns the Errors of equation (9.23)
function Errors=getErrors(Point)
scaleMatrix=[1.0e3 0 0 0 0; 0 1.0e-6 0 0 0; 0 0 1.0e3 0 0;
   0 0 0 1.0e-6 0; 0 0 0 0 1.0e3 ]; %scaling is used resistors in Kohm and
                                    %capacitors in microfarads
scaledParameters=scaleMatrix*Point; %get the scaled parameters
Frequencies=1.0e3*[0.05 0.100 0.150 0.200 0.375 0.400 0.425 0.500]'; % frequencies
Targets=[0 0 0 0 50 50 50 0]'; %these are the targets at the different frequencies
numberOfFrequencies=size(Frequencies,1); %this is the number of frequencies
Errors=zeros(numberOfFrequencies,1); %allocate storage for the errors
for i=1:numberOfFrequencies
  currentFrequency=Frequencies(i); %ith frequency
  currentTarget=Targets(i); %target at ith frequency
  Response=getFilterResponse(scaledParameters, currentFrequency); %get Vo(x, freq)
  Errors(i)=(abs(Response)-currentTarget); %evaluate error
end

%This function evaluates the pth norm of the error
function Objective=getObjective(Point)
Objective=0; %initialize objective
p=5;
Errors=getErrors(Point);
numberOfResponses=size(Errors,1);
for k=1: numberOfResponses%
  Objective=Objective+(abs(Errors(k,1)))^p;
end
Objective=Objective^(1/p);
```

where $s = J2\pi freq$ is the Laplace parameter. The MATLAB listing M9.1 is utilized in this problem after making the necessary changes to the number of parameters. The statistical random point for optimization is given by:

```
oldPoint=[1 1 2.5 1 1.0]'+rand(numberOfParameters,1)*0.5;
```

It follows that random starting points are generated around the point $x_c = [1.0\ \text{K}\Omega\quad 1.0\ \mu\text{F}\quad 2.5\ \text{K}\Omega\quad 1.0\ \mu\text{F}\quad 1.0\ \text{K}\Omega]^T$. Notice that scaling is used within the **getErrors** function to recover the parameters to their physical values. We allowed for 10 random starting points of the steepest descent approach. The output from the code for all 10 iterations and the corresponding final objective function value are given by:

Points =

0.0521	1.5264	2.8939	0.7338	1.0398
0.0500	1.1889	3.0811	0.9144	1.0403
0.0510	1.5320	2.8252	0.7546	1.2647
0.0500	1.3642	3.0274	0.8205	1.3509
0.0500	1.6415	2.8063	0.6930	1.4681
0.0539	1.4629	3.1968	0.6899	1.1280
0.0500	1.4388	3.0400	0.7909	1.2381
0.0500	1.7172	2.9141	0.9086	1.0502
0.0529	1.5766	3.1024	0.6821	1.1274
0.0500	1.6591	2.9389	0.8258	1.0833

Values =

21.8538
23.2981
22.0499
21.3560
23.1248
20.8909
20.9041
32.7748
20.9587
26.0300

It follows that the best point is at $\boldsymbol{x}^* = [0.0539\ \text{K}\Omega\quad 1.4629\ \mu\text{F}\quad 3.1968\ \text{K}\Omega\quad 0.6899\ \mu\text{F}\quad 1.1280\ \text{K}\Omega]^T$. Figures 9.9 and 9.10 show the response at the points $\boldsymbol{x}_c$ and $\boldsymbol{x}^*$. The bandpass of the filter was successfully moved to the desired frequency band even though the bandwidth is larger than it should be.

A9.2 Pattern recognition

Humans are gifted with the ability to recognize patterns and images in a fast way. When we look at a certain picture of a person, we can immediately remember if we have seen this face before. Similarly, when we see a pattern, we can also tell if we have seen it before or not. The reason for our ability to solve this problem in a fast way is probably evolutionary. Humans needed to recognize and respond to threats fast in order to survive. Our brains are able to solve these recognition problems in the fastest possible way. For example, if we consider the two letters shown in Figure 9.11, one can also see that a shift, scaling, and rotation were applied to the original letter as compared to the first one.

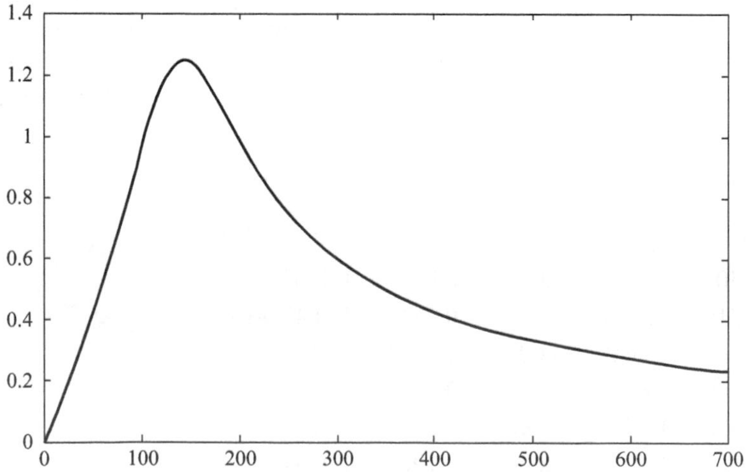

Figure 9.9 The response of the circuit of application A9.1 at the point $x_c = [1.0\ K\Omega\quad 1.0\ \mu F\quad 0.5\ K\Omega\quad 1.0\ \mu F\quad 1.0\ K\Omega]^T$

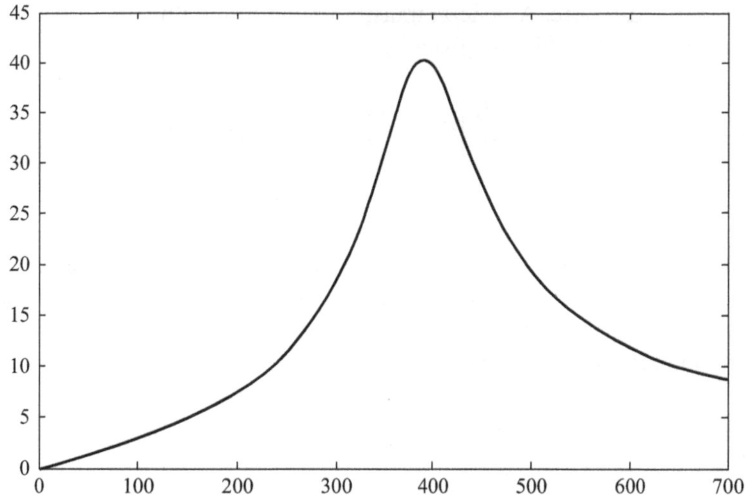

Figure 9.10 The response of the circuit of application A9.1 at the best point $\boldsymbol{x}^* = [0.0539\ K\Omega\ \ 1.1629\ \mu F\ \ 3.1968\ K\Omega\ \ 0.6899\ \mu F\ \ 1.1280\ K\Omega]^T$

Figure 9.11 An illustration of the pattern recognition approach; the second character is immediately recognized as scaled, rotated, and shifted version of the upright A character

The recognition problem can be cast as an optimization problem. By comparing a given image or pattern to all images or patterns in the database and defining some similarity measure, we can determine the image or pattern that has the highest similarity to the given image/pattern.

To illustrate this approach, we consider here a simple point pattern recognition problem [10, 11]. We assume that we have a database of M known patterns. The ith pattern is given by the set of N points:

$$P^{(i)} = \{\boldsymbol{p}_1^{(i)}, \boldsymbol{p}_2^{(i)}, \ldots, \boldsymbol{p}_N^{(i)}\}, \quad i = 1, 2, \ldots, M \tag{9.25}$$

Every one of these points is a two-dimensional vector with x and y components. These sets of points describe how the patterns look like. Our target is to determine if another pattern Q defined by the set of points:

$$Q = \{\boldsymbol{q}_1, \boldsymbol{q}_2, \ldots, \boldsymbol{q}_N\} \tag{9.26}$$

is similar to any of the patterns in the database defined by (9.25). By the word "similar," we mean that Q is very close to a possibly scaled, rotated, and shifted pattern from the database. A general transformation of the *i*th pattern that involves scaling, rotation, and shifting is given by:

$$\tilde{\boldsymbol{p}}_j^{(i)} = \rho \begin{bmatrix} \cos(\theta) & -\sin(\theta) \\ \sin(\theta) & \cos(\theta) \end{bmatrix} \boldsymbol{p}_j^{(i)} + \begin{bmatrix} d_1 \\ d_2 \end{bmatrix}, \quad j = 1, 2, \ldots, N \tag{9.27}$$

where $\tilde{\boldsymbol{p}}_j^{(i)}$ is the transformed *j*th point of the *i*th pattern. The positive scaling parameter ρ allows for stretching ($\rho > 1$) or compressing ($\rho < 1$) the points of the pattern. The rotation angle θ, $0 \leq \theta \leq 2\pi$, allows for rotating the pattern. The shift vector $\boldsymbol{d} = [d_1 \;\; d_2]^T$ allows for shifting the pattern around in the *xy* plane. Multiplying ρ by the transformation matrix, we can put the transformation (9.27) into the more compact but equivalent form:

$$\tilde{\boldsymbol{p}}_j^{(i)} = \begin{bmatrix} a & -b \\ b & a \end{bmatrix} \boldsymbol{p}_j^{(i)} + \begin{bmatrix} d_1 \\ d_2 \end{bmatrix}, \quad j = 1, 2, \ldots, N \tag{9.28}$$

The transformation (9.28) is thus characterized by the four parameters $\boldsymbol{x} = [a \;\; b \;\; d_1 \;\; d_2]^T$.

The target of point pattern recognition is to determine if there is a transformation $\boldsymbol{x}$ that, when applied to one of the database patterns, makes it look similar to the given pattern Q. To cast this as an optimization problem, we define the error between the *j*th point of the given pattern Q and the corresponding point in the transformed pattern. This error is given by:

$$\boldsymbol{E}_j^{(i)}(\boldsymbol{x}) = \begin{bmatrix} a & -b \\ b & a \end{bmatrix} \boldsymbol{p}_j^{(i)} + \begin{bmatrix} d_1 \\ d_2 \end{bmatrix} - \boldsymbol{q}_j, \quad j = 1, 2, \ldots, N \tag{9.29}$$

The error vector for all points of the *i*th database pattern is thus given by:

$$\boldsymbol{E}^{(i)} = \begin{bmatrix} \boldsymbol{E}_1^{(i)T} & \boldsymbol{E}_2^{(i)T} & \cdots & \boldsymbol{E}_N^{(i)T} \end{bmatrix}^T \tag{9.30}$$

The transformation that minimizes this error over all points may be obtained by solving the optimization problem:

$$\underset{\boldsymbol{x}}{\text{minimize}} \; U^{(i)}(\boldsymbol{x}) = \|\boldsymbol{E}^{(i)}(\boldsymbol{x})\| \tag{9.31}$$

where a suitable norm is used in solving (9.31). Notice that the problem (9.31) has to be solved for all patterns in the database. The pattern that gives the lowest possible objective function (best match through the optimal transformation) is the one that looks more similar to the given pattern Q.

We illustrate this approach by considering the database patterns shown in Figure 9.12(a)–(e). The five Greek letters beta, nu, small omega, tau, and capital omega are used as the database patterns. It follows that $M = 5$. Each pattern is defined by $N = 120$ points. The pattern Q to be recognized is shown in Figure 9.12(f). This pattern is also defined by $N = 120$ points. The optimization problem (9.31) is solved five times for all patterns. We utilized the *p*th norm defined in the previous section with $p = 6$.

The genetic algorithm listing M9.3 is utilized in solving this pattern recognition problem with the following initialization part.

```
PopulationSize=1000; %number of chromosomes
NumberOfParameters=4; %number of designable parameters
NumberOfGenerations=30; %number of generations
LowerBounds=[-5 -5 -5  -5]'; %lower bounds on parameters
UpperBounds=[5 5  5   5]'; %lower bounds on parameters
```

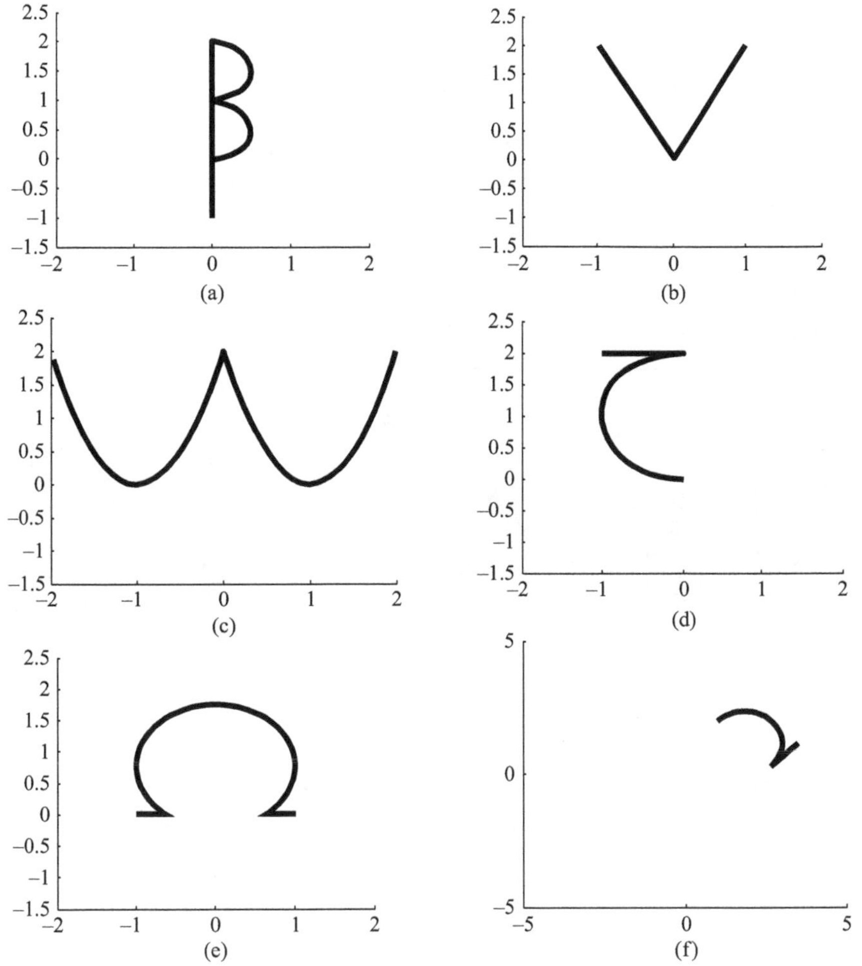

Figure 9.12 The five database patterns of the pattern recognition example (a)–(e) and the given pattern to be recognaized (f)

It follows that we allow for a population of size 1000. Thirty generations are used within GA optimization. The initial population is generated within the region $-5.0 \le x_i \le 5.0$, $i = 1, 2, 3, 4$.

The fitness, as defined earlier, is the negative of the objective function to be minimized. The objective function calculation is shown in the listing M9.8. This function loads the file containing all five patterns. Each pattern is defined by a matrix of two rows and 120 columns. The pattern Q is also loaded. The database pattern to be compared to the given pattern Q is selected by the parameter *patternIndex*. The function evaluates the errors (9.29) between the transformed database pattern and the given pattern and subsequently evaluates its L_6 norm.

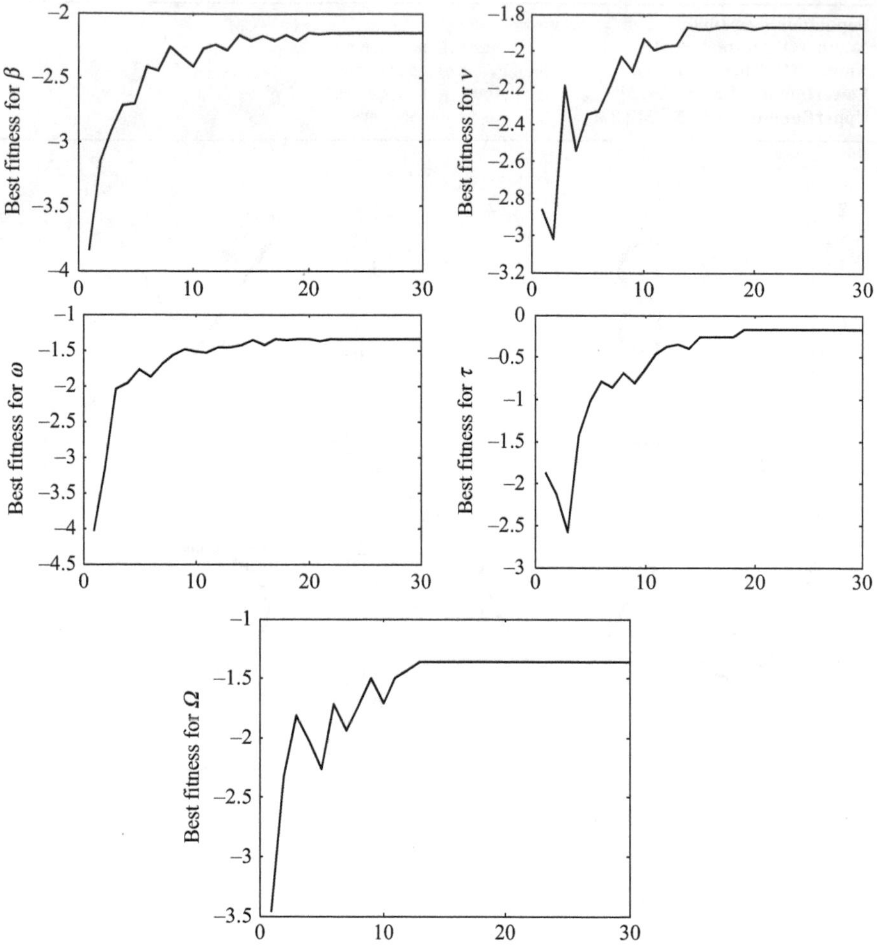

Figure 9.13 *The best fitness achieved in every iteration for all five database patterns; the best fitness is achieved for the τ pattern*

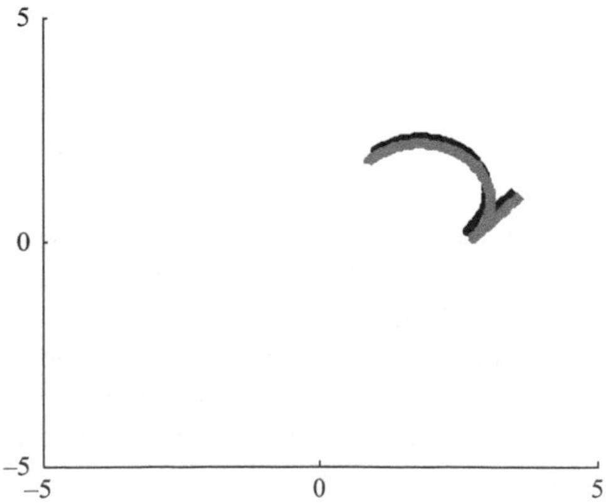

Figure 9.14 *The given pattern and the transformed τ pattern using the optimal transformation. Very good match is observed*

```
%M9.8
% This function evaluates the objective function for the pattern recognition problem
function   Objective=getObjective(Point)
p=6; %exponent of norm
load Patterns.mat %load pattern matrices
load GivenPattern.mat %load the pattern Q
patternIndex=5; %which pattern to match to the given pattern Q
switch patternIndex
   case 1
      targetPattern=Beta;
   case 2
      targetPattern= Nu;
   case 3
      targetPattern= OmegaSmall;
   case 4
      targetPattern= Tau;
   case 5
      targetPattern= OmegaCapital;
end
numberOfPoints=size(targetPattern,2); %get number of points to match
%now we match the target pattern to the given pattern
a=Point(1,1); % a=x(1)
b=Point(2,1); % b=x(2)
d1=Point(3,1); % d1=x(3)
d2=Point(4,1); % d2=x(4)
scaleMatrix=[a   -b; b   a]; shiftVector=[d1   d2]'; %build scale matrix and shift vectors
%now we check all points of the transformed pattern against the given pattern
Objective=0; %initialize objective
%evaluate pth norm
for k=1: numberOfPoints %repeat for all points in the pattern
   %calculate two-dimensional error vector of the kth pattern point
   Error=(scaleMatrix*targetPattern(:,k)+shiftVector)-givenPattern(:,k);
   Objective=Objective+Error(1)^p+Error(2)^p; %raise to power p
end
Objective=Objective^(1/p); %evaluate norm
```

Figure 9.13 shows the best fitness obtained using the GA optimization for each database pattern over all iterations. It is obvious from these figures that the best fitness over all database patterns is obtained for the pattern of the Greek symbol tau. The solution with the optimal fitness over all database patterns is $x^* = [-0.8433 \quad -0.9434 \quad 0.8938 \quad 1.8047]^T$. The transformed tau pattern using this optimal solution is compared to the given pattern in Figure 9.14. Very good match is obtained. It follows, as expected from first glance, that the given pattern is a transformed tau symbol.

References

1. Erwin Kreyszig, *Advanced Engineering Mathematics*, Seventh Edition, John Wiley & Sons Inc., Boston, 1999
2. Eldon Hansen and G. William Walster, *Global Optimization Using Interval Analysis*, Marcel Dekker, New York, 2004
3. John W. Bandler and R.M. Biernacki, *ECE3KB3 Courseware on Optimization Theory*, McMaster University, Canada, 1997
4. R.V. Kacelenga, P.J. Graumann, and L.E. Turner, 'Design of digital filters using simulated annealing', *IEEE International Symposium on Circuits and Systems*, pp. 642–645, 1990

5. D.S. Weile and E. Michielssen, 'Genetic algorithm optimization applied to electromagnetics: A review', *IEEE Transactions on Antennas and Propagation*, vol. 45, no. 3, pp. 343–353, 1997
6. Nanbo Jin and Y. Rahmat-Samii, 'Advances in particle swarm optimization for antenna designs: Real-number, binary, single-objective and multiobjective implementations', *IEEE Transactions on Antennas and Propagation*, vol. 55, no. 3, pp. 556–567, 2007
7. S. Karimkashi and A.A. Kishk, 'Antenna array synthesis using invasive weed optimization: A new optimization technique in electromagnetics', *APSURSI '09: IEEE Antennas and Propagation Society International Symposium*, pp. 1–4, 2009
8. Z. Bayraktar, M. Komurcu, and D.H. Werner, 'Wind driven optimization (WDO): A novel nature-inspired optimization algorithm and its application to electromagnetics', *2010 IEEE Antennas and Propagation Society International Symposium (APSURSI)*, pp. 1–4, 2010
9. Kwang Mong Sim and Weng Hong Sun, 'Ant colony optimization for routing and load-balancing: Survey and new directions', *IEEE Transactions on Systems and Humans*, vol. 33, no. 5, pp. 560–572, 2003
10. Lihua Zhang, Wenli Xu, and Cheng Chang, 'Genetic algorithm for affine point pattern matching', *Pattern Recognition Letters*, vol. 24, pp. 9–19, 2003
11. Andreas Antoniou and Wu-Sheng Lu, *Practical Optimization Algorithms and Engineering Applications*, Springer, 2007

Problems

9.1 Solve the optimization problem:

$$f(x_1, x_2) = x_1^2 + x_2^2 - 3x_1x_2$$

using the MATLAB listing M9.1. Select random starting points in the interval $-3 \le x_1 \le 3$ and $-3 \le x_2 \le 3$. Verify your answers by plotting the contours of this function. How many minima does this problem have?

9.2 Find the global minimum of the optimization problem:

$$\text{minimize}\, f(\boldsymbol{x}) = x_1 \sin(x_1) - x_2 \sin(4x_2)$$

using simulated annealing. Utilize the Simulated Annealing code M9.2 with a starting point $\boldsymbol{x} = [0 \quad 0]^T$. Allow for 100 iterations. Verify your answer by plotting the contours of this objective function.

9.3 Solve the minimax optimization problem:

$$\min_{\boldsymbol{x}} U = \max_{j} f_j(\boldsymbol{x})$$

$$\begin{aligned} \text{with } f_1(\boldsymbol{x}) &= x_1^4 + x_2^2 \\ f_2(\boldsymbol{x}) &= (x_1 - 2)^2 + (x_2 - 2)^2 \\ f_3(\boldsymbol{x}) &= \exp(-x_1 + x_2) \end{aligned}$$

using GAs. Utilize the MATLAB listing M9.3 with initial population of 50 members spread over the interval $-4 \le x_1 \le 4$ and $-4 \le x_2 \le 4$. Allow for 10 generations. Verify your answer by plotting the contours of the objective function U.

9.4 Solve the optimization problem:

$$f(\boldsymbol{x}) = (1.5 - x_1(1 - x_2))^2 + \left(2.25 - x_1\left(1 - x_2^2\right)\right)^2 + \left(2.625 - x_1\left(1 - x_2^3\right)\right)^2$$

using GAs. Utilize the MATLAB listing M9.3 with initial population of 100 members spread over the interval $-6 \le x_1 \le 6$ and $-6 \le x_2 \le 6$. Allow for 10 generations. Verify your answer by plotting the contours of the objective function.

9.5 Minimize the function:

$$f(\boldsymbol{x}) = 2(1 - x_1)^2 \exp\left(-x_1^2 - (x_2 + 1)^2\right) - \left(x_1 - 5x_1^3 - 5x_2^5\right)\exp\left(-x_1^2 - x_2^2\right)$$

using particle swarm optimization (PSO). Utilize the MATLAB listing M9.6 with a swarm of 40 points spread randomly over the range $-6 \le x_1 \le 6$ and $-6 \le x_2 \le 6$. Allow for 50 iterations. Verify your answer by plotting the contours of this objective function.

Chapter 10
Adjoint sensitivity analysis

10.1 Introduction

As was explained in previous chapters, many optimization algorithms utilize sensitivity information to guide the optimization iterations to better points in the parameter space. Sensitivity information is, however, expensive to estimate through classical approaches. These classical approaches require a repeated number of simulations with perturbed values of the parameters. The cost of estimating these sensitivities grows significantly if the algorithm requires also second-order sensitivities. We reviewed in Chapter 1 different formulas of the classical first-order finite difference approaches. We showed that the cost of estimating first-order sensitivities grows linearly with the number of parameters. For problems with time-intensive simulation or with large number of parameters, this computational cost can be extensive. This motivated research for smarter sensitivity estimation approaches.

Adjoint sensitivity analysis offers an efficient approach for sensitivity estimation. Using at most one extra simulation of an adjoint system (or circuit), the first-order sensitivities of the desired response (or objective function) are evaluated with respect to all parameters regardless of their number. This approach applies to both frequency-domain and time-domain responses. It was shown to be applicable to higher-order sensitivities as well. The cost of estimating these higher-order sensitivities using adjoint approaches is also less than that of the corresponding finite difference approach.

There are a number of adjoint sensitivity analysis approaches reported in the literature. These approaches were developed for network theory, frequency-domain sensitivity analysis, and time-domain sensitivity analysis. For every type of analysis, the expression for the adjoint simulation is derived and its excitation is determined. These approaches are applicable in many areas of electrical engineering including the design of high-frequency structures, solution of inverse problems, and control theory. I will give a general overview of some of these approaches using electrical circuits to illustrate different approaches.

This chapter is divided as follows: we first start by introducing Tellegen's theorem, which is the core of adjoint sensitivity analysis of electrical networks in Section 10.2. I then show how this theory is utilized to derive the adjoint electric network corresponding to a given original network in Section 10.3. This approach obtains sensitivities of currents and voltages of arbitrary branches with respect to all circuit parameters. The frequency-domain sensitivity analysis approach is then discussed in Section 10.4. This approach obtains the sensitivities of a general frequency-domain objective function calculated using frequency-domain calculations. I also show in Section 10.5 how an adjoint system can be developed for a general time-domain simulation. This approach is important when a system (or a circuit) utilizes wideband excitation.

10.2 Tellegen's theorem

Any electric network consists of a number of branches. These branches may connect together in a series form, a parallel form, or any other combination to form the electric network. We denote by $\boldsymbol{v}(t)$ the vector of branch voltages and by $\boldsymbol{i}(t)$ the vector of currents flowing through these branches. We assume that in every branch the instantaneous current is considered positive if it flows in the branch from the assumed positive voltage to the negative voltage. Figure 10.1 illustrates the topology of a possible electric circuit and the assumed polarities of the voltages and currents. In this figure, no electric elements are shown but any branch can have any electric component (resistors, capacitors, inductors, or others).

The conservation of instantaneous power of any electric circuit states that:

$$\boldsymbol{v}^T\boldsymbol{i} = v_1 i_1 + v_2 i_2 + \cdots + v_N i_{N_b} = 0 \tag{10.1}$$

where N_b is the number of branches and v_j and i_j are the voltage and current of the jth branch, respectively. Equation (10.1) states that the power supplied by the sources is equal to the power consumed by other components of the circuit.

Tellegen [1, 2] showed that the same relationship applies to two different electric networks that have the same topology. If $\hat{\boldsymbol{v}}(t)$ and $\hat{\boldsymbol{i}}(t)$ are the vectors of currents and voltages of another electric circuit with the same topology, then we have:

$$\hat{\boldsymbol{v}}^T\boldsymbol{i} = 0 \quad \text{and} \quad \hat{\boldsymbol{i}}^T\boldsymbol{v} = 0 \tag{10.2}$$

at any time t. Tellegen also showed that the same relationship applies between the phasors of currents and voltages in two different circuits at the same frequency. It follows that we have:

$$\hat{\boldsymbol{V}}^T\boldsymbol{I} = 0 \quad \text{and} \quad \hat{\boldsymbol{I}}^T\boldsymbol{V} = 0 \tag{10.3}$$

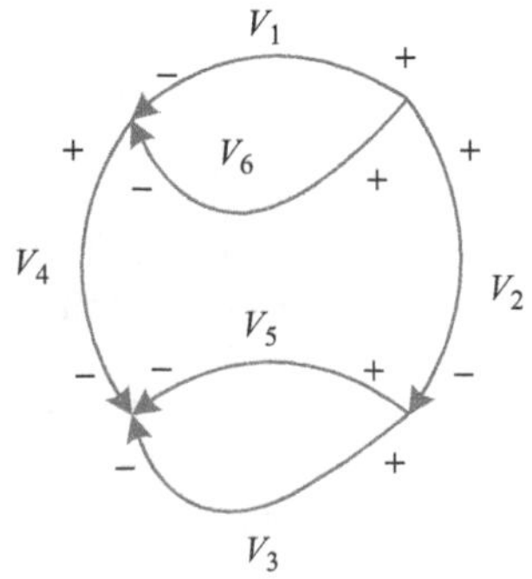

Figure 10.1 An example of the topology of an electric circuit; the directions of the arrows indicate the positive directions of the currents

where capital letters indicate phasors at a certain frequency. The expressions (10.2) and (10.3) show an interesting relationship between any two electric circuits as long as they have the same topology. They represent a generalization of (10.1) to two different electric networks.

The following two examples illustrate Tellegen's theorem.

Example 10.1: Consider the two electric circuits shown in Figure 10.2. Illustrate Tellegen's theorem for these two circuits.

Solution: Both circuits in Figure 10.2 have the same topology. They both have three branches. It follows that for each circuit, we have three branch voltages and three branch currents. Solving the first circuit (Figure 10.2(a)), we get $v_1 = -1.0$ V, $v_2 = 1.0$ V, and $v_3 = 1.0$ V. Notice that the polarities of these voltages are chosen arbitrarily. Solving the second circuit (Figure 10.2(b)), we get the currents $\hat{i}_1 = 1.0$ A, $\hat{i}_2 = 0.5$ A, and $\hat{i}_3 = 0.5$ A. It follows that from the solution of these two circuits we have:

$$\sum_{j=1}^{3} v_j \hat{i}_j = -1.0 \times 1.0 + 1.0 \times 0.5 + 1.0 \times 0.5 = 0 \tag{10.4}$$

The same result can be obtained if we take the inner product between the currents of the first circuit $\boldsymbol{i}(t)$ and the voltages of the second circuit $\hat{\boldsymbol{v}}(t)$. This example is a DC example, where voltages and currents do not change with time. The following example shows that the same theorem applies also for the widely used phasor case.

Example 10.2: Consider the two circuits shown in Figure 10.3. Show that Tellegen's theorem holds for these two circuits in the frequency domain.

Solution: These two circuits have three branches and these branches are connected in parallel. The currents in the second circuit (Figure 10.3(b)) are given by:

$$\hat{I}_1 = \frac{5.0\angle 0}{-J2} = J2.5 \text{ A}$$

$$\hat{I}_3 = \frac{5.0\angle 0}{1+J2} = \frac{5.0(1-J2)}{5.0} = 1.0 - J2 \text{ A} \tag{10.5}$$

$$\hat{I}_2 = -\hat{I}_1 - \hat{I}_3 = -J2.5 - 1.0 + J2 = -1.0 - J0.5 \text{ A}$$

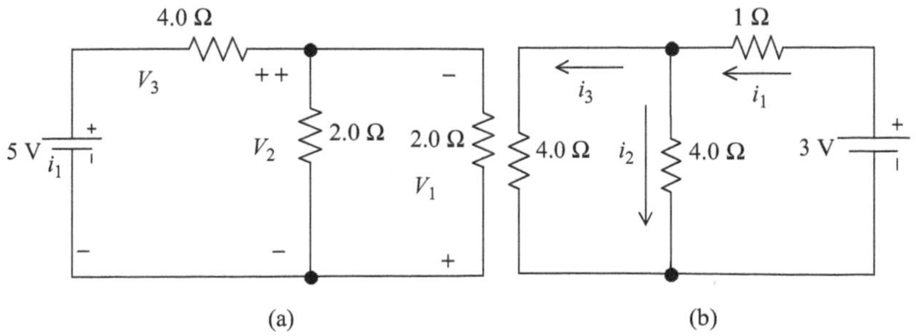

Figure 10.2 The two electrical circuits of Example 10.1

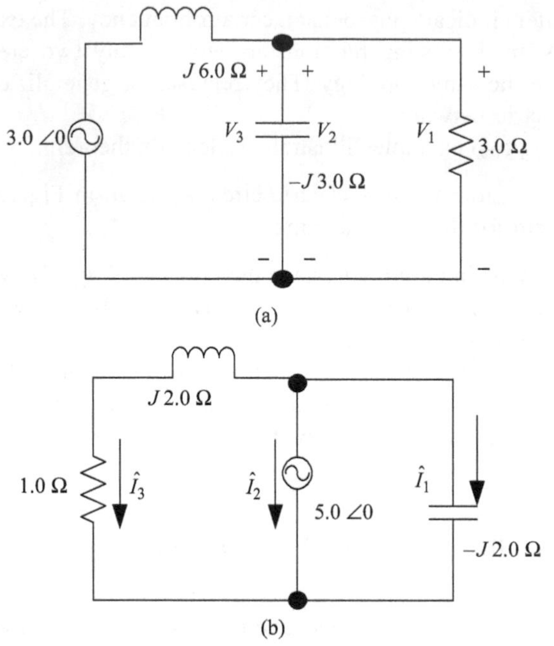

Figure 10.3 The two electric circuits of Example 10.2

where $J=\sqrt{-1}$. Next, we solve for the voltages of the first circuit. The impedance of the two parallel branches is given by:

$$Z = 3//(-J3) = \frac{3 \times (-J3)}{3 - J3} = 6.75 - J2.25\ \Omega \qquad (10.6)$$

The voltages V_1, V_2, and V_3 are given by:

$$V_1 = V_2 = V_3 = \frac{3.0 \times (6.75 - J2.25)}{6.75 - J2.25 + J6} = 2.6415 + J0.509\ \text{V} \qquad (10.7)$$

The inner product between the voltages and currents of the two circuits is given by:

$$V_1\hat{I}_1 + V_2\hat{I}_2 + V_3\hat{I}_3 = (2.6415 + J0.509)(J2.5 - 1.0 - J0.5 + 1.0 - J2.0) = 0 \qquad (10.8)$$

The same result could be obtained by carrying out an inner product between the currents of the first circuit and the voltages of the second circuit.

10.3 Adjoint network method

In the previous examples, Tellegen's theorem would still hold if, in the original circuit, the value of a circuit element such as a resistor, a capacitor, or an inductor was changed. In this case, the new voltages and currents would be $V+\Delta V$ and $I+\Delta I$, respectively. This property is used to derive the adjoint network method.

The derivation of the adjoint system in network theory is instructive and is discussed in detail. This derivation follows closely the one given in Reference 3. We assume that we want to estimate the sensitivities of the current drawn from a voltage source or the voltage across a current source with respect to all desired parameters. This will be shown later not to be a limitation to general elements such as resistors, inductors, and capacitors. Our target is to determine how these sensitivities are estimated. We start the derivation by combining the two expressions in (10.3) into the expression:

$$\hat{\boldsymbol{I}}^T \boldsymbol{V} - \hat{\boldsymbol{V}}^T \boldsymbol{I} = 0 \tag{10.9}$$

The expression (10.9) can be expanded from its vector form to a branch-by-branch form to get:

$$\sum_{k=1}^{N_b} V_k \hat{I}_k - \sum_{k=1}^{N_b} I_k \hat{V}_k = 0 \tag{10.10}$$

As our target is to get the sensitivities of the currents drawn from voltage sources and voltages across current sources, we separate the branches in (10.10) into sources and elements to get:

$$\sum_{\substack{\text{voltage}\\ \text{sources}}} (V_k \hat{I}_k - I_k \hat{V}_k) + \sum_{\substack{\text{current}\\ \text{sources}}} (V_k \hat{I}_k - I_k \hat{V}_k) + \sum_{\text{elements}} (V_k \hat{I}_k - I_k \hat{V}_k) = 0 \tag{10.11}$$

If one of the elements in the original circuits is perturbed, the currents drawn from voltage sources will change and so are the voltages across current sources. The voltage across a voltage source or the current of a current source does not change because these are independent sources. Equation (10.11) can then be rewritten for the perturbed original circuit as:

$$\sum_{\substack{\text{voltage}\\ \text{sources}}} \left(V_k \hat{I}_k - (I_k + \Delta I_k) \hat{V}_k \right) + \sum_{\substack{\text{current}\\ \text{sources}}} \left((V_k + \Delta V_k) \hat{I}_k - I_k \hat{V}_k \right)$$

$$+ \sum_{\text{elements}} \left((V_k + \Delta V_k) \hat{I}_k - (I_k + \Delta I_k) \hat{V}_k \right) = 0 \tag{10.12}$$

A common step in the derivations of adjoint approaches is to subtract the original system of equations from the perturbed system of equations. We thus subtract (10.11) from (10.12) to obtain a system with only differential changes in currents and voltages of the original circuit:

$$-\sum_{\substack{\text{voltage}\\ \text{sources}}} (\Delta I_k \hat{V}_k) + \sum_{\substack{\text{cuurent}\\ \text{sources}}} (\Delta V_k \hat{I}_k) + \sum_{\text{elements}} (\Delta V_k \hat{I}_k - \Delta I_k \hat{V}_k) = 0 \tag{10.13}$$

We assume that only one element x_i is perturbed in the original circuit. This element may be a resistor, a capacitor, and inductor, a parameter of a dependent source, or any other electrical element. We then divide both sides of (10.13) by Δx_i, and take the limit as Δx_i becomes too small to get:

$$\sum_{\substack{\text{voltage}\\ \text{sources}}} \left(\frac{\partial I_k}{\partial x_i}\hat{V}_k\right) - \sum_{\substack{\text{current}\\ \text{sources}}} \left(\frac{\partial V_k}{\partial x_i}\hat{I}_k\right) = \left(\sum_{\text{elements}} \left(\frac{\partial V_k}{\partial x_i}\hat{I}_k - \frac{\partial I_k}{\partial x_i}\hat{V}_k\right)\right) \tag{10.14}$$

The left-hand side of (10.14) contains the desired derivatives of the voltage and current sources. We can construct the excitation of the adjoint circuit such that only one derivative exists on the left-hand side. The derivatives of the currents and voltages of the different elements in the right-hand side of (10.14) must be eliminated because they are unknowns. We satisfy that by imposing certain conditions on the adjoint circuit.

For a general formulation, we assume that the elements of the original circuit are governed by the general hybrid formulation:

$$\begin{bmatrix} \boldsymbol{I}_a \\ \boldsymbol{V}_b \end{bmatrix} = \begin{bmatrix} \boldsymbol{Y} & \boldsymbol{A} \\ \boldsymbol{M} & \boldsymbol{Z} \end{bmatrix} \begin{bmatrix} \boldsymbol{V}_a \\ \boldsymbol{I}_b \end{bmatrix} \tag{10.15}$$

where $\boldsymbol{V}_a$ and $\boldsymbol{V}_b$ are subsets of the branch voltages and $\boldsymbol{I}_a$ and $\boldsymbol{I}_b$ are the corresponding branch currents. The formulation (10.15) takes into account linearized types of transistors, transformers, transmission lines, and other types of linear electrical components. Differentiating both sides of (10.15) with respect to the parameter x_i, we get:

$$\begin{bmatrix} \dfrac{\partial \boldsymbol{I}_a}{\partial x_i} \\ \dfrac{\partial \boldsymbol{V}_b}{\partial x_i} \end{bmatrix} = \begin{bmatrix} \boldsymbol{Y} & \boldsymbol{A} \\ \boldsymbol{M} & \boldsymbol{Z} \end{bmatrix} \begin{bmatrix} \dfrac{\partial \boldsymbol{V}_a}{\partial x_i} \\ \dfrac{\partial \boldsymbol{I}_b}{\partial x_i} \end{bmatrix} + \begin{bmatrix} \dfrac{\partial \boldsymbol{Y}}{\partial x_i} & \dfrac{\partial \boldsymbol{A}}{\partial x_i} \\ \dfrac{\partial \boldsymbol{M}}{\partial x_i} & \dfrac{\partial \boldsymbol{Z}}{\partial x_i} \end{bmatrix} \begin{bmatrix} \boldsymbol{V}_a \\ \boldsymbol{I}_b \end{bmatrix} \tag{10.16}$$

To substitute (10.16) into (10.14), we first write (10.14) into a more organized way that combines similar components together to get:

$$\sum_{\substack{\text{voltage}\\ \text{sources}}} \left(\frac{\partial I_k}{\partial x_i}\hat{V}_k\right) - \sum_{\substack{\text{current}\\ \text{sources}}} \left(\frac{\partial V_k}{\partial x_i}\hat{I}_k\right) = \begin{bmatrix} \dfrac{\partial \boldsymbol{I}_a}{\partial x_i} \\ \dfrac{\partial \boldsymbol{V}_b}{\partial x_i} \end{bmatrix}^T \begin{bmatrix} -\hat{\boldsymbol{V}}_a \\ \hat{\boldsymbol{I}}_b \end{bmatrix} + \begin{bmatrix} \dfrac{\partial \boldsymbol{V}_a}{\partial x_i} \\ \dfrac{\partial \boldsymbol{I}_b}{\partial x_i} \end{bmatrix}^T \begin{bmatrix} \hat{\boldsymbol{I}}_a \\ -\hat{\boldsymbol{V}}_b \end{bmatrix} \tag{10.17}$$

Substituting (10.16) into (10.17), and organizing, we get:

$$\sum_{\substack{\text{voltage}\\\text{sources}}}\left(\frac{\partial I_k}{\partial x_i}\hat{V}_k\right)-\sum_{\substack{\text{current}\\\text{sources}}}\left(\frac{\partial V_k}{\partial x_i}\hat{I}_k\right)$$

$$=\begin{bmatrix}\dfrac{\partial \boldsymbol{V}_a}{\partial x_i}\\[2mm]\dfrac{\partial \boldsymbol{I}_b}{\partial x_i}\end{bmatrix}^T\underbrace{\left(\begin{bmatrix}\boldsymbol{Y}^T & \boldsymbol{M}^T\\ \boldsymbol{A}^T & \boldsymbol{Z}^T\end{bmatrix}\begin{bmatrix}-\hat{\boldsymbol{V}}_a\\ \hat{\boldsymbol{I}}_b\end{bmatrix}+\begin{bmatrix}\hat{\boldsymbol{I}}_a\\ -\hat{\boldsymbol{V}}_b\end{bmatrix}\right)}_{\boldsymbol{0}}+\begin{bmatrix}\boldsymbol{V}_a\\ \boldsymbol{I}_b\end{bmatrix}^T\begin{bmatrix}\dfrac{\partial \boldsymbol{Y}^T}{\partial x_i} & \dfrac{\partial \boldsymbol{Y}^T}{\partial x_i}\\[2mm]\dfrac{\partial \boldsymbol{A}^T}{\partial x_i} & \dfrac{\partial \boldsymbol{Z}^T}{\partial x_i}\end{bmatrix}\begin{bmatrix}-\hat{\boldsymbol{V}}_a\\ \hat{\boldsymbol{I}}_b\end{bmatrix}$$

(10.18)

In (10.18), all terms involving the unknown derivatives of the branch voltages and currents are multiplied by the underlined term. Setting this term to zero enforces the following condition on the adjoint circuit:

$$\begin{bmatrix}\boldsymbol{Y}^T & \boldsymbol{M}^T\\ \boldsymbol{A}^T & \boldsymbol{Z}^T\end{bmatrix}\begin{bmatrix}-\hat{\boldsymbol{V}}_a\\ \hat{\boldsymbol{I}}_b\end{bmatrix}=-\begin{bmatrix}\hat{\boldsymbol{I}}_a\\ -\hat{\boldsymbol{V}}_b\end{bmatrix}\Rightarrow\begin{bmatrix}\boldsymbol{Y}^T & -\boldsymbol{M}^T\\ -\boldsymbol{A}^T & \boldsymbol{Z}^T\end{bmatrix}\begin{bmatrix}\hat{\boldsymbol{V}}_a\\ \hat{\boldsymbol{I}}_b\end{bmatrix}=\begin{bmatrix}\hat{\boldsymbol{I}}_a\\ \hat{\boldsymbol{V}}_b\end{bmatrix}\qquad(10.19)$$

Equation (10.19) gives the relationship between the branch voltages and currents of the adjoint problem. The sensitivity expression (10.18) is now simplified to the efficient one:

$$\sum_{\substack{\text{voltage}\\\text{sources}}}\left(\frac{\partial I_k}{\partial x_i}\hat{V}_k\right)-\sum_{\substack{\text{current}\\\text{sources}}}\left(\frac{\partial V_k}{\partial x_i}\hat{I}_k\right)=\begin{bmatrix}\boldsymbol{V}_a\\ \boldsymbol{I}_b\end{bmatrix}^T\begin{bmatrix}-\dfrac{\partial \boldsymbol{Y}^T}{\partial x_i} & \dfrac{\partial \boldsymbol{M}^T}{\partial x_i}\\[2mm]-\dfrac{\partial \boldsymbol{A}^T}{\partial x_i} & \dfrac{\partial \boldsymbol{Z}^T}{\partial x_i}\end{bmatrix}\begin{bmatrix}\hat{\boldsymbol{V}}_a\\ \hat{\boldsymbol{I}}_b\end{bmatrix}$$

(10.20)

Few things to notice about the expression (10.20): First, we can select the sensitivity of the voltage across a current source or the current drawn from a voltage source by properly selecting the corresponding excitation in the adjoint circuit. For example, to get the sensitivity of the current drawn from the *j*th voltage source ($\partial I_j/\partial x_i$), we set the corresponding excitation in the adjoint problem $\hat{V}_j = 1.0$ V. All other adjoint sources are set to zero ($\hat{V}_k = 0, k \neq j, \hat{I}_k = 0, \forall k$). This implies that in the adjoint circuit, all branches corresponding to voltage sources in the original circuit are shorted out. It also implies that all branches corresponding to current sources in the original circuit are open circuited. It follows that the desired objective function determines the excitation of the adjoint circuit.

The second thing to notice is that the only term depending on the parameter x_i on the right-hand side of (10.20) is the derivative of the circuit matrices $\boldsymbol{Y}$, $\boldsymbol{Z}$, $\boldsymbol{M}$, and $\boldsymbol{A}$. These derivatives are known analytically because the topology of the original circuit/network is known. It follows that by solving for the original circuit responses $[\,\boldsymbol{V}_a \quad \boldsymbol{I}_b\,]^T$ and the adjoint circuit responses $[\,\hat{\boldsymbol{V}}_a \quad \hat{\boldsymbol{I}}_b\,]^T$, we can estimate the sensitivities of the response of interest with respect to x_i, $\forall i$. Using only one adjoint simulation, we can estimate all the sensitivities regardless of the number of parameters n.

The steps for the adjoint theory presented in (10.9) and (10.20) can be summarized as follows:

Step 1: Original simulation: Solve the original circuit to determine the currents and voltages for all branches.
Step 2: Adjoint simulation: Solve the adjoint circuit (10.19) with the excitation used to select the derivatives of the response of interest. Determine all adjoint branch voltages and currents.
Step 3: For every parameter i, calculate the sensitivities by carrying out the product (10.20). The derivatives of the circuit matrices are known analytically.

The following examples illustrate the adjoint network method.

Example 10.3: Consider the circuit shown in Figure 10.4. Evaluate the sensitivities of the output voltage V_o with respect to all the circuit parameters using the adjoint network method.

Solution: It is required to estimate the sensitivities $\partial V_o/\partial R$, $\partial V_o/\partial R_L$, and $\partial V_o/\partial C$. The theory derived earlier estimates the sensitivities of currents drawn from voltage sources or voltages across current sources. To make use of this theory, we modify the original circuit by adding a redundant current source with zero current in parallel with the target voltage V_o. This redundant source does not change the solution of the original circuit but allows us to make use of theory as V_o is now the voltage across a current source. The original circuit with the redundant current source is shown in Figure 10.5. The branch equations governing elements of the original circuit are given by:

$$\begin{bmatrix} V_R \\ V_{R_L} \\ V_C \end{bmatrix} = \begin{bmatrix} R & 0 & 0 \\ 0 & R_L & 0 \\ 0 & 0 & Z_C \end{bmatrix} \begin{bmatrix} I_R \\ I_{R_L} \\ I_C \end{bmatrix} \tag{10.21}$$

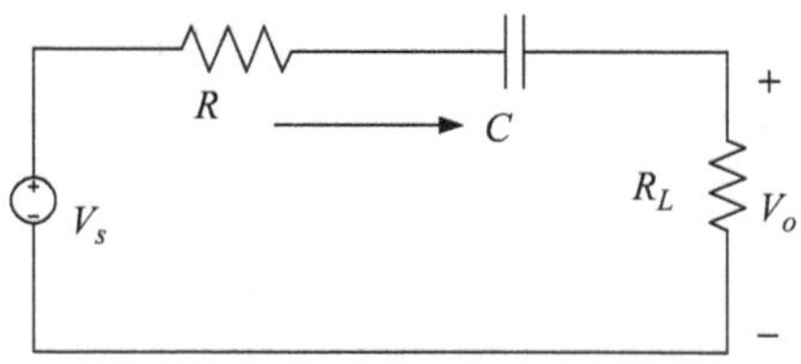

Figure 10.4 The original electric circuit of Example 10.3

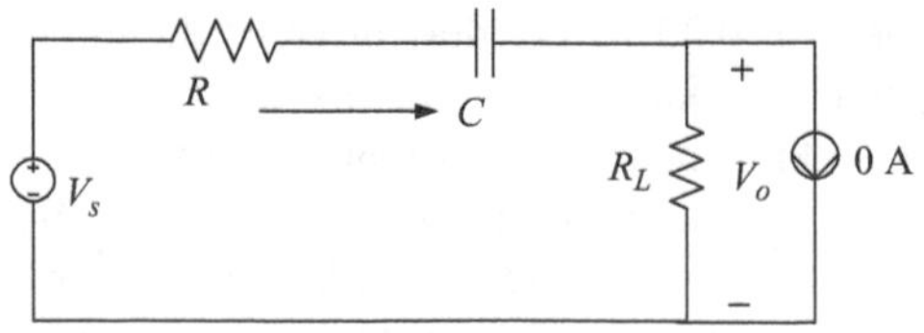

Figure 10.5 The original circuit of Example 10.3 with a redundant current source of 0 A in parallel with the voltage V_o

where $Z_c = 1/J\omega C$. Comparing this impedance matrix with (10.15), we notice that we have only an impedance matrix $\boldsymbol{Z}$. The impedance matrix of the adjoint circuit, according to (10.19), is equal to $\boldsymbol{Z}^T$. The equations governing the branches of the adjoint circuit (10.19) are thus given by:

$$\begin{bmatrix} \hat{V}_R \\ \hat{V}_{R_L} \\ \hat{V}_C \end{bmatrix} = \begin{bmatrix} R & 0 & 0 \\ 0 & R_L & 0 \\ 0 & 0 & Z_C \end{bmatrix} \begin{bmatrix} \hat{I}_R \\ \hat{I}_{R_L} \\ \hat{I}_C \end{bmatrix} \tag{10.22}$$

It follows that in the adjoint circuit, the resistances and capacitance remain unchanged. To evaluate the sensitivities of V_o with respect to all parameters, according to (10.20), we excite the adjoint circuit with a unity current source as shown in Figure 10.6. The original voltage source has been shorted out. Using the formula (10.20), the sensitivities of V_o with respect to any circuit parameter x_i are given by:

$$\frac{\partial V_o}{\partial x_i} = -\begin{bmatrix} I_R \\ I_{R_L} \\ I_C \end{bmatrix}^T \begin{bmatrix} \dfrac{\partial R}{\partial x_i} & 0 & 0 \\ 0 & \dfrac{\partial R_L}{\partial x_i} & 0 \\ 0 & 0 & \dfrac{\partial}{\partial x_i}(Z_C) \end{bmatrix} \begin{bmatrix} \hat{I}_R \\ \hat{I}_{R_L} \\ \hat{I}_C \end{bmatrix} \tag{10.23}$$

Equation (10.23) implies that we have the following sensitivities for each of the parameters:

$$\frac{\partial V_o}{\partial R} = -I_R \hat{I}_R, \quad \frac{\partial V_o}{\partial R_L} = -I_{R_L} \hat{I}_{R_L}, \quad \frac{\partial V_o}{\partial C} = I_C \left(\frac{-1}{J\omega c^2} \right) \hat{I}_C \tag{10.24}$$

It follows that by knowing the currents passing through each element, in the original and adjoint simulations, we determine the sensitivities of the output voltage.

The MATLAB® listing M10.1 solves both the original and adjoint problems for some numerical values of the circuit parameters. The sensitivities of V_o with respect to all parameters are estimated. For $\boldsymbol{x} = [R \quad R_L \quad C]^T$, the output of this code is given by:

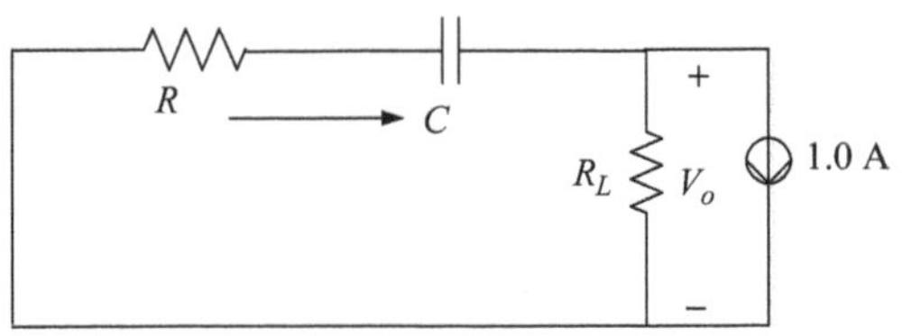

Figure 10.6 The adjoint circuit of Example 10.3 with the original excitation shorted out and an adjoint excitation of 1.0 A is placed in parallel with the target voltage V_o

```
%M10.1
%This example illustrates the adjoint network method for a simple circuit
i=sqrt(-1);
Vs=1.0; %source value of original circuit
R=1.0e3; %resistor
C=1.0e-6; %capacitor
RL=2.0e3; %load resistor
Frequency=5.0e3; %frequency of the circuit
Omega=2*pi*Frequency; %angular frequency
Zc=1/(i*Omega*C); %impedance of capacitor
%solve original circuit
I=Vs/(R+RL+Zc); %current in original circuit
IR=I; IC=I; IRL=I; %all currents are equal in original circuit
%solve adjoint circuit
I1=1.0*RL/(RL+R+Zc); %current through R in adjoint problem
I2=1.0*(R+Zc)/(RL+R+Zc); %current through RL in adjoin problem
IRAdjoint=I1; ICAdjoint=I1;
IRLAdjoint=-I2;
%estimate all sensitivities
DerivativeR=-1.0*IR*IRAdjoint; %derivative of Vo w.r.t. R
DerivativeRL=-1.0*IRL*IRLAdjoint; %derivative of Vo w.r.t. RL
DerivativeC=IC*(1/(i*Omega*C*C))*ICAdjoint; %derivative w.r.t. C
GradientAdjoint=[DerivativeR;
          DerivativeRL;
          DerivativeC]
```

```
%M10.2
%This example estimates the sensitivities using central differences
Vs=1.0; %source value of original circuit
Frequency=5.0e3; %frequency of the circuit
Omega=2*pi*Frequency; %angular frequency
x=[1.0e3 2.0e3 1.0e-6]'; %values of R, RL, and C
perturbations=[20 40 0.05e-6]';%perturbations used in CFD
IMatrix=eye(3); %identity matrix of 3 columns
J=sqrt(-1);
gradientDirect=zeros(3,1); %storage for gradient
for i=1:3
  dx=perturbations(i); %get perturbation of the ith parameter
  xBack=x-dx*IMatrix(:,i); %make a backward perturbation in the ith parameter
  R=xBack(1); RL=xBack(2); C=xBack(3); %get perturbed parameters
  ZC=1/(J*Omega*C); %get impedance
  %calculate Vo;
  backwardPerturbedObjective=1.0* RL/(R+RL+ZC); %get Vo
  %perturbed simulation
  xForward=x+dx*IMatrix(:,i);
  R=xForward(1); RL=xForward(2); C=xForward(3); %get perturbed parameters
  ZC=1/(J*Omega*C); %get impedance
  forwardPerturbedObjective=1.0* RL/(R+RL+ZC); %get Vo for forward perturbation
  gradientDirect(i)=(forwardPerturbedObjective-
                backwardPerturbedObjective)/(2*dx); %use CFD
end
gradientDirect
```

GradientAdjoint =

    1.0e+03 *
  −0.000000222147184−0.000000004714640i
    0.000000111148628−0.000000001178262i
    0.150071665481941−7.071164471938115i

The result obtained by the MATLAB listing M10.1 is checked through the more expensive central finite differences (CFDs). The MATLAB listing M10.2 shows how CFD was used to estimate the sensitivities. The results obtained using the MATLAB listing M10.2 is given by:

```
gradientDirect =
      1.0e+03 *
   -0.000000222157049-0.000000004715059i
    0.000000111168389-0.000000001178052i
    0.150824591350629-7.088872669601789i
```

These sensitivities match well those obtained using the adjoint network method. The accurate CFD requires six extra simulations ($n = 3$), while adjoint sensitivities are estimated using only one extra simulation.

Example 10.4: Consider the circuit in Figure 10.7. Find the sensitivities of the voltage V_{C_2} with respect to all parameters exploiting adjoint sensitivity analysis.

Solution: The circuit in Figure 10.7 is an active circuit with the active element replaced by its equivalent model. Because our target is to find the sensitivities of the voltage V_{C_2}, a redundant current source of zero value is added in parallel with this branch in the original circuit. Our target is to find the sensitivities of the voltage across this redundant current source. The relationship between the voltages and currents of the different branches in the original circuit may be formulated as:

$$\begin{bmatrix} I_a \\ I_{R_1} \\ I_{C_1} \\ I_{C_2} \\ I_{R_2} \end{bmatrix} = \begin{bmatrix} 0 & 0 & g_m & 0 & 0 \\ 0 & G_1 & 0 & 0 & 0 \\ 0 & 0 & J\omega C_1 & 0 & 0 \\ 0 & 0 & 0 & J\omega C_2 & 0 \\ 0 & 0 & 0 & 0 & G_2 \end{bmatrix} \begin{bmatrix} V_a \\ V_{R_1} \\ V_{C_1} \\ V_{C_2} \\ V_{R_2} \end{bmatrix} \tag{10.25}$$

where $G_i = 1/R_i$, $i = 1$, and 2 is the corresponding conductance. Comparing (10.25) with (10.15), we see that the formulation (10.25) has only an admittance matrix $\boldsymbol{Y}$. Using (10.19), the corresponding adjoint circuit is governed by the transpose of $\boldsymbol{Y}$ and is given by:

$$\begin{bmatrix} \hat{I}_a \\ \hat{I}_{R_1} \\ \hat{I}_{C_1} \\ \hat{I}_{C_2} \\ \hat{I}_{R_2} \end{bmatrix} = \begin{bmatrix} 0 & 0 & 0 & 0 & 0 \\ 0 & G_1 & 0 & 0 & 0 \\ g_m & 0 & J\omega C_1 & 0 & 0 \\ 0 & 0 & 0 & J\omega C_2 & 0 \\ 0 & 0 & 0 & 0 & G_2 \end{bmatrix} \begin{bmatrix} \hat{V}_a \\ \hat{V}_{R_1} \\ \hat{V}_{C_1} \\ \hat{V}_{C_2} \\ \hat{V}_{R_2} \end{bmatrix} \tag{10.26}$$

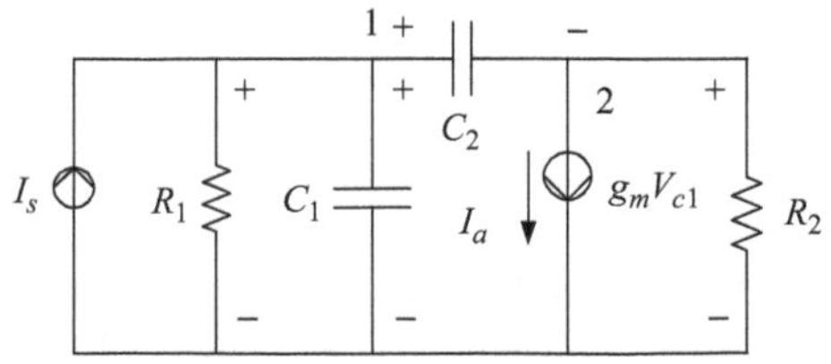

Figure 10.7 The original circuit of Example 10.4

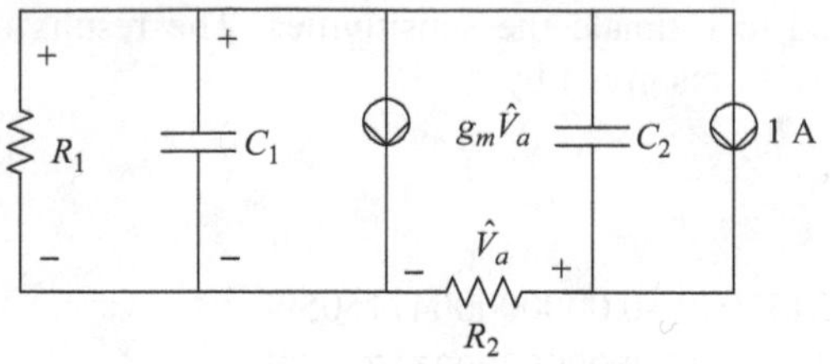

Figure 10.8 The adjoint circuit of Example 10.4 with V_{C_2} as the target response

Equation (10.26) indicates that the voltage-controlled current source is open circuited in the adjoint circuit (first row is all zeros). Also, it indicates that another voltage-controlled current source is added in parallel with the capacitor C_1 in the adjoint circuit. The adjoint circuit is thus as shown in Figure 10.8. The expression (10.20) states that the sensitivity of the voltage V_{C_1} is given by:

$$\frac{\partial V_{C_2}}{\partial x_i} = \begin{bmatrix} V_a \\ V_{R_1} \\ V_{C_1} \\ V_{C_2} \\ V_{R_2} \end{bmatrix}^T \begin{bmatrix} 0 & 0 & 0 & 0 & 0 \\ 0 & \dfrac{\partial G_1}{\partial x_i} & 0 & 0 & 0 \\ \dfrac{\partial g_m}{\partial x_i} & 0 & J\omega\dfrac{\partial C_1}{\partial x_i} & 0 & 0 \\ 0 & 0 & 0 & J\omega\dfrac{\partial C_2}{\partial x_i} & 0 \\ 0 & 0 & 0 & 0 & \dfrac{\partial G_2}{\partial x_i} \end{bmatrix} \begin{bmatrix} \hat{V}_a \\ \hat{V}_{R_1} \\ \hat{V}_{C_1} \\ \hat{V}_{C_2} \\ \hat{V}_{R_2} \end{bmatrix} \tag{10.27}$$

where x_i is any of the circuit parameters. Notice that the negative sign multiplying the derivative of the admittance matrix cancels out with the negative sign multiplying the source on the left-hand side in (10.20). Putting (10.27) in a scalar form, we get:

$$\begin{aligned} \frac{\partial V_{C_2}}{\partial x_i} &= V_{R_1}\frac{\partial G_1}{\partial x_i}\hat{V}_{R_1} + V_{C_1}\frac{\partial g_m}{\partial x_i}\hat{V}_a + V_{C_1}\left(J\omega\frac{\partial C_1}{\partial x_i}\right)\hat{V}_{C_1} \\ &\quad + V_{C_2}\left(J\omega\frac{\partial C_2}{\partial x_i}\right)\hat{V}_{C_2} + V_{R_2}\frac{\partial G_2}{\partial x_i}\hat{V}_{R_2} \end{aligned} \tag{10.28}$$

For $\boldsymbol{x} = [\,G_1 \quad g_m \quad C_1 \quad C_2 \quad G_2\,]^T$, we have the following gradient:

$$\frac{\partial V_{C_2}}{\partial \boldsymbol{x}} = \left[\,V_{R_1}\hat{V}_{R_1} \quad V_{C_1}\hat{V}_a \quad J\omega V_{C_1}\hat{V}_{C_1} \quad J\omega V_{C_2}\hat{V}_{C_2} \quad V_{R_2}\hat{V}_{R_2}\,\right]^T \tag{10.29}$$

It follows that by only solving the voltages of the original and the adjoint circuits, we can estimate the sensitivities of the voltage V_{C_2} with respect to all parameters regardless of their number.

To illustrate our gradient evaluation with some numbers, we assume the following values for the circuit parameters: $\boldsymbol{x} = [\,1.0\text{e-}3\ \Omega^{-1} \quad 2.0\text{e-}3\ \Omega^{-1} \quad 1.0\text{e-}6\ \text{F} \quad 2.0\text{e-}6\ \text{F} \quad 0.5\text{e-}3\ \Omega^{-1}\,]^T$. The MATLAB listing M10.3 is used to illustrate our results:

```
%M10.3
%This file calculates the sensitivities for example 10.4
i=sqrt(-1);
Is=0.5; %current source
R1=1.0e3; %first resistor
gm= 2.0e-3; %current source coefficient
C1= 1.0e-6; %first capacitor
C2= 2.0e-6 %second capacitor
R2=2.0e3; %second resistor
Frequency=3.0e3; %this is the frequency
Omega=2*pi*Frequency;
%call original solver to get [Vc1 Va];
OriginalResponse=getOriginalResponse(R1,C1, R2,C2, gm, Is, Frequency);
VR1=OriginalResponse(1,1); VC1=OriginalResponse(1,1);
VR2=OriginalResponse(2,1); Vgm=OriginalResponse(1,1);
VC2=OriginalResponse(1,1)-OriginalResponse(2,1);
%call adjoint solver to get [Vc1Adjoint VaAdjoint]
AdjointResponse=getAdjointResponse(R1,C1, R2,C2, gm, Frequency);
VR1Adjoint=AdjointResponse(1); VC1Adjoint=AdjointResponse(1);
VR2Adjoint=AdjointResponse(2); VgmAdjoint=AdjointResponse(2);
VC2Adjoint=AdjointResponse(1)-AdjointResponse(2); %Vc2=Vc1-Va
Gradient=[(VR1*VR1Adjoint) (Vgm*VgmAdjoint) (i*Omega*VC1*VC1Adjoint)...
                        (i*Omega*VC2*VC2Adjoint) (VR2*VR2Adjoint)]
```

The function **getOriginalResponse** solves the nodal equations of the original circuit for the two voltages V_1 and V_2 as shown in Figure 10.7. The voltages and currents for all other branches are deduced from these two values. The function **getAdjointResponse** does exactly the same function for the adjoint circuit. The original response supplied by the MATLAB code is $[V_{C_1} \quad V_a]^T = [4.7445 - J25.6000 \quad 6.4377 - J25.2629]^T$. The corresponding adjoint response is evaluated to be $[3.3864 + J0.6742 \quad 3.7286 - J25.8022]^T$. Using these values the gradient is evaluated to be:

Gradient =

1.0e+06 *

0.000033325187754−0.000083492019023i

−0.000642859773−0.0002178706979i

1.573787481582732+0.628164990160438i

0.842836769595989+0.179149931631100i

−0.000627834411124−0.000260300091010i

To check this answer, we also evaluated the gradient using CFDs. A simple MATLAB code similar to the listing M10.2 was implemented. Using this code, the gradient was calculated to be:

1.0e+06 *

0.000033342−0.000083523i

−0.000642831231343−0.000217857058307i

1.576788392477067+0.630944618571618i

0.844929305763995+0.179651714100970i

−0.0006280738393−0.00026042184066i

The CFD results match very well the ones obtained using adjoint sensitivities. CFD requires 10 extra simulations ($n = 5$), while the adjoint approach requires only one adjoint simulation.

Example 10.5: Consider the circuit of Figure 10.9. Utilize the adjoint network method to estimate the sensitivities of the response I_{R_3} with respect to all parameters.

Solution: Because the target response $F = I_{R_3}$ in this example is a current, a redundant voltage source of zero value is connected in series with R_3 in the original circuit. In the adjoint circuit this voltage source has a unity value to estimate the sensitivities of the current passing through it, I_{R_3}.

The branch voltages and currents of the original circuit are governed by the matrix:

$$\begin{bmatrix} I_{C_1} \\ I_{C_2} \\ I_{C_3} \\ I_{R_4} \\ V_{R_1} \\ V_{R_2} \\ V_{R_3} \\ V_o \end{bmatrix} = \begin{bmatrix} X_{C_1} & 0 & 0 & 0 & 0 & 0 & 0 & 0 \\ 0 & X_{C_2} & 0 & 0 & 0 & 0 & 0 & 0 \\ 0 & 0 & X_{C_3} & 0 & 0 & 0 & 0 & 0 \\ 0 & 0 & 0 & G_4 & 0 & 0 & 0 & 0 \\ 0 & 0 & 0 & 0 & R_1 & 0 & 0 & 0 \\ 0 & 0 & 0 & 0 & 0 & R_2 & 0 & 0 \\ 0 & 0 & 0 & 0 & 0 & 0 & R_3 & 0 \\ 0 & 0 & 0 & A_v & 0 & 0 & 0 & 0 \end{bmatrix} \begin{bmatrix} V_{C_1} \\ V_{C_2} \\ V_{C_3} \\ V_{R_4} \\ I_{R_1} \\ I_{R_2} \\ I_{R_3} \\ I_o \end{bmatrix} \quad (10.30)$$

where $X_{C_i} = J\omega C_i$, $i = 1, 2, 3$ and $G_4 = (1/R_4)$. Comparing with (10.15), we notice that (10.30) has the following four submatrices:

$$\boldsymbol{Y} = \begin{bmatrix} X_{C_1} & 0 & 0 & 0 \\ 0 & X_{C_2} & 0 & 0 \\ 0 & 0 & X_{C_3} & 0 \\ 0 & 0 & 0 & G_4 \end{bmatrix}, \quad \boldsymbol{M} = \begin{bmatrix} 0 & 0 & 0 & 0 \\ 0 & 0 & 0 & 0 \\ 0 & 0 & 0 & 0 \\ 0 & 0 & 0 & A_v \end{bmatrix},$$
$$\boldsymbol{Z} = \begin{bmatrix} R_1 & 0 & 0 & 0 \\ 0 & R_2 & 0 & 0 \\ 0 & 0 & R_3 & 0 \\ 0 & 0 & 0 & 0 \end{bmatrix}, \quad \text{and} \quad \boldsymbol{A} = \boldsymbol{0} \quad (10.31)$$

Using (10.19), the corresponding adjoint circuit has the following branch voltages and currents:

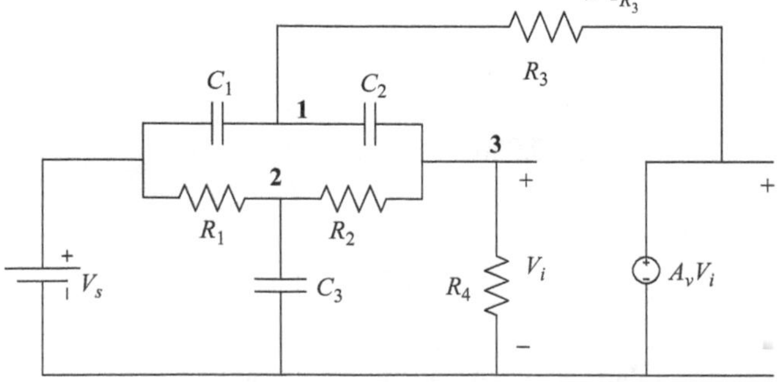

Figure 10.9 The original circuit of Example 10.5

$$\begin{bmatrix} \hat{I}_{C_1} \\ \hat{I}_{C_2} \\ \hat{I}_{C_3} \\ \hat{I}_{R_4} \\ \hat{V}_{R_1} \\ \hat{V}_{R_2} \\ \hat{V}_{R_3} \\ \hat{V}_o \end{bmatrix} = \begin{bmatrix} X_{C_1} & 0 & 0 & 0 & 0 & 0 & 0 & 0 \\ 0 & X_{C_2} & 0 & 0 & 0 & 0 & 0 & 0 \\ 0 & 0 & X_{C_3} & 0 & 0 & 0 & 0 & 0 \\ 0 & 0 & 0 & G_4 & 0 & 0 & 0 & -A_v \\ 0 & 0 & 0 & 0 & R_1 & 0 & 0 & 0 \\ 0 & 0 & 0 & 0 & 0 & R_2 & 0 & 0 \\ 0 & 0 & 0 & 0 & 0 & 0 & R_3 & 0 \\ 0 & 0 & 0 & 0 & 0 & 0 & 0 & 0 \end{bmatrix} \begin{bmatrix} \hat{V}_{C_1} \\ \hat{V}_{C_2} \\ \hat{V}_{C_3} \\ \hat{V}_{R_4} \\ \hat{I}_{R_1} \\ \hat{I}_{R_2} \\ \hat{I}_{R_3} \\ \hat{I}_o \end{bmatrix} \tag{10.32}$$

Few things to notice about (10.32): The output voltage V_o is shorted out in the adjoint circuit because of the last row in (10.32). The resistors and capacitors in the adjoint circuit remain the same as in the original circuit. The fourth equation in (10.36) indicates that the voltage-controlled voltage source in the original circuit has been replaced by a current-controlled current source in parallel with R_4. The adjoint circuit for estimating the sensitivities of I_{R_3} is thus as shown in Figure 10.10.

Once the two circuits have been solved for the currents and voltages in each branch, and using (10.20), the sensitivities with respect to the ith parameter are given by:

$$\frac{\partial I_{R_3}}{\partial x_i} = \begin{bmatrix} V_{C_1} \\ V_{C_2} \\ V_{C_3} \\ V_{R_4} \\ I_{R_1} \\ I_{R_2} \\ I_{R_3} \\ I_o \end{bmatrix}^T \begin{bmatrix} -\frac{\partial X_{C_1}}{\partial x_i} & 0 & 0 & 0 & 0 & 0 & 0 & 0 \\ 0 & -\frac{\partial X_{C_2}}{\partial x_i} & 0 & 0 & 0 & 0 & 0 & 0 \\ 0 & 0 & -\frac{\partial X_{C_3}}{\partial x_i} & 0 & 0 & 0 & 0 & 0 \\ 0 & 0 & 0 & -\frac{\partial G}{\partial x_i} & 0 & 0 & 0 & \frac{\partial A_v}{\partial x_i} \\ 0 & 0 & 0 & 0 & \frac{\partial R_1}{\partial x_i} & 0 & 0 & 0 \\ 0 & 0 & 0 & 0 & 0 & \frac{\partial R_2}{\partial x_i} & 0 & 0 \\ 0 & 0 & 0 & 0 & 0 & 0 & \frac{\partial R_3}{\partial x_i} & 0 \\ 0 & 0 & 0 & 0 & 0 & 0 & 0 & 0 \end{bmatrix} \begin{bmatrix} \hat{V}_{C_1} \\ \hat{V}_{C_2} \\ \hat{V}_{C_3} \\ \hat{V}_{R_4} \\ \hat{I}_{R_1} \\ \hat{I}_{R_2} \\ \hat{I}_{R_3} \\ \hat{I}_o \end{bmatrix} \tag{10.33}$$

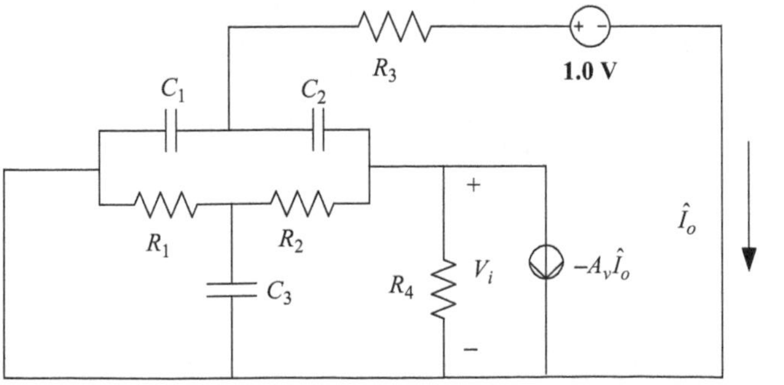

Figure 10.10 The adjoint circuit corresponding to the original circuit in Figure 10.5 used to estimate the sensitivities of I_{R_3}

Using (10.33), the sensitivities with respect to all parameters are given by:

$$\frac{\partial F}{\partial C_1} = J\omega V_{C_1}\hat{V}_{C_1}, \quad \frac{\partial F}{\partial C_2} = J\omega V_{C_2}\hat{V}_{C_2}, \quad \frac{\partial F}{\partial C_3} = J\omega V_{C_3}\hat{V}_{C_3}$$
$$\frac{\partial F}{\partial R_1} = I_{R_1}\hat{I}_{R_1}, \quad \frac{\partial F}{\partial R_2} = I_{R_2}\hat{I}_{R_2}, \quad \frac{\partial F}{\partial R_3} = I_{R_3}\hat{I}_{R_3}, \quad \frac{\partial F}{\partial R_4} = \frac{1}{R_4^2}V_{R_4}\hat{V}_{R_4}, \quad (10.34)$$
$$\frac{\partial F}{\partial A_v} = V_{R_4}\hat{I}_o$$

It follows that the sensitivities with respect to all parameters are estimated using the original and adjoint simulations. The MATLAB listing M10.4 shows an implementation of the adjoint network method to obtain the sensitivities of the response I_{R_3} with respect to all parameters for specific numerical values of the circuit parameters. The function **getOriginalResponse** evaluates the voltages $[V_1 \quad V_2 \quad V_3]^T$. All branch voltages and currents can be evaluated using these three voltages. The function **getAdjointResponse** does the same functionality for the adjoint problem. The output of this MATLAB listing for the given parameters values is given by:

```
%M10.4
%Adjoint sensitivity analysis of the circuit in example 10.5
Vs=2.0; %current source
R1=1.0e3; %first resistor
A= 3; %voltage source coefficient
C1= 1.0e-6; %first capacitor
C2= 2.0e-6; %second capacitor
C3=1.0e-6; %third capacitor
R2=2.0e3; %second resistor
R3=3e3; %third resistor
R4=2.0e3; %fourth resistor
Frequency=3.0e3; %this is the frequency
Omega=2*pi*Frequency; %angular frequency
i=sqrt(-1);
%call original solver to get [V1 V2 V3];
OriginalResponse=getOriginalResponse(R1,C1, R2,C2, R3, C3, R4, A, Vs, Frequency);
%by knowing the original nodal voltages V1, V2, and V3, we calculate all branch
%voltages and currents
VR1=OriginalResponse(2,1)-Vs; VC1=OriginalResponse(1,1)-Vs;
VR2=OriginalResponse(2,1)-OriginalResponse(3,1); VC2=OriginalResponse(1,1)-OriginalResponse(3,1);
VR3=OriginalResponse(1,1)-A*OriginalResponse(3,1); VC3= OriginalResponse(2,1);
VR4=OriginalResponse(3,1); Vo=A*OriginalResponse(3,1);
IR1=VR1/R1; IR2=VR2/R2; IR3=VR3/R3; IR4=VR4/R4; Io= VR3/R3;%get currents
%call adjoint solver to get [V1Adjoint V2Adjoint V3Adjoint]
AdjointResponse1=getAdjointResponse(R1,C1, R2,C2, R3, C3, R4, A, Frequency);
%by knowing the adjoint nodal voltages V1Adjoint, V2Adjoint, and V3Adjoint, we calculate all branch
%voltages and currents for the adjoint circuit
VR1Adjoint=AdjointResponse1(2,1); VC1Adjoint=AdjointResponse1(1,1);
VR2Adjoint=AdjointResponse1(2,1)-AdjointResponse1(3,1); VC2Adjoint=AdjointResponse1(1,1)-
AdjointResponse1(3,1);
VR3Adjoint=AdjointResponse1(1,1)-1.0; VC3Adjoint= AdjointResponse1(2,1);
VR4Adjoint=AdjointResponse1(3,1); VoAdjoint=0; IoAdjoint=VR3Adjoint/R3;
IR1Adjoint=VR1Adjoint/R1; IR2Adjoint=VR2Adjoint/R2; IR3Adjoint=VR3Adjoint/R3;
IR4Adjoint=VR4Adjoint/R4; %get currents
%Using original and adjoint response, we estimate all sensitivities
%regardless of their number
GradientAdjoint=[(IR1*IR1Adjoint); (-i*Omega*VC1*VC1Adjoint); (IR2*IR2Adjoint); (-
i*Omega*VC2*VC2Adjoint); (IR3*IR3Adjoint); (-i*Omega*VC3*VC3Adjoint);...
        (IR4*IR4Adjoint); (VR4*IoAdjoint)]
```

GradientAdjoint =

−0.000000003187383−0.000000000655332i
−4.520537011903178+22.999592827777899i
−0.000000005918742+0.000000030131259i
−2.897169921930644+26.207041140587766i
0.000000443692193+0.000000009762776i
−4.764798027960165−1.053015568184380i
−0.000000002705737+0.000000030667024i
−0.000665966159050−0.000005832165183i

These results were checked using CFDs. The code used to generate the CFD estimates is shown in the MATLAB listing M10.5.

```
%M10.5
%Evaluation of the gradient of example 10.5 using CFD
Vs=2.0; %current source
x=[1.0e3 1.0e-6 2.0e3 2.0e-6 3e3 1.0e-6 2.0e3 3]'; %vector of nominal parameters
perturbations=[20 0.05e-6 30 0.05e-6 50 0.05e-6 30 0.05]'; %perturbation for each parameter
IMatrix=eye(8);%identity matrix
GradientDirect=zeros(8,1); %storage for gradient
Frequency=3.0e3; %this is the frequency
Omega=2*pi*Frequency;
for i=1:8
  dx=perturbations(i); %get the perturbation of the ith parameter
  xBack=x-dx*IMatrix(:,i); %backward perturb the ith parameter
  R1=xBack(1); C1=xBack(2); R2=xBack(3); C2=xBack(4);
  R3=xBack(5); C3=xBack(6); R4=xBack(7); A=xBack(8);
  %call original solver to get [V1 V2 V3];
  OriginalResponse=getOriginalResponse(R1,C1, R2,C2, R3, C3, R4, A, Vs, Frequency);
  backwardPerturbedObjective=(OriginalResponse(1)-A*OriginalResponse(3))/R3; %get IR3
  %perturbed simulation
  xForward=x+dx*IMatrix(:,i); %forward perturb the ith parameter
  R1=xForward(1); C1=xForward(2); R2=xForward(3); C2=xForward(4);
  R3=xForward(5); C3=xForward(6); R4=xForward(7); A=xForward(8);
  %call original solver to get [V1 V2 V3]
  OriginalResponse=getOriginalResponse(R1,C1, R2,C2, R3, C3, R4, A, Vs, Frequency);
  forwardPerturbedObjective=(OriginalResponse(1)-A*OriginalResponse(3))/R3; %get IR3
  GradientDirect(i)=(forwardPerturbedObjective-backwardPerturbedObjective)/(2*dx); %use CFD
end
format long
GradientDirect
```

The output of the CFD code is given by:

GradientDirect =

−0.000000003188617−0.000000000655725i
−4.533870019694017+23.056702746979674i
−0.000000005920940+0.000000030137760i
−2.899843119864073+26.223289173490624i
0.000000443816013+0.000000009756785i
−4.776075812659275−1.057473948017280i
−0.000000002706891+0.000000030673845i
−0.000665965642428−0.000005832114958i

The two gradients estimated using the adjoint network method and using CFD are almost identical. The accurate CFD sensitivities requires 16 extra simulations

($n = 8$), while the adjoint approach requires only 1 extra simulation. It is obvious from previous examples that the higher the number of parameters, the more efficient the adjoint network method becomes.

10.4 Adjoint sensitivity analysis of a linear system of equations

In the previous section, I showed how adjoint sensitivity analysis can be applied to electric networks. The same concept of adjoint analysis can also be applied to systems of linear equations. Consider a system of linear equations of the form:

$$\boldsymbol{Z}(\boldsymbol{x})\boldsymbol{I} = \boldsymbol{V} \tag{10.35}$$

where $\boldsymbol{Z} \in C^{m\times m}$ is the system matrix, which is, in general, a complex matrix. The components of this matrix are functions of the parameters $\boldsymbol{x}$. The vector $\boldsymbol{I}$ is the vector of state variables. These state variables may represent currents in some applications or electric field values at certain spatial positions in other applications. The vector $\boldsymbol{V}$ is the vector of excitation of the problem. It may represent voltage sources in some problems or vector of incident electric fields at certain positions in space in other applications. Both the method of moments (MoM) and the finite element method (FEM), utilized in computational electromagnetics, result in a linear system similar to (10.35).

Our target is to find the gradient of the response $f(\boldsymbol{x}, \boldsymbol{I})$ with respect to $\boldsymbol{x}$. The classical way of estimating this gradient is to perturb each parameter x_i, and then solve the linear system (10.35) for the perturbed parameters. Finite differences are then used to estimate the sensitivities of the response as explained in Chapter 1. This approach requires constructing the matrix $\boldsymbol{Z}$ for all perturbed parameters and then solving the system (10.35) at least n times. It follows that the classical approach to gradient estimation requires at least n matrix constructions and n matrix inversions.

The adjoint variable method can estimate the required sensitivities in a more efficient way. The derivation of this method starts by differentiating equation (10.35) with respect to the ith parameter x_i to obtain:

$$\frac{\partial(\boldsymbol{Z}\bar{\boldsymbol{I}})}{\partial x_i} + \boldsymbol{Z}\frac{\partial \boldsymbol{I}}{\partial x_i} = \frac{\partial \boldsymbol{V}}{\partial x_i} \tag{10.36}$$

The first term in (10.36) implies differentiating the matrix $\boldsymbol{Z}$ while keeping $\boldsymbol{I}$ fixed at its nominal value $\bar{\boldsymbol{I}}$. Solving for the state variables' derivatives, we get:

$$\frac{\partial \boldsymbol{I}}{\partial x_i} = \boldsymbol{Z}^{-1}\left(\frac{\partial \boldsymbol{V}}{\partial x_i} - \frac{\partial(\boldsymbol{Z}\bar{\boldsymbol{I}})}{\partial x_i}\right) \tag{10.37}$$

The derivative of the response $f(\boldsymbol{x}, \boldsymbol{I})$ with respect to ith parameter x_i is given by:

$$\frac{\partial f}{\partial x_i} = \frac{\partial^e f}{\partial x_i} + \left(\frac{\partial f}{\partial \boldsymbol{I}}\right)^T \left(\frac{\partial \boldsymbol{I}}{\partial x_i}\right) \tag{10.38}$$

where the first derivative on the right-hand side is the explicit dependence of f on the parameter x_i, which is usually known. The second term denotes the implicit dependence on $\boldsymbol{x}$ through the state variables, which is unknown.

The derivative of the state variables in (10.38) is eliminated using (10.37) to obtain:

$$\frac{\partial f}{\partial x_i} = \frac{\partial^e f}{\partial x_i} + \underbrace{\left(\frac{\partial f}{\partial \boldsymbol{I}}\right)^T \boldsymbol{Z}^{-1}}\left(\frac{\partial \boldsymbol{V}}{\partial x_i} - \frac{\partial(\boldsymbol{Z}\bar{\boldsymbol{I}})}{\partial x_i}\right) \tag{10.39}$$

The parameter-dependent part can be used to define the adjoint state variables using:

$$\hat{\boldsymbol{I}}^T = \left(\frac{\partial f}{\partial \boldsymbol{I}}\right)^T \boldsymbol{Z}^{-1} \Rightarrow \hat{\boldsymbol{I}} = \boldsymbol{Z}^{T^{-1}}\left(\frac{\partial f}{\partial \boldsymbol{I}}\right) \Rightarrow \boldsymbol{Z}^T\hat{\boldsymbol{I}} = \left(\frac{\partial f}{\partial \boldsymbol{I}}\right) \tag{10.40}$$

It follows that the vector of adjoint state variables $\hat{\boldsymbol{I}}$ is obtained by solving the system (10.40). The system matrix of (10.40) is the transpose of the system (10.35). The same LU factorization used in (10.35) can be reused to solve the adjoint system. The excitation of the adjoint system (10.40) depends on the response *f* and its derivatives with respect to the state variables. Once the adjoint system is solved, the sensitivities of the response with respect to the *i*th parameter, $\forall i$, are given by:

$$\frac{\partial f}{\partial x_i} = \frac{\partial^e f}{\partial x_i} + \hat{\boldsymbol{I}}^T\left(\frac{\partial \boldsymbol{V}}{\partial x_i} - \frac{\partial(\boldsymbol{Z}\bar{\boldsymbol{I}})}{\partial x_i}\right) \tag{10.41}$$

Once the solutions of the original system (10.35) given by $\bar{\boldsymbol{I}}$ and the adjoint system (10.40) are obtained, the sensitivities with respect to all the parameters $\boldsymbol{x}$ can be estimated using (10.41).

The excitation vector $\boldsymbol{V}$ is usually given, so its derivatives on the parameters are known. In many practical applications $\boldsymbol{V}$ is not dependent on $\boldsymbol{x}$ and the excitation derivatives vanish. The derivative of the system matrix term is sometimes available analytically. In this case, the derivatives are available through closed expressions. In many cases, however, this is not the case and the matrix is only available numerically. The most applicable approach for estimating these sensitivities is to utilize finite differences. The *j*th column of the system matrix derivative in (10.41) is given by:

$$\frac{\partial(\boldsymbol{Z}\bar{\boldsymbol{I}})}{\partial x_j} = \left(\frac{\boldsymbol{Z}(\boldsymbol{x} + \Delta x_j \boldsymbol{e}_j) - \boldsymbol{Z}}{\Delta x_j}\right)\bar{\boldsymbol{I}}, \quad j = 1, 2, \ldots, n \tag{10.42}$$

The approximation (10.42) requires extra n matrix fills. The following example illustrates the application of this approach to a practical problem.

Example 10.6: Consider the circuit shown in Figure 10.11. Using loop analysis, write the equations governing this circuit. Utilize frequency-domain adjoint sensitivity analysis to estimate the sensitivities of the input impedance with respect to all parameters.

Solution: As shown in Figure 10.11, we have four loops. The loop equations for these loops are given by:

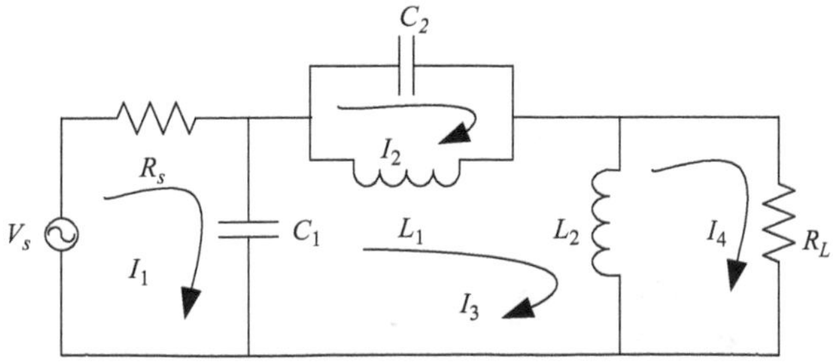

Figure 10.11 The circuit of Example 10.6

$$\begin{bmatrix} R_s + Z_{c_1} & 0 & -Z_{c_1} & 0 \\ 0 & Z_{c_2} + Z_{L_1} & -Z_{L_1} & 0 \\ -Z_{c_1} & -Z_{L_1} & Z_{c_1} + Z_{L_1} + Z_{L_2} & -Z_{L_2} \\ 0 & 0 & -Z_{L_2} & R_L + Z_{L_2} \end{bmatrix} \begin{bmatrix} I_1 \\ I_2 \\ I_3 \\ I_4 \end{bmatrix} = \begin{bmatrix} V_s \\ 0 \\ 0 \\ 0 \end{bmatrix} \tag{10.43}$$

The target response for this example is Z_{in}, the input impedance seen by the source. This input impedance is given by $Z_{in} = V_s/I_1$. According to (10.40), the corresponding adjoint problem is given by:

$$\begin{bmatrix} R_s + Z_{c_1} & 0 & -Z_{c_1} & 0 \\ 0 & Z_{c_2} + Z_{L_1} & -Z_{L_1} & 0 \\ -Z_{c_1} & -Z_{L_1} & Z_{c_1} + Z_{L_1} + Z_{L_2} & -Z_{L_2} \\ 0 & 0 & -Z_{L_2} & R_L + Z_{L_2} \end{bmatrix} \begin{bmatrix} \hat{I}_1 \\ \hat{I}_2 \\ \hat{I}_3 \\ \hat{I}_4 \end{bmatrix} = \begin{bmatrix} -\dfrac{V_s}{I_1^2} \\ 0 \\ 0 \\ 0 \end{bmatrix} \tag{10.44}$$

Notice that the system matrix of the original problem (10.43) is symmetric and this is why the adjoint problem has the same system matrix. The same LU decomposition used in solving (10.43) can be reused in solving (10.44) for time-intensive problems. For this simple problem, we solve (10.44) directly. Once the solutions of (10.43) and (10.44) are known, the sensitivities relative to any parameter x_i are given by (10.41). For the current example, this is given by:

$$\frac{\partial f}{\partial x_i} = \hat{\boldsymbol{I}}^T \left(-\frac{\partial (\boldsymbol{Z})}{\partial x_i} \right) \bar{\boldsymbol{I}} \tag{10.45}$$

where $\bar{\boldsymbol{I}}$ is the solution of (10.43). Notice that in this case, the excitation vector in (10.43) is not dependent on any of the circuit parameters. Also, the explicit derivative in (10.41) vanishes. Also, the system matrix $\boldsymbol{Z}$ is known analytically and its analytic derivatives can be used to estimate the sensitivities.

The MATLAB M10.6 listing solves this problem for a set of values of the circuit parameters. Notice that we used the full matrix sensitivities in this code even though a more concise scalar form could have been obtained. The gradient of the objective function ($\partial Z_{in}/\partial \boldsymbol{x}$) with $\boldsymbol{x} = [R_S \quad R_1 \quad L_1 \quad L_2 \quad C_1 \quad C_2]^T$ is estimated using this code using adjoint sensitivities. The result is given by:

```
gradientAdjoint =
    1.0e+05 *

     0.000010000000000-0.000000000000000i
    -0.000000044376474+0.000000005717472i
     0.000222109330675+0.084274776428242i
     0.009099638733091+0.070627352723206i
     0.298403998207694+7.075886110985558i
     0.008768524910207+3.327034817318012i
```

```
%M10.6
%Adjoint sensitivity analysis of the circuit of example 10.6
Vs=1.0;%source voltage
Rs=50; %source resistance
RL=50; %load resistance
C1=1.0e-6; %first capacitor
C2=2.0e-6; %second capacitor
L1=1.0e-3; %first inductor
L2=0.5e-3; %second inductor
Frequency=1.0e3; %frequency
Omega=2*pi*Frequency; %anular frequency
J=sqrt(-1);
ZL1=J*Omega*L1; %impedance of first inductor
ZL2=J*Omega*L2; %impedance of second inductor
ZC1=1/(J*Omega*C1); %impedance of first capacitor
ZC2=1/(J*Omega*C2); %impedance of second capacitor
Z=[Rs+ZC1  0       -ZC1      0;
 0  (ZC2+ZL1)   -ZL1      0;
 -ZC1   -ZL1   (ZC1+ZL1+ZL2)  -ZL2;
 0   0       -ZL2       (RL+ZL2)];
OriginalExcitation=[Vs; 0; 0; 0]; %original excitation
ZInverse=inv(Z); %get inverse of Z matrix
%solve for the original circuit
IOriginal=ZInverse*OriginalExcitation; %get the original currents
%now we determine the adjoint excitation
AdjointExcitation=[-Vs/(IOriginal(1)^2);  0;  0;  0];
IAdjoint=ZInverse*AdjointExcitation; %get the adjoint vector
%build analytical sensitivities of the impedance matrix relative to all parameters
dZBydRs=[1  0  0  0; 0  0  0  0;  0  0  0  0; 0  0  0  0];
dZBydRL=[0  0  0  0; 0  0  0  0;0  0  0  0; 0 0 0  1];
dZBydL1=[0   0  0  0; 0  J*Omega  -J*Omega  0; 0  -J*Omega   J*Omega  0; 0 0  0  0];
dZBydL2=[0   0   0  0; 0   0  0  0; 0  0  J*Omega  -J*Omega; 0  0 -J*Omega  J*Omega];
dZBydC1=[-1/(J*Omega*C1^2)  0  1/(J*Omega*C1^2)  0; 0   0   0  0;1/(J*Omega*C1^2)  0  -
                                                 1/(J*Omega*C1^2)  0;  0  0  0  0];
dZBydC2=[0  0  0  0; 0  -1/(J*Omega*C2^2)  0  0;  0  0  0  0;  0  0  0  0];
%evaluate the adjoint Sensitivities
DerivativeRs=transpose(IAdjoint)*-1.0*dZBydRs*IOriginal;
DerivativeRL=transpose(IAdjoint)*-1.0*dZBydRL*IOriginal;
DerivativeL1=transpose(IAdjoint)*-1.0*dZBydL1*IOriginal;
DerivativeL2=transpose(IAdjoint)*-1.0*dZBydL2*IOriginal;
DerivativeC1=transpose(IAdjoint)*-1.0*dZBydC1*IOriginal;
DerivativeC2=transpose(IAdjoint)*-1.0*dZBydC2*IOriginal;
%print gradient
gradientAdjoint=[DerivativeRs; DerivativeRL; DerivativeL1;
                              DerivativeL2; DerivativeC1; DerivativeC2]
```

The same sensitivities are also estimated using CFD using a MATLAB code using a code similar to 10.6. These results are given by:

1.0e+05 *

$$
\begin{array}{r}
0.000010000000000-0.000000000000000i \\
-0.000000044445547+0.000000005735586i \\
0.000222112807251+0.084275741594925i \\
0.009099937079422+0.070626601453266i \\
0.298405656559453+7.075905737796618i \\
0.008768570876327+3.327051527901581i
\end{array}
$$

The two results match very well. The CFD approach is evaluated using 10 extra circuit simulations while adjoint sensitivities are estimated with minor computational overhead. The same matrix inversion used to solve the original system is reused in solving the adjoint system.

10.5 Time-domain adjoint sensitivity analysis

In the previous two sections, we discussed the adjoint network method and the adjoint method for sensitivity analysis of linear systems of equations. These approaches are usually applicable for problems where responses do not change with time (DC) or if the system is solved for one frequency at a time. Many applications, however, obtain a wideband solution of the electric system or circuit. Through solving the differential equations governing the system in the time domain, all transient responses are obtained. These responses are time-varying quantities whose values change by changing different parameters of problem. Iterative numerical approaches are usually used for solving such problems. We discuss in this section adjoint techniques applicable to these problems.

Without loss of generality, we assume that the considered problem is governed by the following system of differential equations:

$$\boldsymbol{M}\ddot{\boldsymbol{E}} + \boldsymbol{N}\dot{\boldsymbol{E}} + \boldsymbol{K}\boldsymbol{E} = \boldsymbol{G} \tag{10.46}$$

where $\boldsymbol{E} \in \Re^m$ is the vector of state responses. $\dot{\boldsymbol{E}}$ and $\ddot{\boldsymbol{E}}$ are its first- and second-order time derivatives. The square matrices $\boldsymbol{M}$, $\boldsymbol{N}$, and $\boldsymbol{K}$ are the system matrices. These matrices are assumed to be functions of the parameters $\boldsymbol{x}$. The vector $\boldsymbol{G} \in \Re^m$ is the vector of time-varying excitation of the system.

Our target is to find the sensitivities of a given objective function that depends on the temporal response $\boldsymbol{E}$. This objective function may be given by the integral:

$$f = \int_0^{T_m} \psi(\boldsymbol{x}, \boldsymbol{E})\, dt \tag{10.47}$$

where ψ is the kernel of the integral. The kernel function may have an explicit dependence on $\boldsymbol{x}$ and it may have also an implicit dependence through the temporal vector $\boldsymbol{E}$. T_m is the time over which the system response is observed. Our target is to find the gradient of f with respect to the parameters $\boldsymbol{x}$. From (10.47), the ith component of the gradient is given by:

$$\frac{\partial f}{\partial x_i} = \int_0^{T_m} \frac{\partial^e \psi}{\partial x_i} dt + \int_0^{T_m} \left(\frac{\partial \psi}{\partial \boldsymbol{E}}\right)^T \frac{\partial \boldsymbol{E}}{\partial x_i} dt \tag{10.48}$$

The first integral in (10.48) calculates the explicit derivative, which is assumed to be known. The second integral requires the derivative of $\boldsymbol{E}$ with respect to $\boldsymbol{x}$, which is unknown. The classical way of calculating (10.48) involves solving (10.46) for perturbed values of the parameters $\boldsymbol{x}$. At least n extra solutions of the system (10.46) are required to estimate all sensitivities of the function f.

The adjoint variable method estimates the gradient in (10.48) using at most one extra simulation. The derivation of the method is simple and it is instructive to go over it. This derivation starts by differentiating both sides of (10.46) with respect to the ith parameter x_i to get:

$$\frac{\partial \boldsymbol{M}}{\partial x_i}\ddot{\boldsymbol{E}} + \boldsymbol{M}\frac{\partial \ddot{\boldsymbol{E}}}{\partial x_i} + \frac{\partial \boldsymbol{N}}{\partial x_i}\dot{\boldsymbol{E}} + \boldsymbol{N}\frac{\partial \dot{\boldsymbol{E}}}{\partial x_i} + \frac{\partial \boldsymbol{K}}{\partial x_i}\boldsymbol{E} + \boldsymbol{K}\frac{\partial \boldsymbol{E}}{\partial x_i} = \frac{\partial \boldsymbol{G}}{\partial x_i} \tag{10.49}$$

In (10.49), we assume that the system matrices are analytical functions of the parameters. If these matrices are only available numerically, as in the case in many applications, then finite differences can be used to approximate the system matrices' derivatives. Multiplying both sides of (10.49) by the transpose of the temporal adjoint variable $\boldsymbol{\lambda}$, whose value is to be determined later, and integrating, we obtain the expression:

$$\int_0^{T_m}\left(\boldsymbol{\lambda}^T\boldsymbol{M}\frac{\partial\ddot{\boldsymbol{E}}}{\partial x_i}+\boldsymbol{\lambda}^T\boldsymbol{N}\frac{\partial\dot{\boldsymbol{E}}}{\partial x_i}+\boldsymbol{\lambda}^T\boldsymbol{K}\frac{\partial\boldsymbol{E}}{\partial x_i}\right)dt=\int_0^{T_m}\boldsymbol{\lambda}^T\boldsymbol{R}_i\,dt, \tag{10.50}$$

where $\boldsymbol{R}_i=\dfrac{\partial\boldsymbol{G}}{\partial x_i}-\dfrac{\partial\boldsymbol{M}}{\partial x_i}\ddot{\boldsymbol{E}}-\dfrac{\partial\boldsymbol{N}}{\partial x_i}\dot{\boldsymbol{E}}-\dfrac{\partial\boldsymbol{K}}{\partial x_i}\boldsymbol{E}$

Notice that the residue vector $\boldsymbol{R}_i$ is known $\forall i$. This is because the vector $\boldsymbol{E}$ and its derivatives are known once the original system (10.46) is solved. Also, the derivatives of the system matrices with respect to parameters are known either analytically or numerically. The derivation proceeds by integrating the second-order term by parts twice and integrating the first-order derivatives once to get:

$$\boldsymbol{\lambda}^T\boldsymbol{M}\frac{\partial\dot{\boldsymbol{E}}}{\partial x_i}\bigg|_0^{T_m}-\dot{\boldsymbol{\lambda}}^T\boldsymbol{M}\frac{\partial\boldsymbol{E}}{\partial x_i}\bigg|_0^{T_m}+\int_0^{T_m}\ddot{\boldsymbol{\lambda}}^T\boldsymbol{M}\frac{\partial\boldsymbol{E}}{\partial x_i}\,dt+\boldsymbol{\lambda}^T\boldsymbol{N}\frac{\partial\boldsymbol{E}}{\partial x_i}\bigg|_0^{T_m}$$

$$-\int_0^{T_m}\dot{\boldsymbol{\lambda}}^T\boldsymbol{N}\frac{\partial\boldsymbol{E}}{\partial x_i}dt+\int_0^{T_m}\boldsymbol{\lambda}^T\boldsymbol{K}\frac{\partial\boldsymbol{E}}{\partial x_i}dt=\int_0^{T_m}\boldsymbol{\lambda}^T\boldsymbol{R}_i\,dt \tag{10.51}$$

The system (10.46) is assumed to start with zero initial conditions. This implies that energy-storage elements such as capacitors and inductors have initial zero energy. This implies that $\boldsymbol{E}(0)=\dot{\boldsymbol{E}}(0)=\boldsymbol{0}$. Also, to simplify (10.51), we impose zero terminal values on the adjoint variable, i.e., $\boldsymbol{\lambda}(T_m)=\dot{\boldsymbol{\lambda}}(T_m)=\boldsymbol{0}$. Equation (10.51) is thus simplified into:

$$\int_0^{T_m}(\ddot{\boldsymbol{\lambda}}^T\boldsymbol{M}-\dot{\boldsymbol{\lambda}}^T\boldsymbol{N}+\boldsymbol{\lambda}^T\boldsymbol{K})\frac{\partial\boldsymbol{E}}{\partial x_i}dt=\int_0^{T_m}\boldsymbol{\lambda}^T\boldsymbol{R}_i\,dt \tag{10.52}$$

Comparing (10.52) to the implicit integral in (10.48), we see that if the adjoint vector satisfies:

$$\ddot{\boldsymbol{\lambda}}^T\boldsymbol{M}-\dot{\boldsymbol{\lambda}}^T\boldsymbol{N}+\boldsymbol{\lambda}^T\boldsymbol{K}=\left(\frac{\partial\psi}{\partial\boldsymbol{E}}\right)^T\Rightarrow\boldsymbol{M}^T\ddot{\boldsymbol{\lambda}}-\boldsymbol{N}^T\dot{\boldsymbol{\lambda}}+\boldsymbol{K}^T\boldsymbol{\lambda}=\frac{\partial\psi}{\partial\boldsymbol{E}} \tag{10.53}$$

then the sensitivities (10.48) are given by:

$$\frac{\partial f}{\partial x_i}=\int_0^{T_m}\frac{\partial^e\psi}{\partial x_i}\,dt+\int_0^{T_m}\boldsymbol{\lambda}^T\boldsymbol{R}_i\,dt \tag{10.54}$$

It follows that by solving the original system (10.46) with zero initial values and then solving the adjoint system (10.53) with zero terminal values, we can estimate the sensitivities with respect to all the parameters regardless of their number. Notice that the residue $\boldsymbol{R}_i$ is known using the original simulation for all parameters.

Few things to notice about this procedure: First, the excitation of the adjoint problem (10.53) is the temporal vector $\partial\psi/\partial\boldsymbol{E}$, which is estimated during the original system. Second, the adjoint system in (10.53) is similar to the original system with the adjoint matrices given by:

$$\hat{\boldsymbol{M}} = \boldsymbol{M}^T, \quad \hat{\boldsymbol{N}} = -\boldsymbol{N}, \quad \text{and} \quad \hat{\boldsymbol{K}} = \boldsymbol{K} \tag{10.55}$$

The same numerical solver used for solving (10.46) can thus be used for solving (10.53) with the appropriate matrices. Actually, if we make a change of variables in (10.53) to use the adjoint time coordinate $\tau = T_m - t$, the sign of the first-order derivative is reversed and (10.53) is written as:

$$\boldsymbol{M}^T\ddot{\boldsymbol{\lambda}} + \boldsymbol{N}^T\dot{\boldsymbol{\lambda}} + \boldsymbol{K}^T\boldsymbol{\lambda} = \frac{\partial\psi(T_m - \tau)}{\partial \boldsymbol{E}} \tag{10.56}$$

where all temporal derivatives in (10.56) are now with respect to τ. The same solver used to solve (10.46) with respect to t can also solve (10.56) with respect to τ, the backward time coordinate of the adjoint problem.

Finally, the inner product in (10.54) is usually approximated by the summation:

$$\int_0^{T_m} \boldsymbol{\lambda}^T \boldsymbol{R}_i \, dt \approx \Delta t \sum_k \boldsymbol{\lambda}^T(k\Delta t)\boldsymbol{R}_i(k\Delta t) \tag{10.57}$$

The values of the original and adjoint responses are usually obtained through numerical techniques at discrete multiples of the time step Δt. The following example illustrates the time-domain adjoint sensitivity analysis.

Example 10.7: Consider the circuit shown in Figure 10.12. Find the derivatives of the energy dissipated in the resistor with respect to all parameters $\boldsymbol{x} = [R \quad C_1 \quad C_2 \quad L \quad \alpha]^T$ using time-domain adjoint sensitivity analysis.

Solution: First, we write the differential equations governing this circuit using nodal analysis to obtain:

$$\begin{aligned} &\frac{(v_1 - v_s)}{R} + C_1\frac{\partial v_1}{\partial t} + \frac{1}{L}\int_0^t (v_1 - v_2)d\tau = 0 \\ &\frac{1}{L}\int_0^t (v_2 - v_1)d\tau + C_2\frac{\partial v_2}{\partial t} + \alpha v_1 = 0 \end{aligned} \tag{10.58}$$

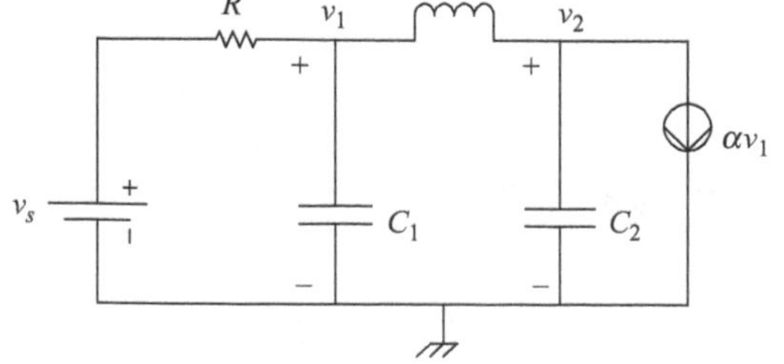

Figure 10.12 The circuit of the Example 10.7

Differentiating both equations with respect to time and organizing terms, we obtain the system:

$$\begin{bmatrix} C_1 & 0 \\ 0 & C_2 \end{bmatrix} \begin{bmatrix} \dfrac{d^2v_1}{dt^2} \\ \dfrac{d^2v_2}{dt^2} \end{bmatrix} + \begin{bmatrix} \dfrac{1}{R} & 0 \\ \alpha & 0 \end{bmatrix} \begin{bmatrix} \dfrac{dv_1}{dt} \\ \dfrac{dv_2}{dt} \end{bmatrix} + \begin{bmatrix} \dfrac{1}{L} & -\dfrac{1}{L} \\ -\dfrac{1}{L} & \dfrac{1}{L} \end{bmatrix} \begin{bmatrix} v_1 \\ v_2 \end{bmatrix} = \begin{bmatrix} \dfrac{1}{R}\dfrac{dv_s}{dt} \\ 0 \end{bmatrix} \tag{10.59}$$

This system has the form (10.46) with the state variables $\boldsymbol{v} = [v_1 \quad v_2]^T$ and the parameter-dependent system matrices:

$$\boldsymbol{M} = \begin{bmatrix} C_1 & 0 \\ 0 & C_2 \end{bmatrix}, \quad \boldsymbol{N} = \begin{bmatrix} \dfrac{1}{R} & 0 \\ \alpha & 0 \end{bmatrix}, \quad \boldsymbol{K} = \begin{bmatrix} \dfrac{1}{L} & -\dfrac{1}{L} \\ -\dfrac{1}{L} & \dfrac{1}{L} \end{bmatrix}, \quad \boldsymbol{G} = \begin{bmatrix} \dfrac{1}{R}\dfrac{dv_s}{dt} \\ 0 \end{bmatrix} \tag{10.60}$$

The objective function of this problem is given by:

$$f = \int_0^{T_m} \psi(R, \boldsymbol{v})\, dt = \int_0^{T_m} \frac{1}{R}(v_1 - v_s)^2\, dt \tag{10.61}$$

According to the theory presented in (10.46)–(10.57), the corresponding adjoint problem is given by:

$$\begin{bmatrix} C_1 & 0 \\ 0 & C_2 \end{bmatrix} \begin{bmatrix} \dfrac{d^2\lambda_1}{dt^2} \\ \dfrac{d^2\lambda_2}{dt^2} \end{bmatrix} - \begin{bmatrix} \dfrac{1}{R} & \alpha \\ 0 & 0 \end{bmatrix} \begin{bmatrix} \dfrac{d\lambda_1}{dt} \\ \dfrac{d\lambda_2}{dt} \end{bmatrix} + \begin{bmatrix} \dfrac{1}{L} & -\dfrac{1}{L} \\ -\dfrac{1}{L} & \dfrac{1}{L} \end{bmatrix} \begin{bmatrix} \lambda_1 \\ \lambda_2 \end{bmatrix} = \hat{\boldsymbol{G}} \tag{10.62}$$

Using (10.53) and (10.61), the adjoint excitation for this problem $\hat{\boldsymbol{G}}$ is given by:

$$\hat{\boldsymbol{G}}(t) = \frac{\partial \psi}{\partial \boldsymbol{v}} = \begin{bmatrix} \dfrac{2}{R}(v_1(t) - v_s(t)) \\ 0 \end{bmatrix} \tag{10.63}$$

The value of this excitation is determined during the original simulation. The two systems (10.59) and (10.62) are then solved for the original responses $\boldsymbol{v}(t)$ and the adjoint response $\boldsymbol{\lambda}(t)$. The residues with respect to all parameters in (10.50) are calculated and stored during the original simulation. They are given by:

$$\boldsymbol{R}_1 = \begin{bmatrix} -\dfrac{1}{R^2}\dfrac{dv_s}{dt} \\ 0 \end{bmatrix} - \begin{bmatrix} -\dfrac{1}{R^2} & 0 \\ 0 & 0 \end{bmatrix} \begin{bmatrix} \dfrac{dv_1}{dt} \\ \dfrac{dv_2}{dt} \end{bmatrix}, \quad \boldsymbol{R}_2 = -\begin{bmatrix} 1 & 0 \\ 0 & 0 \end{bmatrix} \begin{bmatrix} \dfrac{d^2v_1}{dt^2} \\ \dfrac{d^2v_2}{dt^2} \end{bmatrix}$$

$$R_3 = -\begin{bmatrix} 0 & 0 \\ 0 & 1 \end{bmatrix}\begin{bmatrix} \dfrac{d^2v_1}{dt^2} \\ \dfrac{d^2v_2}{dt^2} \end{bmatrix}, \quad R_4 = -\begin{bmatrix} -\dfrac{1}{L^2} & \dfrac{1}{L^2} \\ \dfrac{1}{L^2} & -\dfrac{1}{L^2} \end{bmatrix}\begin{bmatrix} v_1 \\ v_2 \end{bmatrix},$$

$$R_5 = -\begin{bmatrix} 0 & 0 \\ 1 & 0 \end{bmatrix}\begin{bmatrix} \dfrac{dv_1}{dt} \\ \dfrac{dv_2}{dt} \end{bmatrix} \tag{10.64}$$

where R_i is the residue vector related to the ith parameter where $x = [R \quad C_1 \quad C_2 \quad L \quad \alpha]^T$. Once these residues are determined equation (10.57) is used to estimate the sensitivities relative to all parameters.

A point that remains is how to solve the two systems (10.59) and (10.62) with one of them running forward in time and the other one running backward in time. There are actually many approaches for solving this problem. In this example, we utilize the second-order accurate finite differences to obtain a time marching scheme. We illustrate this approach for the original system (10.59). The second-order derivatives of the state variables are replaced by their second-order accurate finite difference approximations to get:

$$M\frac{(v^+ - 2v + v^-)}{\Delta t^2} + N\frac{(v^+ - v^-)}{2\Delta t} + Kv = G \tag{10.65}$$

where $v^+ = v((k+1)\Delta t)$, $v^- = v((k-1)\Delta t)$, and $v = v(k\Delta t)$. Multiplying both sides by $2\Delta t^2$ and organizing, we get:

$$(2M + \Delta tN)v^+ = (\Delta tN - 2M)v^- + (4M - 2\Delta t^2 K)v + 2\Delta t^2 G \tag{10.66}$$

Using (10.66), the values of the voltages at time $t = (k+1)\Delta t$ are evaluated using the two previous values in a marching forward scheme. Both sides of (10.66) are multiplied by the inverse of the time-invariant matrix $(2M + \Delta tN)$ to solve for the new values for the voltages. The system (10.66) is solved with zero initial conditions.

Similarly, for the adjoint system (10.62), we can write the backward running steps:

$$(2M^T + \Delta tN^T)\lambda^+ = (\Delta tN^T - 2M^T)\lambda^- + (4M^T - 2\Delta t^2 K^T)\lambda + 2\Delta t^2 \hat{G}(N_t - k) \tag{10.67}$$

where N_t is the total number of time steps. Notice that in (10.67) the adjoint excitation is applied backward. Using this system, the value of the adjoint vector at $(k+1)\Delta\tau$ is calculated using the two previous values. This simulation runs backward in time with zero terminal values.

The procedure illustrated in this example is implemented in the following MATLAB listing which uses specific values for the circuit components.

```
%M10.7
%This example illustrates the adjoint variable method for time-varying responses
C1=1.0e-6; %first capacitor
C2=2.0e-6; %second capacitor
R=15; %resistor
L=1.0e-3; %inductor
Alpha=1.0e-3; %coefficient of current controlled current surce
startTime=0; %start of simulation
endTime=3.0e-3; %end of simulation
numberOfTimeSteps=3000; %number of simulation time steps
simulationTime=linspace(startTime,endTime, numberOfTimeSteps); %total simulation time
dt=(endTime-startTime)/(numberOfTimeSteps-1); %this is the time step
excitationCentre=(endTime+startTime)/2; %parameter for excitation
excitationSpread=(endTime+startTime)/10;%spread of excitation waveform
M=[C1  0; 0  C2]; %first system matrix
N=[1/R      0; Alpha  0]; %second system matrix
K=[1/L  -1/L; -1/L  1/L]; %third system matrix
%initialization
V=[0   0]'; %current vector of voltages
VMinus=[0  0]'; %previous vector of voltages
MatrixVMinus=inv(2*M+dt*N)*(dt*N-2*M); %matrix multiplying Vminus in time marching
MatrixV=inv(2*M+dt*N)*(4*M-2*dt*dt*K); %matrix multiplying V
MatrixExcitation=inv(2*M+dt*N)*2*dt*dt; %matrix multiplying excitation
%initialize needed storage
AdjointExcitation=zeros(2,numberOfTimeSteps); %storage of adjoint exciation
ResidueR=zeros(2,numberOfTimeSteps); %residue vector for the parameter R
ResidueC1=zeros(2,numberOfTimeSteps); %residue vector for the parameter C1
ResidueC2=zeros(2,numberOfTimeSteps);%residue vector for the parameter C2
ResidueL=zeros(2,numberOfTimeSteps);%residue vector for the parameter L
ResidueAlpha=zeros(2,numberOfTimeSteps);%reside vector for the parameter Alpha
%initialize excitation
SourceWaveForm=2.0*exp(-(simulationTime-excitationCentre).^2/(2.0*
                                                    excitationSpread*excitationSpread));
excitationTerm=(1/R)*diff(SourceWaveForm)/dt; %excitation term is the derivative of the voltage
                                              % waveform divided by R
explicitDerivativeR=0;
%solving original system
for k=1:(numberOfTimeSteps-1)
   VPlus=MatrixVMinus*VMinus+MatrixV*V+MatrixExcitation*[excitationTerm(k)  0]'; %Main
                                                                                % update equation
  %estimate second-order derivatives
  SecondDerivativeApproximation=(VPlus-2*V+VMinus)/(dt*dt);
  %estimate first-order derivatives
  FirstDerivativeApproximation=(VPlus-VMinus)/(2*dt);
  AdjointExcitation(:,k)=[(2/R)*(V(1)-SourceWaveForm(k))   0]'; %store adjoint excitation
  % update explicit part of resistance derivative at current time step.
   explicitDerivativeR=explicitDerivativeR+dt*((-1/(R^2))*(V(1)-SourceWaveForm(k))^2);
  %store residues at the current time step
  ResidueR(:,k)=[-excitationTerm(k)/R  0]'-[-1/(R*R)  0; 0  0]*FirstDerivativeApproximation;
  ResidueC1(:,k)=-1.0*[1  0; 0  0]*SecondDerivativeApproximation; %residue for C1
  ResidueC2(:,k)=-1.0*[0  0; 0  1]*SecondDerivativeApproximation; %residue for C2
  ResidueL(:,k)=-1.0*[-1/(L*L)  1/(L*L); 1/(L*L)  -1/(L*L)]*V;
  ResidueAlpha(:,k)=-1.*[0  0; 1  0]*FirstDerivativeApproximation; %residue for Alpha
  %move in time by updating the values
  VMinus=V;
  V=VPlus;
end
```

```
%solving the adjoint system
MatrixVMinusAdjoint=inv(2*M'+dt*N')*(dt*N'-2*M'); %matrix multiplying Vminus in the time marching
                                                   % scheme
MatrixVAdjoint=inv(2*M'+dt*N')*(4*M'-2*dt*dt*K'); %matrix multiplying V
MatrixExcitationAdjoint=inv(2*M'+dt*N')*2*dt*dt; %matrix multiplying excitation
VMinusAdjoint=[0 0]';
VAdjoint=[0 0]';
%initialize derivatives
DerivativeR=0; DerivativeC1=0;
DerivativeC2=0; DerivativeL=0;
DerivativeAlpha=0;
%now we run the adjoint simulation
for k=1:(numberOfTimeSteps-1)
  %get new value of the vector VPlusAdjoint using previous values

VPlusAdjoint=MatrixVMinusAdjoint*VMinusAdjoint+MatrixVAdjoint*VAdjoint+MatrixExcitationAdjoint*
                                   AdjointExcitation(:,numberOfTimeSteps-k+1);

 SecondDerivativeApproximation=(VPlusAdjoint-2*VAdjoint+VMinusAdjoint)/(dt*dt);
 FirstDerivativeApproximation=(VPlusAdjoint-VMinusAdjoint)/(2*dt);
 (M'*SecondDerivativeApproximation)+(N'*FirstDerivativeApproximation)+(K'*VAdjoint)-
                                                          AdjointExcitation(:,k)
 %we evaluate sensitivity expressions on the fly
 % notice how residues are inverted in time because adjoint responses are running backward in time
 DerivativeR=DerivativeR+dt*VPlusAdjoint'*ResidueR(:,numberOfTimeSteps-k);
 DerivativeC1=DerivativeC1+dt*VPlusAdjoint'*ResidueC1(:,numberOfTimeSteps-k);
 DerivativeC2=DerivativeC2+dt*VPlusAdjoint'*ResidueC2(:,numberOfTimeSteps-k);
 DerivativeL=DerivativeL+dt*VPlusAdjoint'*ResidueL(:,numberOfTimeSteps-k);
 DerivativeAlpha=DerivativeAlpha+dt*VPlusAdjoint'*ResidueAlpha(:,numberOfTimeSteps-k);
 VMinusAdjoint=VAdjoint; %move backward in time by updating the values
 VAdjoint=VPlusAdjoint;
end
DerivativeR=DerivativeR+explicitDerivativeR
```

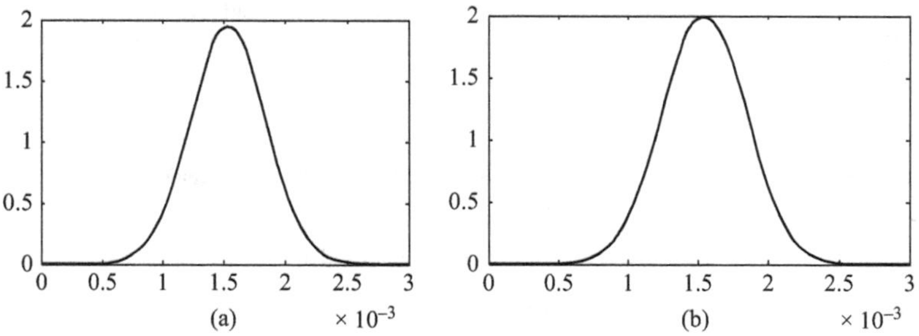

Figure 10.13 The original responses of the circuit of Example 10.7; (a) the voltage $v_1(t)$ and (b) the voltage $v_2(t)$

The MATLAB listing M10.7 is intentionally detailed and has some redundancy to illustrate the basic steps. The original and adjoint responses of this circuit are shown in Figures 10.13 and 10.14, respectively. The output of this code for the given parameter values and source is given by:

```
DerivativeR =

  9.64376e-08

DerivativeC1 =

  0.9754
DerivativeC2 =

  1.0436
```

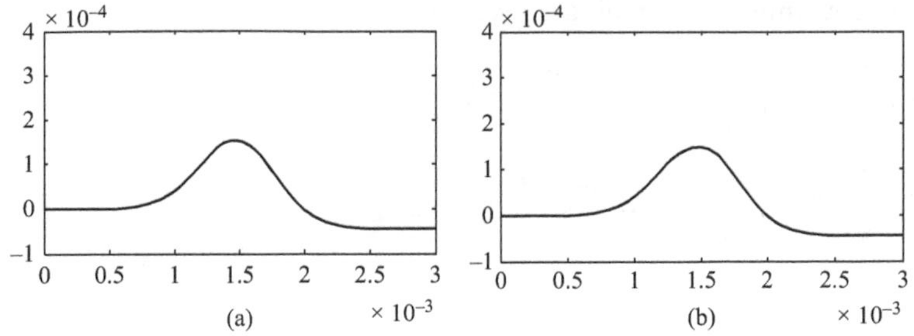

Figure 10.14 The adjoint responses of the circuit in Example 10.7; (a) the response $\lambda_1(\tau)$ and (b) the response $\lambda_2(\tau)$. The time axis for this figure is τ

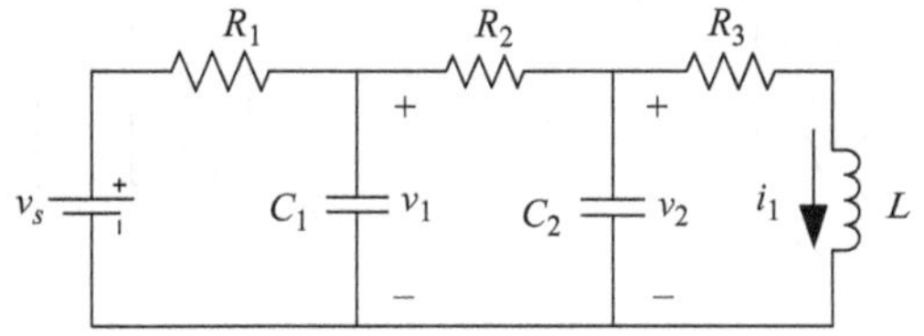

Figure 10.15 The circuit topology of Example 10.8

DerivativeL =

6.9759e-05

DerivativeAlpha =

1.6488e-05

The gradient obtained using the expensive central differences is given by $\nabla f = [9.6756\text{e-}08 \quad 0.9546 \quad 1.0215 \quad 6.8394\text{e-}05 \quad 1.8473\text{e-}05]^T$. Very good match is obtained for all sensitivities. Notice that the derivative with respect to R has both an explicit and an implicit part. The total derivative is the sum of both parts.

The following example utilizes the state space representation to solve for the adjoint sensitivities.

Example 10.8: Consider the circuit shown in Figure 10.15. The parameters of this circuit are $\boldsymbol{x} = [R_1 \quad R_2 \quad R_3 \quad C_1 \quad C_2 \quad L]^T$. Utilize the adjoint variable method to estimate the sensitivities of the objective function:

$$F = \int_0^{T_m} p_s\, dt = \int_0^{T_m} v_s(t) i_s(t)\, dt = \int_0^{T_m} v_s(t)\left(\frac{v_s(t) - v_1}{R_1}\right) dt \tag{10.68}$$

Solution: The objective function in (10.68) measures the energy supplied by the voltage source throughout the simulation time. This circuit contains three energy-storage elements (two capacitors and one inductor). The voltages across the capacitors and the current through the inductor are independent of each other and thus can be considered state variables. We thus express the differential equations governing the circuit using the state variables $\boldsymbol{v} = [v_1 \quad v_2 \quad i_1]^T$. The three nodal equations governing this circuit are given by:

$$
\begin{aligned}
&C_1\frac{dv_1}{dt}+\frac{(v_1-v_s)}{R_1}+\frac{(v_1-v_2)}{R_2}=0\\
&C_2\frac{dv_2}{dt}+\frac{(v_2-v_1)}{R_2}+i_1=0\\
&v_2-i_1R_3-L\frac{di_1}{dt}=0
\end{aligned}
\tag{10.69}
$$

In matrix form, (10.69) is written as:

$$
\begin{bmatrix} C_1 & 0 & 0\\ 0 & C_2 & 0\\ 0 & 0 & L \end{bmatrix}
\begin{bmatrix} \frac{dv_1}{dt}\\ \frac{dv_2}{dt}\\ \frac{di_1}{dt} \end{bmatrix}
+
\begin{bmatrix} \frac{1}{R_1}+\frac{1}{R_2} & -\frac{1}{R_2} & 0\\ -\frac{1}{R_2} & \frac{1}{R_2} & 1\\ 0 & -1 & R_3 \end{bmatrix}
\begin{bmatrix} v_1\\ v_2\\ i_1 \end{bmatrix}
=
\begin{bmatrix} \frac{v_s(t)}{R_1}\\ 0\\ 0 \end{bmatrix}
\tag{10.70}
$$

We can simplify this equation further by writing:

$$
\begin{bmatrix} \frac{dv_1}{dt}\\ \frac{dv_2}{dt}\\ \frac{di_1}{dt} \end{bmatrix}
+
\begin{bmatrix} \frac{1}{C_1R_1}+\frac{1}{C_1R_2} & -\frac{1}{C_1R_2} & 0\\ -\frac{1}{C_2R_2} & \frac{1}{C_2R_2} & \frac{1}{C_2}\\ 0 & \frac{-1}{L} & \frac{R_3}{L} \end{bmatrix}
\begin{bmatrix} v_1\\ v_2\\ i_1 \end{bmatrix}
=
\begin{bmatrix} \frac{v_s(t)}{C_1R_1}\\ 0\\ 0 \end{bmatrix}
\tag{10.71}
$$

Comparing (10.71) with (10.46), we see that in this example we have:

$$
\boldsymbol{M}=\boldsymbol{0},\quad \boldsymbol{N}=\boldsymbol{I},\quad \boldsymbol{K}=
\begin{bmatrix} \frac{1}{C_1R_1}+\frac{1}{C_1R_2} & -\frac{1}{C_1R_2} & 0\\ -\frac{1}{C_2R_2} & \frac{1}{C_2R_2} & \frac{1}{C_2}\\ 0 & \frac{-1}{L} & \frac{R_3}{L} \end{bmatrix},\quad \text{and}\quad \boldsymbol{G}=
\begin{bmatrix} \frac{v_s(t)}{C_1R_1}\\ 0\\ 0 \end{bmatrix}
\tag{10.72}
$$

Using (10.53), the adjoint system corresponding to (10.70) is given by:

$$
-\begin{bmatrix} C_1 & 0 & 0\\ 0 & C_2 & 0\\ 0 & 0 & L \end{bmatrix}
\begin{bmatrix} \frac{d\lambda_1}{dt}\\ \frac{d\lambda_2}{dt}\\ \frac{d\lambda_3}{dt} \end{bmatrix}
+
\begin{bmatrix} \frac{1}{R_1}+\frac{1}{R_2} & -\frac{1}{R_2} & 0\\ -\frac{1}{R_2} & \frac{1}{R_2} & -1\\ 0 & 1 & R_3 \end{bmatrix}
\begin{bmatrix} \lambda_1\\ \lambda_2\\ \lambda_3 \end{bmatrix}
=\frac{\partial\psi}{\partial \boldsymbol{v}}
\tag{10.73}
$$

where the adjoint excitation is given by:

$$\frac{\partial \psi}{\partial v} = \begin{bmatrix} -\frac{v_s(t)}{R_1} & 0 & 0 \end{bmatrix}^T \tag{10.74}$$

By defining $\tau = T_m - t$, the time axis of the adjoint simulation, the adjoint problem (10.74) can be rewritten as:

$$\begin{bmatrix} \frac{d\lambda_1}{d\tau} \\ \frac{d\lambda_2}{d\tau} \\ \frac{d\lambda_3}{d\tau} \end{bmatrix} + \begin{bmatrix} \frac{1}{C_1R_1} + \frac{1}{C_1R_2} & -\frac{1}{C_1R_2} & 0 \\ -\frac{1}{C_2R_2} & \frac{1}{C_2R_2} & \frac{-1}{C_2} \\ 0 & \frac{1}{L} & \frac{R_3}{L} \end{bmatrix} \begin{bmatrix} \lambda_1 \\ \lambda_2 \\ \lambda_3 \end{bmatrix} = \begin{bmatrix} -\frac{v_s(\tau)}{C_1R_1} \\ 0 \\ 0 \end{bmatrix} \tag{10.75}$$

The residues with respect to all parameters in (10.51) are given by:

$$\boldsymbol{R}_1 = \begin{bmatrix} -\frac{v_s(t)}{R_1^2} \\ 0 \\ 0 \end{bmatrix} - \begin{bmatrix} \frac{-1}{R_1^2} & 0 & 0 \\ 0 & 0 & 0 \\ 0 & 0 & 0 \end{bmatrix} \begin{bmatrix} v_1 \\ v_2 \\ i_1 \end{bmatrix}, \quad \boldsymbol{R}_2 = -\begin{bmatrix} \frac{-1}{R_2^2} & \frac{1}{R_2^2} & 0 \\ \frac{1}{R_2^2} & \frac{-1}{R_2^2} & 0 \\ 0 & 0 & 0 \end{bmatrix} \begin{bmatrix} v_1 \\ v_2 \\ i_1 \end{bmatrix},$$

$$\boldsymbol{R}_3 = -\begin{bmatrix} 0 & 0 & 0 \\ 0 & 0 & 0 \\ 0 & 0 & 1 \end{bmatrix} \begin{bmatrix} v_1 \\ v_2 \\ i_1 \end{bmatrix}, \quad \boldsymbol{R}_4 = -\begin{bmatrix} 1 & 0 & 0 \\ 0 & 0 & 0 \\ 0 & 0 & 0 \end{bmatrix} \begin{bmatrix} \frac{dv_1}{dt} \\ \frac{dv_2}{dt} \\ \frac{di_1}{dt} \end{bmatrix},$$

$$\boldsymbol{R}_5 = -\begin{bmatrix} 0 & 0 & 0 \\ 0 & 1 & 0 \\ 0 & 0 & 0 \end{bmatrix} \begin{bmatrix} \frac{dv_1}{dt} \\ \frac{dv_2}{dt} \\ \frac{di_1}{dt} \end{bmatrix}, \quad \text{and } \boldsymbol{R}_6 = -\begin{bmatrix} 0 & 0 & 0 \\ 0 & 0 & 0 \\ 0 & 0 & 1 \end{bmatrix} \begin{bmatrix} \frac{dv_1}{dt} \\ \frac{dv_2}{dt} \\ \frac{di_1}{dt} \end{bmatrix} \tag{10.76}$$

As explained earlier, these residues are stored during the original simulation and then formula (10.57) is used to estimate the sensitivities. We solve the original simulation and the adjoint problem by expanding the first-order derivatives using finite differences. The MATLAB code used for this problem is shown in the MATLAB listing M10.8.

The responses of the original and adjoint circuits for this example are shown in Figures 10.16 and 10.17, respectively. The output of this code is given by:

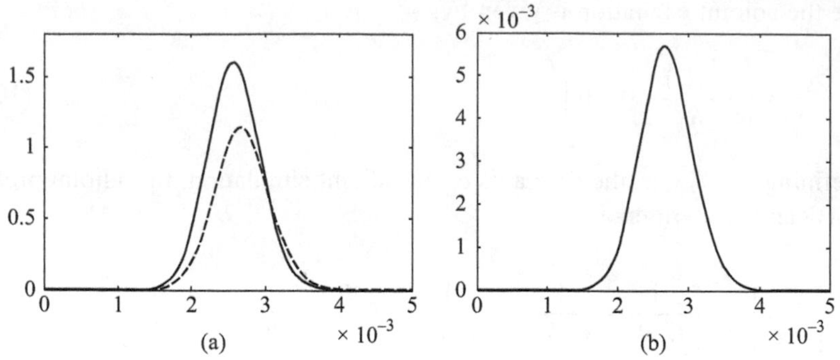

Figure 10.16 The responses of the original circuit of Example 10.8; (a) the voltages $v_1(t)$ (–) and $v_2(t)$ (- - -) and (b) the current $i_1(t)$

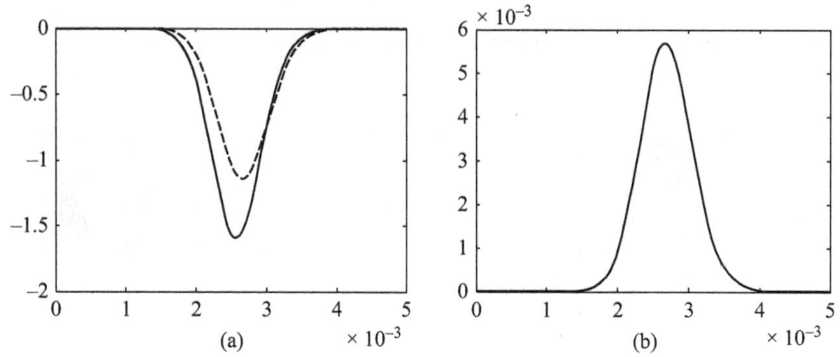

Figure 10.17 The responses of the adjoint circuit for Example 10.8; (a) the voltages $\lambda_1(t)$ (–) and $\lambda_2(t)$ (- - -) and (b) the current $\lambda_3(t)$

DerivativeR1 =

−2.141352428419419e-08

DerivativeR2 =

−2.877441026685108e-08

DerivativeR3 =

−1.622423656568843e-08

DerivativeC1 =

0.955413562516277

DerivativeC2 =

0.895909328093876

DerivativeL =

−2.272452612983195e-05

```
%M10.8
%Adjoint sensitivity analysis of example 10.7
R1=0.05e3; R2=0.07e3; R3=0.2e3; C1=1.0e-6; C2=2.0e-6; L=1.0e-3; %These are the simulation parameters
startTime=0; %start of simulation
endTime=5.0e-3; %end of simulation
numberOfTimeSteps=25000; %number of simulation time steps
simulationTime=linspace(startTime,endTime, numberOfTimeSteps); %total simulation time
dt=(endTime-startTime)/(numberOfTimeSteps-1); %this is the time step
excitationCentre=(endTime+startTime)/2; %parameter for excitation
excitationSpread=(endTime+startTime)/15;%spread of excitation waveform
%we use Gaussian waveform with amplitude of 2 volts
sourceWaveForm=2.0*exp(-(simulationTime-excitationCentre).^2/(2.0*excitationSpread*excitationSpread));
%build the K matrix.
K=[1/(C1*R1)+1/(C1*R2) -1/(C1*R2) 0; -1/(C2*R2) 1/(C2*R2) 1/C2; 0 -1/L R3/L];
OriginalResponse=Integrate(K ,sourceWaveForm/(C1*R1),dt,[0 0 0]');% get original response for all t
%Now we evaluate the explicit part for R1
explicitPartR1=-(((sourceWaveForm-OriginalResponse(1,:))*sourceWaveForm'/(R1*R1))*dt);
%we then evaluate the adjoint excitation
forwardAdjointWaveForm=-1.0*sourceWaveForm/(C1*R1); %This is the adjoint excitation
backwardAdjointWaveForm=flipdim(forwardAdjointWaveForm,2); %backward adjoint excitation
%K matrix of adjoint system
KAdjoint=[1/(C1*R1)+1/(C1*R2) -1/(C1*R2) 0; -1/(C2*R2) 1/(C2*R2) -1/C2; 0 1/L R3/L];
%call adjoint simulator. The original slver is used with different matrix and excitation
AdjointResponse=Integrate(KAdjoint, backwardAdjointWaveForm,dt,[0 0 0]');
%now we evaluate the sensitivities by integrating the product of the adjoint response
%and the residues. Integrals are replaced by dot product of temporal values
DerivativeR1=explicitPartR1+(dt*((OriginalResponse(1,:)-sourceWaveForm)/(R1*R1))
                                         *flipdim(AdjointResponse(1,:),2)')
DerivativeR2=dt*(((OriginalResponse(1,:)-OriginalResponse(2,:))/(R2*R2))*
flipdim(AdjointResponse(1,:),2)'+ (((OriginalResponse(2,:)-OriginalResponse(1,:))/(R2*R2))*
                                             flipdim(AdjointResponse(2,:),2)'))
DerivativeR3= dt* (-1.0*OriginalResponse(3,:))*flipdim(AdjointResponse(3,:),2)'
DerivativeC1= dt*(-1.0*diff(OriginalResponse(1,:))/dt)*
                      (flipdim(AdjointResponse(1,1:numberOfTimeSteps-1),2))'
DerivativeC2= dt*(-1.0*diff(OriginalResponse(2,:))/dt)*
                      (flipdim(AdjointResponse(2,1:numberOfTimeSteps-1),2))'
DerivativeL= dt*(-1.0*diff(OriginalResponse(3,:))/dt)*
                      (flipdim(AdjointResponse(3,1:numberOfTimeSteps-1),2))'

%this function carries out an integration of the state space equation
function OriginalResults=Integrate(K,sourceValues,dt,VInitial)
numberOfTimeSteps=size(sourceValues,2); %get number of time steps
OriginalResults=zeros(3,numberOfTimeSteps); %get original storage
OriginalResults(:,1)=VInitial; %store the initial voltage
V=VInitial;
IMatrix=eye(3);
for k=2:numberOfTimeSteps
 VNew=(IMatrix-dt*K)*V+dt*[sourceValues(k) 0 0]'; %the time marching scheme
 OriginalResults(:,k)=VNew; %store new voltage value
 V=VNew; %make the new vector of values the current values
end
```

Notice that in this listing, the function **Integrate** returns the original response for all time steps as a matrix. The residues (10.76) are not stored in every time step. Rather, they are evaluated from the returned original field and multiplied by the adjoint response.

To confirm the adjoint sensitivity estimates, we also estimated these sensitivities using CFDs. The CFD estimate is given by $\nabla F = [-2.1326179\text{e-}8 \; -2.8783061\text{e-}8 \; -1.6240022\text{e-}8 \; 0.955397625322147 \; 0.895842728659003 \; -2.2724519669\text{e-}5]^T$. Very good agreement is observed between the two estimates. The CFD approach requires 12 extra simulations, while the adjoint variable method requires only one adjoint simulation.

The responses of the original circuit and the adjoint circuit in the previous example have some similarity, yet they are not identical. The question then arises if, for some cases, we do not have to carry out an adjoint simulation at all. In some problems, in circuit theory and other fields, this is indeed the case. Adjoint sensitivity

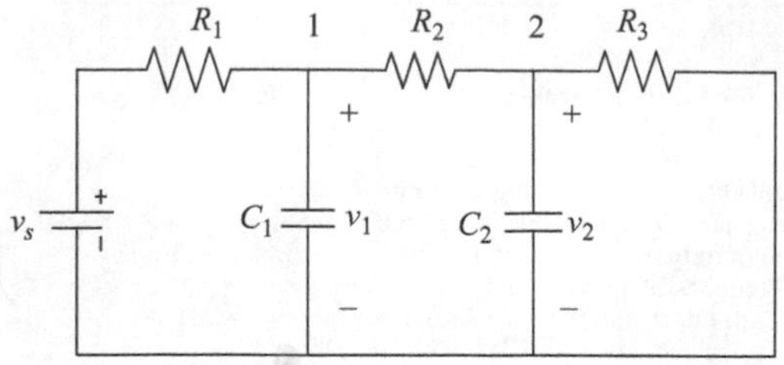

Figure 10.18 The circuit of the self-adjoint problem

analysis that does not require any adjoint simulation is called self-adjoint analysis. For these problems, the adjoint responses are deducible from the original responses. Only the original simulation is needed to estimate the sensitivities of the objective function with respect to all the parameters regardless of their number.

To illustrate this concept, we consider the circuit shown in Figure 10.18. The target objective function for this problem has the form (10.68) as in the previous example. The equations governing the original circuit for this example are given by:

$$\begin{bmatrix} C_1 & 0 \\ 0 & C_2 \end{bmatrix} \begin{bmatrix} \dfrac{dv_1}{dt} \\ \dfrac{dv_2}{dt} \end{bmatrix} + \begin{bmatrix} \dfrac{1}{R_1} + \dfrac{1}{R_2} & -\dfrac{1}{R_2} \\ -\dfrac{1}{R_2} & \dfrac{1}{R_2} + \dfrac{1}{R_3} \end{bmatrix} \begin{bmatrix} v_1 \\ v_2 \end{bmatrix} = \begin{bmatrix} \dfrac{v_s(t)}{R_1} \\ 0 \end{bmatrix} \tag{10.77}$$

The system matrices for this problem are all symmetric. It follows that the adjoint system is given by:

$$-\begin{bmatrix} C_1 & 0 \\ 0 & C_2 \end{bmatrix} \begin{bmatrix} \dfrac{d\lambda_1}{dt} \\ \dfrac{d\lambda_2}{dt} \end{bmatrix} + \begin{bmatrix} \dfrac{1}{R_1} + \dfrac{1}{R_2} & -\dfrac{1}{R_2} \\ -\dfrac{1}{R_2} & \dfrac{1}{R_2} + \dfrac{1}{R_3} \end{bmatrix} \begin{bmatrix} \lambda_1 \\ \lambda_2 \end{bmatrix} = \begin{bmatrix} -\dfrac{v_s(t)}{R_1} \\ 0 \end{bmatrix} \tag{10.78}$$

By changing from the time coordinate of the original problem t to the time axis of the adjoint problem ($\tau = T_m - t$), the adjoint system can be cast in the form:

$$\begin{bmatrix} C_1 & 0 \\ 0 & C_2 \end{bmatrix} \begin{bmatrix} \dfrac{d\lambda_1}{d\tau} \\ \dfrac{d\lambda_2}{d\tau} \end{bmatrix} + \begin{bmatrix} \dfrac{1}{R_1} + \dfrac{1}{R_2} & -\dfrac{1}{R_2} \\ -\dfrac{1}{R_2} & \dfrac{1}{R_2} + \dfrac{1}{R_3} \end{bmatrix} \begin{bmatrix} \lambda_1(\tau) \\ \lambda_2(\tau) \end{bmatrix} = \begin{bmatrix} -\dfrac{v_s(T_m - \tau)}{R_1} \\ 0 \end{bmatrix} \tag{10.79}$$

The adjoint system (10.79) has the same matrices as the original system except that the excitation is reversed in time with a negative sign. It follows that we do not need to solve the adjoint system as the adjoint responses are deducible by:

$$\begin{bmatrix} \lambda_1(\tau) \\ \lambda_2(\tau) \end{bmatrix} = \begin{bmatrix} \lambda_1(T_m - t) \\ \lambda_2(T_m - t) \end{bmatrix} = -\begin{bmatrix} v_1(t) \\ v_2(t) \end{bmatrix} \tag{10.80}$$

Equation (10.80) implies that the adjoint responses are an inverted version of the original response in both time and sign. They are thus known for any time t. For a simple circuit that can be solved in a fast way, this does not make a big difference.

However, if the original and adjoint problems each take long hours of simulation, this approach can save significant simulation time.

A10.1 Sensitivity analysis of high-frequency structures

In the previous sections we utilized three different adjoint sensitivity analysis techniques to simple circuit examples. I showed that using at most one extra simulation of an adjoint system (or circuit), the sensitivities of the target response with respect to all parameters can be estimated. This can be contrasted with finite difference approaches that require at least n extra simulations, where n is the number of parameters.

The real power of this technique manifests itself in problems where the simulation time is excessive. This is the case in modeling of high-frequency structures. The target of the numerical modeling of these structures is to solve Maxwell's equations, the equations governing electric and magnetic fields. A numerical mesh is usually used to solve these problems. The number of field unknowns can reach hundreds of millions or even more. Large electromagnetic (EM) problems may require several days to calculate just one response. In such a case, adjoint sensitivities come very handy. Also, in some of these problems, the parameters are the electric properties of each computational cell. We may have hundreds of millions of cells resulting in unthinkable number of parameters. Estimating these sensitivities would require a formidable time. However, using adjoint sensitivities, the cost is reduced to at most one extra EM simulation.

One of the techniques most commonly used in simulating EM problems is the transmission line method (TLM). This method models the propagation of EM waves in the computational domain with a network of electric transmission lines. The propagation of electric voltage and current impulses on these transmission lines models the propagation of EM fields in different media. The properties of a transmission line (characteristic impedance and wave velocity) are determined by the corresponding values of local permittivity and conductivity.

Figure 10.19 shows a basic computational TLM cell for a two-dimensional problem. At any time step, four incident voltage impulses are incident from the four transmission lines. They give rise to scattering impulses depending on the

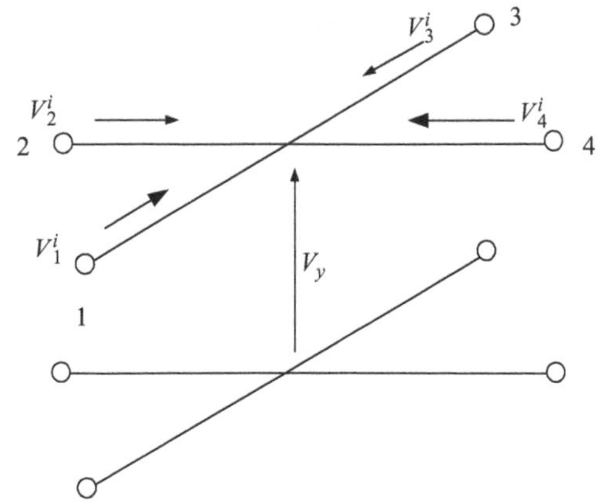

Figure 10.19 A 2D transmission line node

properties of the transmission lines. These impulses then travel to neighboring nodes and become incident at these nodes at the next time step. The scattering equation of one cell is given by:

$$\boldsymbol{V}^r_{j,k} = \boldsymbol{S}_j \boldsymbol{V}^i_{j,k} \tag{10.81}$$

where $\boldsymbol{V}^i_{j,k}$ is the vector of incident voltage impulses of the *j*th cell at the *k*th time step. The size of this vector depends on the TLM computational node used and whether or not it allows for variable permittivity and conductivity. $\boldsymbol{S}_j$ is the scattering matrix of the *j*th node. Its components are known analytically and they are functions of the local material properties and the discretization used. $\boldsymbol{V}^r_{j,k}$ is the vector of reflected impulses of the *j*th node at the *k*th time step. These reflected impulses travel to neighboring nodes and become incident on these nodes at the next time step.

Figure 10.20 shows a two-dimensional computational domain. A large number of TLM cells are connected to model the propagation of EM waves. For problems with non dispersive boundaries, the equation governing the computational domain is given by:

$$\boldsymbol{V}_{k+1} = \boldsymbol{CS}\ \boldsymbol{V}_k + \boldsymbol{V}^s_k \tag{10.82}$$

where $\boldsymbol{S}$ is the scattering matrix of the whole computational domain. It is a block diagonal matrix whose *j*th diagonal matrix is $\boldsymbol{S}_j$. The vector $\boldsymbol{V}_k = [\boldsymbol{V}^T_{1,k}\ \ \boldsymbol{V}^T_{2,k}\ \ \cdots\ \ \boldsymbol{V}^T_{N,k}]^T$ includes the incident impulses on all nodes at the *k*th time step. The matrix $\boldsymbol{C}$ is the connection matrix of the whole domain. It describes how reflected impulses are connected to neighboring nodes. $\boldsymbol{V}^s_k$ is the vector of source impulses at the *k*th time. The time marching equation (10.83) is carried out for all time steps. The electric and magnetic fields at the *k*th time step are linear functions of the impulses incident on the node at this time step.

The system (10.82) can be cast in a form similar to (10.46) by utilizing a first-order Taylor expansion to obtain:

$$\boldsymbol{V}_k + \frac{\partial \boldsymbol{V}_k}{\partial t}\Delta t = \boldsymbol{CS}\ \boldsymbol{V}_k + \boldsymbol{V}^s_k \Rightarrow \frac{\partial \boldsymbol{V}}{\partial t} = \frac{1}{\Delta t}(\boldsymbol{CS} - \boldsymbol{I})\boldsymbol{V} + \frac{1}{\Delta t}\boldsymbol{V}(t) \tag{10.83}$$

where the subscript k has been dropped to indicate an arbitrary time. Comparing with (10.46), we see that the corresponding system matrices are given by:

$$\boldsymbol{M} = \boldsymbol{0}, \quad \boldsymbol{N} = \boldsymbol{I}, \quad \boldsymbol{K} = -\frac{1}{\Delta t}(\boldsymbol{CS} - \boldsymbol{I}) \tag{10.84}$$

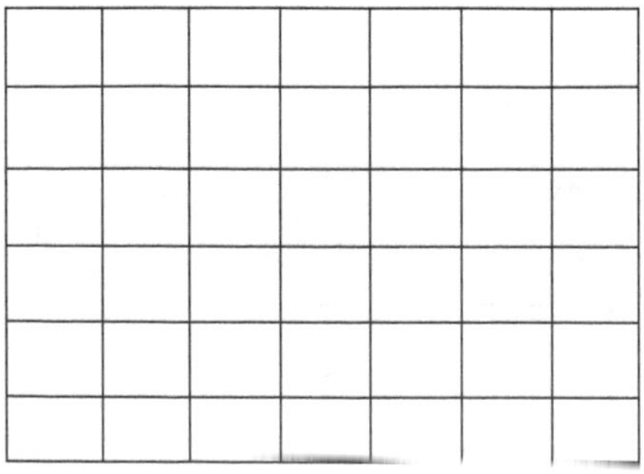

Figure 10.20 A 2D computational domain formed by a number of 2D TLM nodes

For an objective function of the form:

$$f = \int_0^{T_m} \psi(\boldsymbol{x}, \boldsymbol{V})\, dt \tag{10.85}$$

the corresponding adjoint system, according to (10.56), is given by:

$$-\frac{\partial \boldsymbol{\lambda}}{\partial t} = \frac{1}{\Delta t}(\boldsymbol{S}^T \boldsymbol{C}^T - \boldsymbol{I})\boldsymbol{\lambda} + \frac{\partial \psi}{\partial \boldsymbol{V}} \Rightarrow \boldsymbol{\lambda} - \Delta t \frac{\partial \boldsymbol{\lambda}}{\partial t} = \boldsymbol{S}^T \boldsymbol{C}^T \boldsymbol{\lambda} + \Delta t \frac{\partial \psi}{\partial \boldsymbol{V}} \tag{10.86}$$

which can be put in the discrete form [4]:

$$\boldsymbol{\lambda}_{k-1} = \boldsymbol{S}^T \boldsymbol{C}^T \boldsymbol{\lambda}_k + \Delta t \left(\frac{\partial \psi}{\partial \boldsymbol{V}} \right)_{k\Delta t} \tag{10.87}$$

The theory for TLM-based adjoint sensitivity has been developed for different types of responses including the S-parameters that are widely used in microwave engineering. It has been also developed for dispersive materials, which require some modification to the theory presented in this section. Similar techniques were also developed for other numerical EM techniques, see, e.g., References 5–7.

To illustrate this theory, consider Figure 10.21, which shows a schematic of a gold nanoantenna [8]. These nanoantennas are used in many applications including energy harvesting. A light plane wave is incident on this structure. The target is to estimate the sensitivities of the reflection coefficient ($|S_{11}|$) with respect to all parameters. There are 15 geometrical parameters in this example as shown in Figure 10.21. Figures 10.22 and 10.23 show two of the estimated sensitivities using time-domain adjoint sensitivities as compared to CFD. Very good match is observed over a wide frequency range for all parameters. This wideband analysis is one of the main strengthes of time-domain sensitivity analysis. One adjoint simulation is needed for this simulation, while the CFD requires 30 extra simulations.

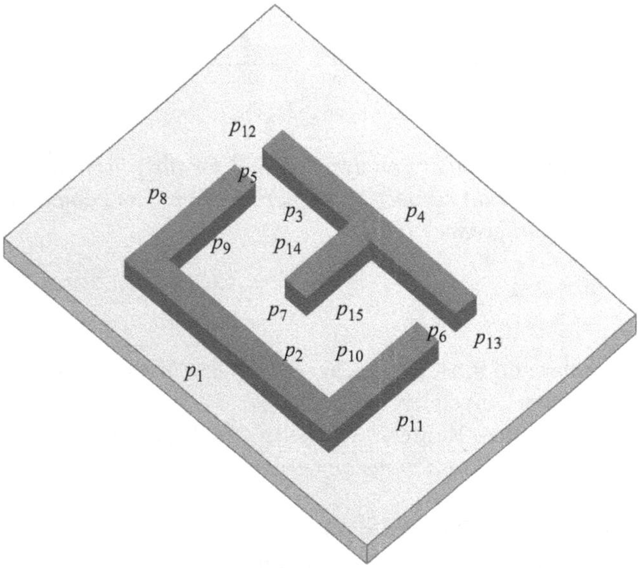

Figure 10.21 A 3D view of the gold nanoplasmonic resonating antenna. There are 15 geometrical parameters for this antenna

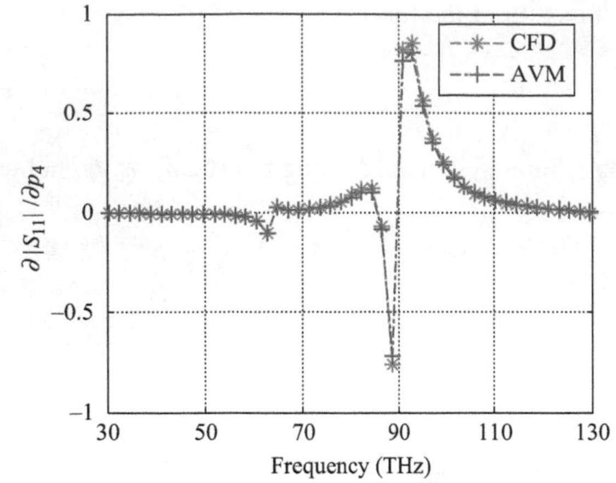

Figure 10.22 Adjoint sensitivity analysis of $|S_{11}|$ for the plasmonic resonator antenna with respect to the parameter p_4 as compared to central finite differences (CFDs)

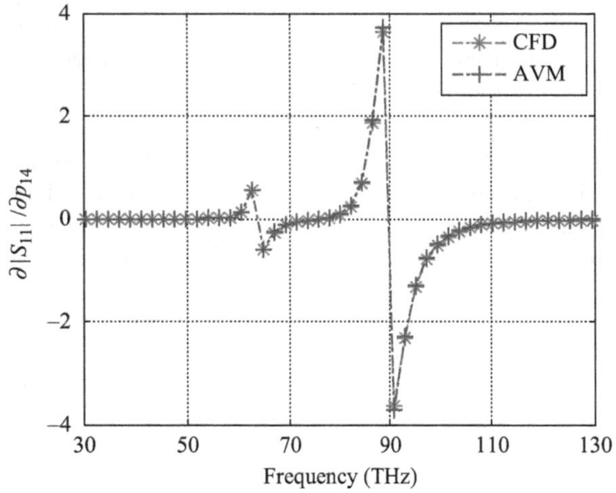

Figure 10.23 Adjoint sensitivity analysis of $|S_{11}|$ for the plasmonic resonator antenna with respect to the parameter p_{14} as compared to central finite differences (CFDs)

References

1. John W. Bandler and R.M. Biernacki, *ECE3KB3 Courseware on Optimization Theory*, McMaster University, 1997
2. S.W. Director and R.A. Rohrer, "The generalized adjoint network and network sensitivities", *IEEE Transactions on Circuit Theory*, vol. CT-16, pp. 318–323, 1969
3. C.A. Desoer and E.S. Kuh, *Basic Circuit Theory*, McGraw Hill, 1969
4. M.H. Bakr and N.K. Nikolova, "An Adjoint variable method for time domain TLM with fixed structured grids", *IEEE Transactions on Microwave Theory Techniques*, vol. 52, pp. 554–559, February 2004

5. N.K. Nikolova, H.W. Tam, and M.H. Bakr, "Sensitivity analysis with the FDTD method on structured grids", *IEEE Transactions on Microwave Theory Techniques*, vol. 52, no. 4, pp. 1207–1216, April 2004
6. J.P. Webb, "Design sensitivity of frequency response in 3-D finite-element analysis of microwave devices", *IEEE Transactions on Magnetics*, vol. 38, pp. 1109–1112, March 2002
7. N.K. Georgieva, S. Glavic, M.H. Bakr, and J.W. Bandler, "Feasible adjoint sensitivity technique for EM design optimization", *IEEE Transactions on Microwave Theory Techniques*, vol. 50, no. 12, pp. 2751–2758, December 2002
8. O.S. Ahmed, M.H. Bakr, X. Li, and T. Nomura, "A time-domain adjoint variable method for materials with dispersive constitutive parameters", *IEEE Transactions on Microwave Theory Techniques*, vol. 60, no. 10, pp. 2959–2971, October 2012
9. Richard C. Dorf and James A. Svoboda, *Introduction to Electrical Circuits*, 8th edn., Wiley, 2010

Problems

10.1 Consider the two circuits shown in Figure 10.24. By solving for the currents and voltages in all branches in both circuits, verify Tellegen's theorem for these two circuits.

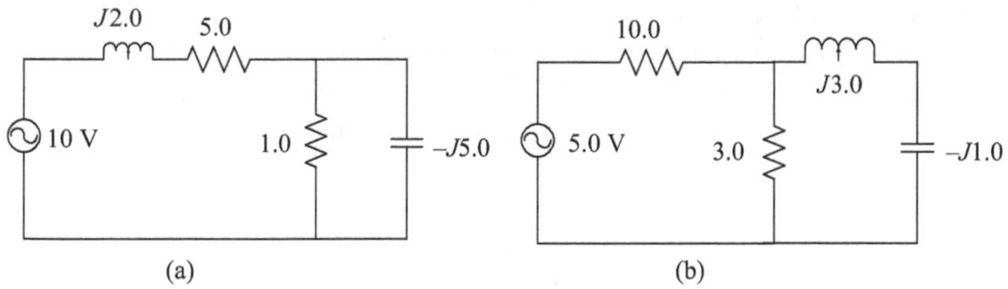

Figure 10.24 The two circuits of Problem 10.1

10.2 Consider the circuit shown in Figure 10.25. Utilize adjoint network method to estimate the sensitivities of the voltage across the resistor R_2 relative to all circuit parameters $\boldsymbol{x} = [R_1 \quad R_2 \quad R_3 \quad R_4]^T$. Take $V_s = 5$ V, $R_1 = 2.5$ kΩ, $R_2 = 5$ kΩ, $R_3 = 4$ kΩ, and $R_4 = 1.0$ kΩ. Verify your answer by analytically deriving the gradient of this voltage.

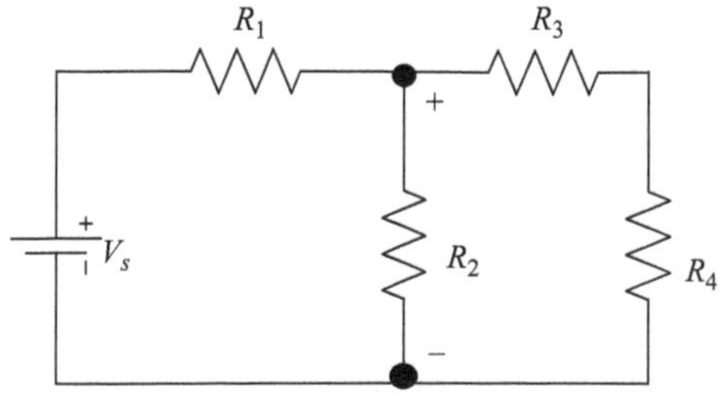

Figure 10.25 The circuit of Problem 10.2

10.3 Consider the circuit shown in Figure 10.26.
   (a) Construct the adjoint circuit to evaluate the sensitivity of the voltage V_L.
   (b) Estimate the gradient $\partial V_L/\partial \boldsymbol{x}$, where $\boldsymbol{x} = [R_s \quad C \quad R \quad \omega]^T$ using the original and adjoint responses. Take $V_s = 1.0 \angle 0$ V, $R_s = 10\ \Omega$, $C = 100\ \mu\text{F}$, $R = 200\ \Omega$, and $\omega = 2000\pi$.
   (c) Verify your answer in (b) by analytically calculating the gradient $\partial V_L/\partial \boldsymbol{x}$.

Hint: Frequency can be treated as a circuit parameter and all matrices can be differentiated relative to frequency.

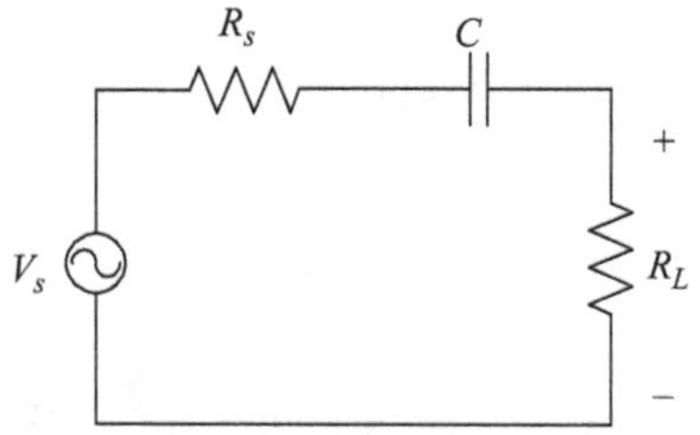

Figure 10.26 The circuit of Problem 10.3

10.4 A resistive network is shown in Figure 10.27.
   (a) Write the nodal equations to solve for the voltages V_1, V_2, and V_3.
   (b) Utilize adjoint analysis of this linear system of equations to evaluate $\partial i_3/\partial \boldsymbol{x}$ with $\boldsymbol{x} = [R_1 \quad R_2 \quad R_3]^T$. Use the values $I_1 = 2.0$ A, $I_2 = 1.0$ A, $R_1 = R_2 = 5\ \Omega$, and $R_3 = 10\ \Omega$.
   (c) Verify your answer in (b) by repeatedly solving the system of equations with perturbed parameter values.

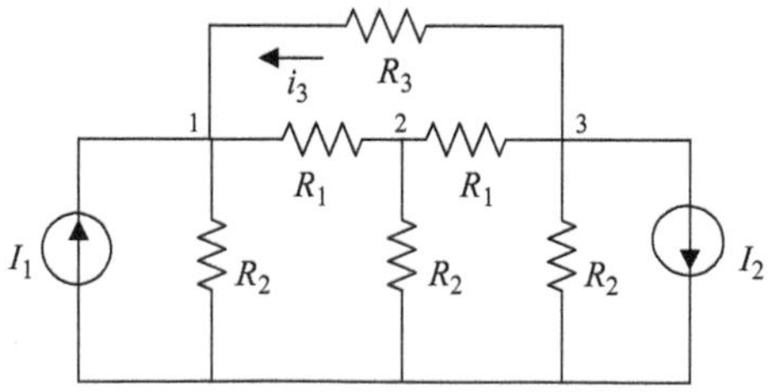

Figure 10.27 The resistive network of Problem 10.4

10.5 Consider the RC filter shown in Figure 10.28.
   (a) Write the system of linear equations to solve for the voltages V_1 and V_2.
   (b) Use adjoint sensitivity analysis of linear system to estimate the gradient $\partial V_3/\partial \boldsymbol{x}$, with $\boldsymbol{x} = [R_1 \quad R_2 \quad R_L \quad C_1 \quad C_2 \quad C_3]^T$. Use the values $V_s = 1.0\angle 0$, $R_1 = R_2 = R_L = 100\ \Omega$, $C_1 = C_2 = C_3 = 10\ \mu\text{F}$, and $\omega = 2000\pi$ rad/s.
   (c) Verify your answers in (b) by using finite differences.

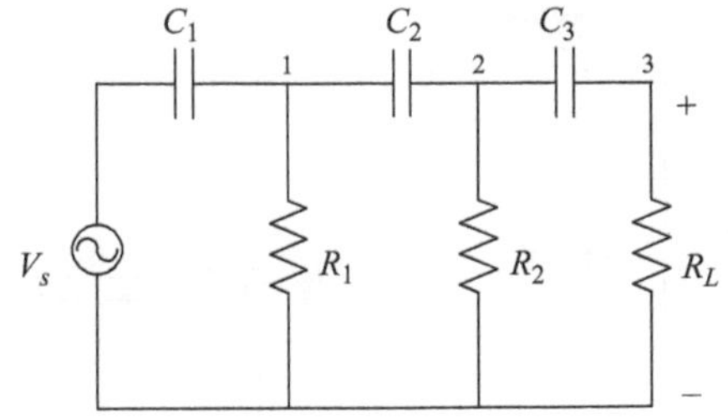

Figure 10.28 The circuit of Problem 10.5

10.6 An RC circuit is shown in Figure 10.29. The switch is closed at time $t=0$.

(a) Write the first-order differential equation governing the capacitor voltage $v_c(t)$.

(b) Solve this differential equation through a numerical technique for the circuit parameter values shown. Take the maximum simulation time as $T_m=0.1$ s.

(c) Construct the corresponding adjoint system to estimate the sensitivities of the energy dissipated in the resistor:

$$f = \int_0^{T_m} \frac{(v_s(t) - v_c(t))^2}{R} dt$$

(d) Solve the adjoint differential equation in (c) to obtain the adjoint voltage $\hat{v}_c(t)$ using the same numerical technique in (b).

(e) Estimate the gradient $\partial f/\partial \mathbf{x}$, where $\boldsymbol{x} = [R \quad C]^T$ using the original and adjoint responses of parts (b) and (d).

(f) Confirm your answer in (e) using finite difference approximation by repeating (b) for perturbed parameter values.

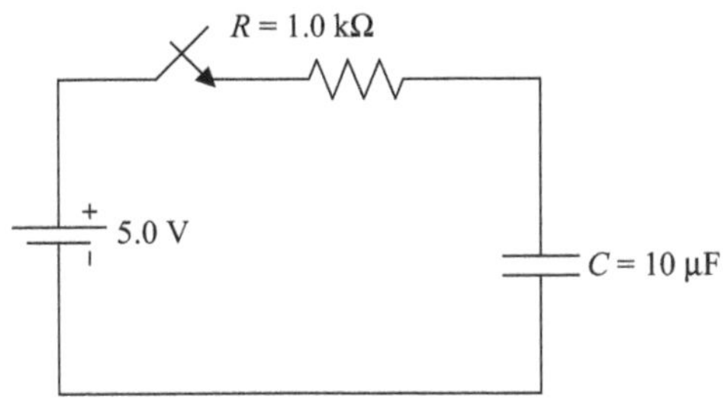

Figure 10.29 The circuit of Problem 10.6

10.7 Consider the RLC circuit shown in Figure 10.30. The circuit elements have all zero energy at time $t=0$.

(a) Write the state differential equations of this circuit in terms of the capacitor voltage v_c and inductor current i_L.

(b) Solve the state equations for $v_c(t)$ and $i_L(t)$ using a numerical technique for the values $R_1=1.0$ kΩ, $R_2=10$ kΩ, $L=1.0$ mH, and $C=10$ μF. Take the simulation time as $T_m=0.5$ s.

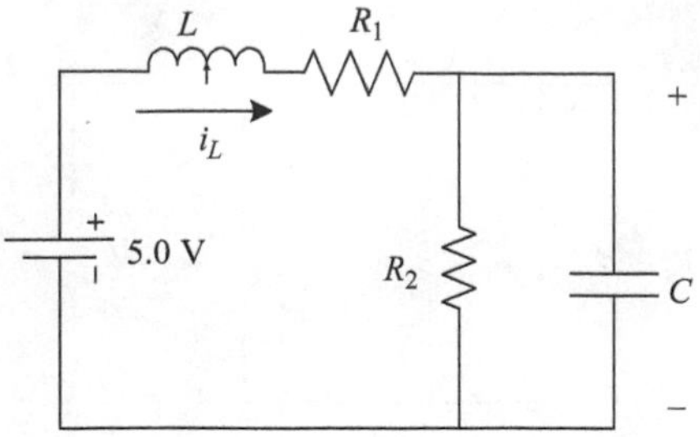

Figure 10.30 The circuit of Problem 10.7

(c) Construct the corresponding adjoint system to estimate the sensitivities of the energy dissipated in R_1:

$$f = \int_0^{T_m} i_L^2 R_1 \, dt$$

(d) Solve the adjoint system in (c) for $\hat{v}_c(t)$ and $\hat{i}_L(t)$.

(e) Utilize your answers in (b) and (d) to estimate the gradient $\partial f/\partial \boldsymbol{x}$, where $\boldsymbol{x} = [R_1 \quad R_2 \quad L \quad C]^T$.

(f) Verify your answer in (e) using finite differences.

Index

Printed in the USA
CPSIA information can be obtained
at www.ICGtesting.com
JSHW051410221024
72173JS00006B/1334

9 781849 195430